Acanthamoeba
Biology and Pathogenesis

Naveed Ahmed Khan

School of Veterinary Medicine and Science,
University of Nottingham, Nottingham, UK

www.caister.com

British Library Cataloguing-in-Publication Data
A catalogue record for this book is available from the British Library

ISBN: 978-1-904455-43-1

Printed and bound in Great Britain

Dedication

This volume is dedicated to my beloved late father, Muhammad Hafeez Khan, and my dear mother, Khalida Khanum – you are the best parents ever.

Naveed Khan

Many of the figures in this book can be viewed in colour at:
http://www.caister.com/acanthamoeba

Contents

Preface

This book provides the first comprehensive review of *Acanthamoeba* research to be published. Everything that is known about *Acanthamoeba* is critically reviewed and divided into easy to follow sections in a thought-provoking manner. This book presents the current state of research on every aspect of this organism, detailing major advances in areas such as genomics, molecular and cellular biology, life cycles, geographical distribution, role in ecosystem, morphology, motility, phylogenetics, genotyping, metabolism, regulation of morphogenesis, host-parasite interactions, the molecular and immunological basis of pathogenesis, methods of transmission, epidemiology, clinical manifestation, diagnosis, treatment, new target development and drug resistance, as well as its role as a Trojan horse of the microbial world, including viral, bacterial, protozoal and fungal pathogens, and much more. There is a significant emphasis on our knowledge of *Acanthamoeba* infections that has grown in the molecular era. In addition, this book provides a historical perspective on *Acanthamoeba* research that will be of considerable interest.

This compilation will serve as an essential reference for microbiologists, immunologists, and physicians in the field of basic and medical microbiology, as well as an invaluable reference for new and experienced researchers who wish to understand this organism better. This book is the definitive guide to current research in this increasingly important organism.

Naveed Khan
Nottingham, 2009

Other Books of Interest

From: *Acanthamoeba*: Biology and Pathogenesis. By: Naveed A. Khan.

Section A

Biology and Phylogeny

1. Introduction

Acanthamoeba is an opportunistic protozoan that is widely distributed in the environment. The organism has two stages in its life cycle, an active trophozoite stage during which *Acanthamoeba* reproduces, and a dormant cyst stage during which it remains inactive with little metabolic activity, but viable, for years. During the last few decades, *Acanthamoeba* has become increasingly appreciated as an important microbe and now is well-recognized to produce serious human infections, including a vision-threatening keratitis (called *Acanthamoeba* keratitis) and a rare but fatal encephalitis, known as granulomatous amoebic encephalitis. Initially the term "granulomatous amoebic encephalitis" was coined specifically to describe brain infection due to *Acanthamoeba*. However, with the discovery of a number of amoebae that can produce granulomatous encephalitis, including *Acanthamoeba*, *Balamuthia mandrillaris*, *Sappinia diploidea*, and perhaps other unidentified amoebae, it is necessary to differentiate the disease according to its causative agent. So, I have referred to "granulomatous amoebic encephalitis due to *Acanthamoeba*" as "*Acanthamoeba* granulomatous encephalitis" throughout this book.

The true burden of keratitis and encephalitis due to *Acanthamoeba* on human health is not known. Furthermore, the pathogenesis and pathophysiology associated with *Acanthamoeba* infections, as well as the molecular identification of virulence traits of *Acanthamoeba*, which may be potential targets for therapeutic interventions and/or the development of preventative measures remain incompletely understood. In recent years, there has been a tremendous interest in this pathogen by the scientific and the medical community. This is due to (i) an increasing number of *Acanthamoeba* infections, associated with a rise in the number of immunocompromised individuals and contact lens wearers, (ii) the potential role in ecosystems, and (iii) the ability to act as a host or reservoir for microbial pathogens, including viruses, prokaryotes, other protozoa and fungi.

1.1. *Acanthamoeba* is a eukaryotic organism

Eukaryotic organisms include animals, plants, fungi, and protists. They are made of cells organized into complex structures enclosed within membranes.

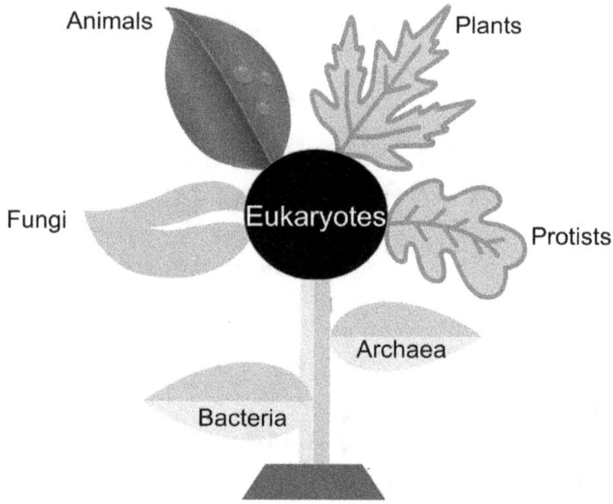

Figure 1. Tree of life.

The defining structure which differentiates eukaryotic cells from prokaryotic cells is the nucleus (Greek "eu" means "true", and "nucleus" means "core". The nucleus is enclosed in a double membrane, called the nuclear envelope, which contains pores (i.e., nuclear pores) to allow the transport of molecules such as ribonucleic acid (RNA) and proteins. Eukaryotic cells contain other membrane-bound organelles such as mitochondria, chloroplasts (plants) and Golgi bodies. Phylogeny studies suggest that eukaryotes are a monophyletic group, and make up one of the three domains of life (Fig. 1). The two other domains, bacteria and archaea, are prokaryotes, and have none of the above features. But eukaryotes do share some aspects of their biochemistry with archaea, and so are grouped close to archaea.

In general, eukaryotic cells are larger than prokaryotes, contain various organelles, and a cytoskeleton composed of microtubules, microfilaments, and intermediate filaments. Furthermore, deoxyribonucleic acid (DNA) in eukaryotes is divided into several linear bundles called chromosomes, which are separated by a microtubular spindle during the nuclear division.

1.2. Protists and protozoa

Protists (Greek "protiston" means "the first of all ones"), are a diverse group of unicellular organisms, comprising those eukaryotes that cannot be classified in any of the other eukaryotic kingdoms such as fungi, animals, or plants. They are usually treated as the kingdom Protista, a term described by Ernst Haeckel in 1866. Protists were traditionally subdivided into several groups, based on similarities to the "higher" kingdoms: the one-celled animal-like protozoa, the plant-like protophyta (mostly one-celled algae), and the fungus-like slime molds and water molds. Because these groups often overlap, they have been replaced by phylogenetic-based classifications. However, they are still useful as informal names for describing the morphology and ecology of protists. The taxonomy of protists is still changing. Newer classification schemes

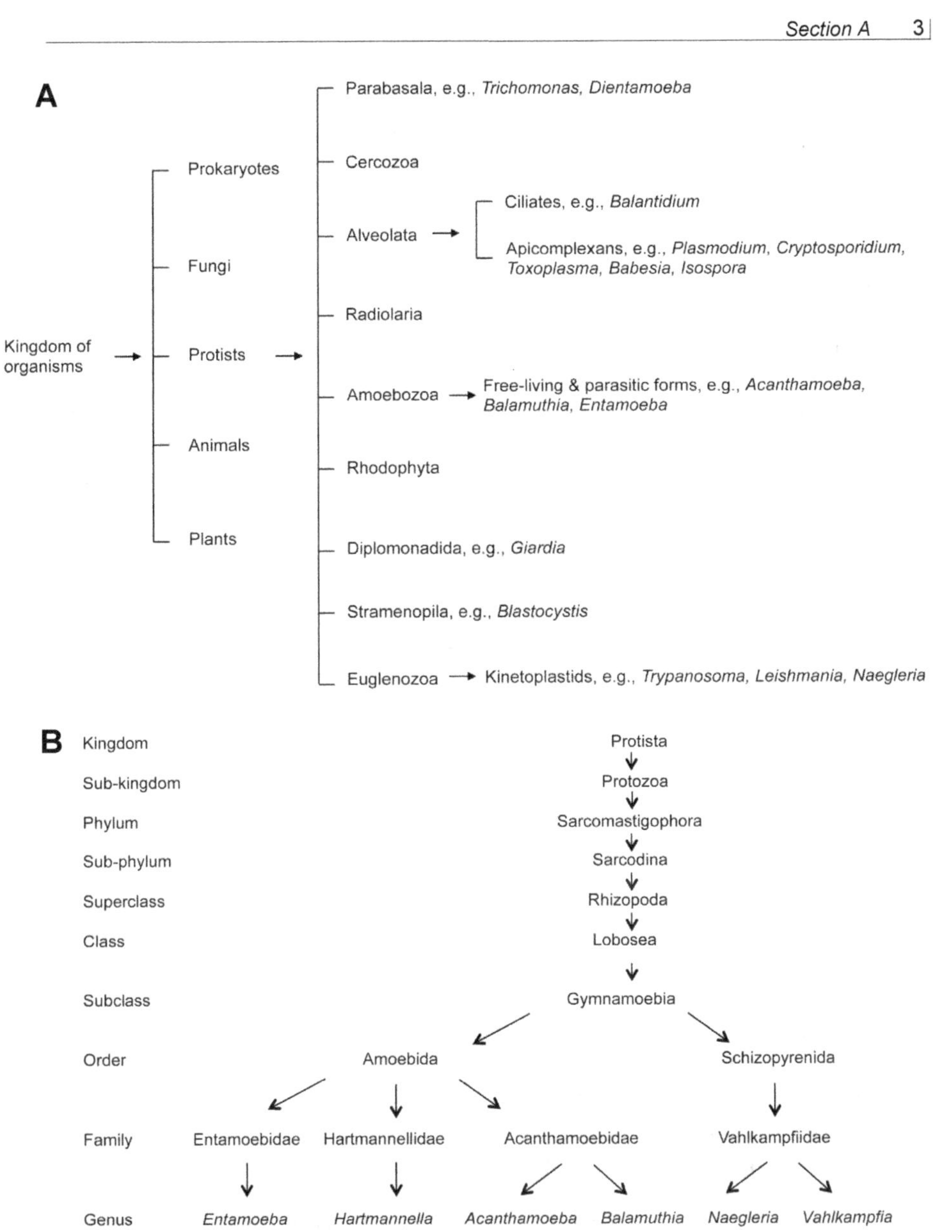

Figure 2. A) The present classification scheme of protists, based largely on their genetic relatedness. **B)** The traditional classification scheme of protists, based largely on morphological characteristics and it is no longer valid.

attempt to present monophyletic groups based on ultrastructure features as well as biochemistry, and genetics information. As a whole, the protists are paraphyletic, rather than a monophyletic group as they do not have much in common besides a relatively simple organization.

Among protists, the animal-like single-celled eukaryotes are called "Protozoa" (Greek "proto" means first and "zoa" means "animal"; singular protozoan) (Fig. 2). The term, "Protozoa" is generally used to refer to a unicellular heterotrophic protist. Protozoa are the largest single-cell, non-photosynthetic eukaryotes that lack cell walls. The study of protozoa, invisible to the naked eye, was initiated with the discovery of the microscope in 1600s by Antonio van Leeuwenhoek (1632 – 1723). Protozoa are generally around 10 – 100 μm, but can grow up to 1 mm and can easily be seen under a microscope or can be even large, e.g., *Xenophyophores* are giant amoeboid single-celled protists between 10 and 25 cm across found on the abyssal plains of the deep oceans. Protozoa exist throughout the aqueous environments and the soil. Protozoa occupy a range of trophic levels. As predators, they prey upon unicellular or filamentous algae, bacteria, and yeast. Protozoa feed by pinocytosis (engulfing liquids/particles by invagination of the plasma membrane) and/or phagocytosis (engulfing large particles, which may require specific cell membrane interactions). Protozoa are known to reproduce using various strategies as follows:

1. Asexual reproduction occurs by binary fission (parent cell mitotically divides into two daughter cells), multiple fission (parent cell divides into several daughter cells), budding (formation of a new organism by the protrusion of part of another organism).
2. Sexual reproduction (two cells join, exchange genetic material and produce progeny by budding or fission).

Some protozoa use both sexual and asexual reproduction during their life cycle. To produce disease, protozoa access their hosts via direct transmission through the oral cavity, the respiratory tract, the genito-urinary tract and the skin, or by indirect transmission through insects, rodents as well as by inanimate objects such as towels, contact lenses and surgical instruments. Once the host tissue is invaded, protozoa multiply to establish themselves in the host, and this may be followed by physical damage to the host tissue or depriving it of nutrients, and/or by the induction of an excessive host immune response resulting in the disease. Among the microbial pathogens, protozoan pathogens alone produce millions of deaths, annually and represent a significant burden on human health. For example, malaria parasite (*Plasmodium* spp.) alone kills more than one million people every year, mostly infants, young children and pregnant women and most of them in Africa. To name a few protozoa of major importance in human health, include the malaria parasite, *Plasmodium* spp., sleeping sickness and the Chagas' disease parasite, *Trypanosoma* spp., the Leishmaniasis disease parasite, *Leishmania* spp. the amoebiasis parasite, *Entamoeba histolytica*, the primary amoebic meningoencephalitis pathogen, *Naegleria fowleri*, and the keratitis and encephalitis pathogen, *Acanthamoeba* spp. In general, the term "parasite" is described for potent parasites such as *Plasmodium* spp., *Trypanosoma* spp., *Leishmania* spp., *Entamoeba histolytica*. In contrast, the free-living amoebae are described as "opportunistic pathogens", however for simplicity they have also been referred to as parasites in some places.

1.3. Discovery of pathogenic free-living and parasitic amoebae

The term "amoebae" (singular is amoeba) encompasses the largest diverse group of organisms in protists that have been studied since the discovery of the early microscope, e.g., *Amoeba proteus*, or closely related *Chaos* that is a genus of giant amoebae, varying from 1 – 5 mm in length. Although these organisms have a common amoeboid motion, i.e., crawling-like movement, they have been classified into several different groups, some of which are described below including both potent parasites and opportunistic pathogens.

- *Entamoeba histolytica*: this is a potent parasitic organism that was discovered in 1873 from a patient suffering from bloody dysentery and named *Entamoeba histolytica* in 1903.
- *Naegleria* spp.: this is a free-living amoeba that was first discovered by Schardinger in 1899, who named it "*Amoeba gruberi*". In 1912, Alexeieff suggested its genus name as *Naegleria*, and much later in the 1970, Carter identified *Naegleria fowleri* as the causative agent of fatal human infections involving the central nervous system (reviewed in De Jonckheere, 2002).
- *Sappinia diploidea* is a free-living amoeba that was isolated from the faeces of lizards and from the soil in 1908-09, and then described as a causative agent of granulomatous amoebic encephalitis in 2001 (Gelman *et al.,* 2001).
- In 1930, *Acanthamoeba* was discovered as a eukaryotic cell culture contaminant and was placed in the genus *Acanthamoeba*, and then described as a causative agent of granulomatous encephalitis in the 1960s (Castellanii, 1930; Douglas, 1930; Volkonsky, 1931).
- *Balamuthia mandrillaris* has been discovered relatively recently, in 1986, from the brain of a baboon that died of meningoencephalitis and was described as a new genus, i.e., *Balamuthia* (Visvesvara *et al.,* 1990; 1993).

Over the years, the aforementioned free-living amoebae have gained increasing attention from the scientific and the medical community, due to their diverse roles in the ecosystem and in particular, their role in causing serious and sometimes fatal human infections (Fig. 3).

2. Discovery of *Acanthamoeba* spp.

It is estimated that single-celled microorganisms emerged on earth, approximately 3-4 billion years ago. Based on the ribosomal RNA (rRNA) sequences, it is estimated that *Acanthamoeba* has diverged from the main line of eukaryotic descent, sometimes between the divergence of yeast (~1.2 x 10^9 years ago) and the divergence of plants and animals (~1 x 10^9 years ago). As an organism, Castellani discovered an amoeba in 1930 in a culture of the fungus, *Cryptococcus pararoseus*. This amoeba was round or oval in shape with a diameter of 13.5 to 22.5 μm. The trophozoite stage (Greek "tropho" means "to nourish") exhibited the presence of pseudopodia (Fig. 4). In addition, the encysted form of this amoeba exhibited a double wall with an

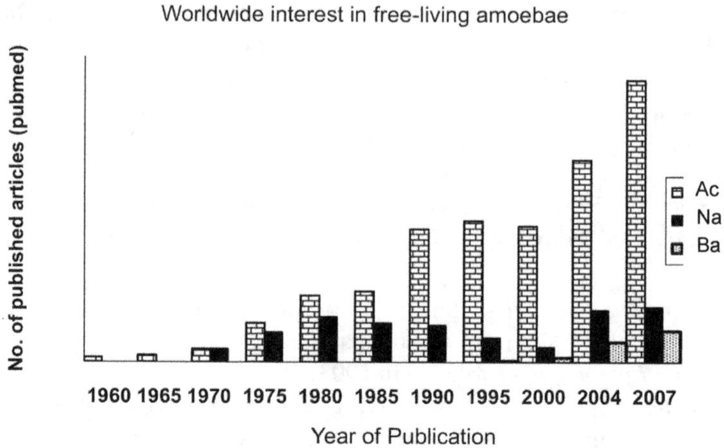

Figure 3. The number of published articles in free-living amoebae of major medical importance. Data for *Acanthamoeba*, *Naegleria* and *Balamuthia* were collected from pubmed, i.e., http://www.ncbi.nlm.nih.gov/entrez/query.fcgi

average diameter of between 9 to 12 µm. Based on these characteristics, this amoeba was placed in the genus *Hartmannella*, and named *Hartmannella castellanii*. A year later in 1931, Volkonsky sub-divided the *Hartmannella* genus into three genera (Volkonsky, 1931), based on the following characteristics:

1. *Hartmannella*: amoebae characterized by round, smooth walled cysts were placed in this genus.
2. *Glaeseria*: amoebae characterized by nuclear division in the cysts were included in this genus.
3. *Acanthamoeba*: amoebae characterized by the appearance of pointed spindles at mitosis, and double-walled cysts and an irregular outer layer, were placed in this genus.

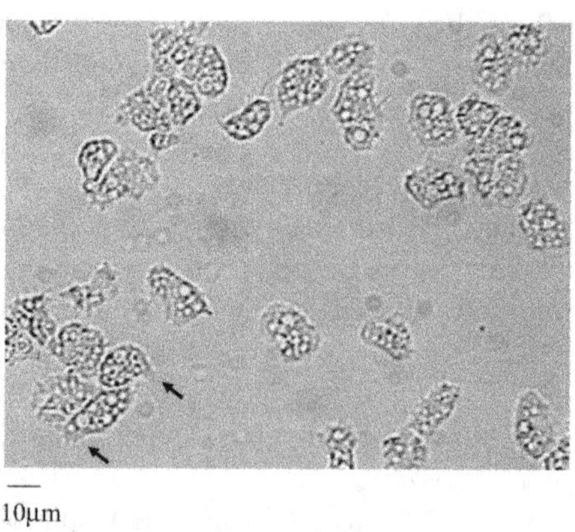

Figure 4. *Acanthamoeba* spp. The infective form of *Acanthamoeba* spp., also known as trophozoites, as observed under phase-contrast microscope. They exhibit round or oval shape, and pseudopodia as shown by arrows. Bar = 10 µm.

In 1952, Singh argued that the classification of amoebae by morphology, locomotion and appearance of cysts was of limited phylogenetic value, and that these characteristics were not diagnostic. It was concluded that the shape of the mitotic spindle was inadequate as a generic character and the genus *Acanthamoeba* was discarded. In 1966, Pussard agreed with Singh (1952) that the spindle shape was an unsatisfactory feature for species differentiation but considered the distinctive morphology of the cyst to be a decisive character at the generic-level and recognized the genus *Acanthamoeba*. After studying several strains of *Hartmannella* and *Acanthamoeba,* Page (1967) also concluded that the shape of the spindle was a doubtful criterion for the species differentiation. He considered the presence of acanthopodia and the structure of the cyst to be sufficiently distinctive to justify the generic designations of *Hartmannella* and *Acanthamoeba*. He also stated that the genus *Hartmannella* had nothing in common with *Acanthamoeba*, except for a general mitotic pattern, which is a property shared with many other amoebae.

In 1975, Sawyer and Griffin created the family Acanthamoebidae and Page (1988) placed *Hartmannella* in the family Hartmannellidae. The current position of *Acanthamoeba* in relation to *Hartmannella*, *Naegleria* and some other amoebae is shown in Fig. 2A. The prefix acanth (Greek "acanth" means "spikes") was added to the term "amoeba" to indicate the presence of spine-like structures (now known as acanthopodia) on the surface of this organism and now it is called "*Acanthamoeba*". After the initial discovery in 1930, *Acanthamoeba* was largely ignored for the next three decades. However, in the late 1950s, *Acanthamoeba* was discovered as a tissue culture contaminant (Culbertson *et al.,* 1958; Jahnes *et al.,* 1957). Later, Culbertson and colleagues (1958-59) demonstrated, for the first time, the pathogenic potential of *Acanthamoeba* by exhibiting its ability to produce cytopathic effects on monkey kidney cells, *in vitro*, and to kill laboratory animals, *in vivo*. The first clearly identified *Acanthamoeba* granulomatous encephalitis case in humans was observed in 1972 (Jager and Stamm, 1972). Later in 1973, *Acanthamoeba* keratitis cases were reported. *Acanthamoeba* was first shown to be infected with bacteria in 1954 (Drozanski, 1956); and then demonstrated to harbour bacteria as endosymbionts, (Proca-Ciobanu *et al.,* 1975) and later as a reservoir for pathogenic facultative mycobacteria (Krishna-Prasad and Gupta, 1978). *Acanthamoeba* was first linked with Legionnaires' disease by Rowbotham in 1980 (Rowbotham, 1980). Since then the worldwide research interest in the field of *Acanthamoeba* has increased dramatically and continues to do so (Fig. 3).

3. Ecology

Strictly speaking, ecology is the study of the distribution and abundance of life and the interactions between organisms and their environment. However, this term is used loosely to describe the interrelations between organisms and their environments, as well as to simply describe the presence of an organism in a given environment. To this end, *Acanthamoeba* has been isolated from

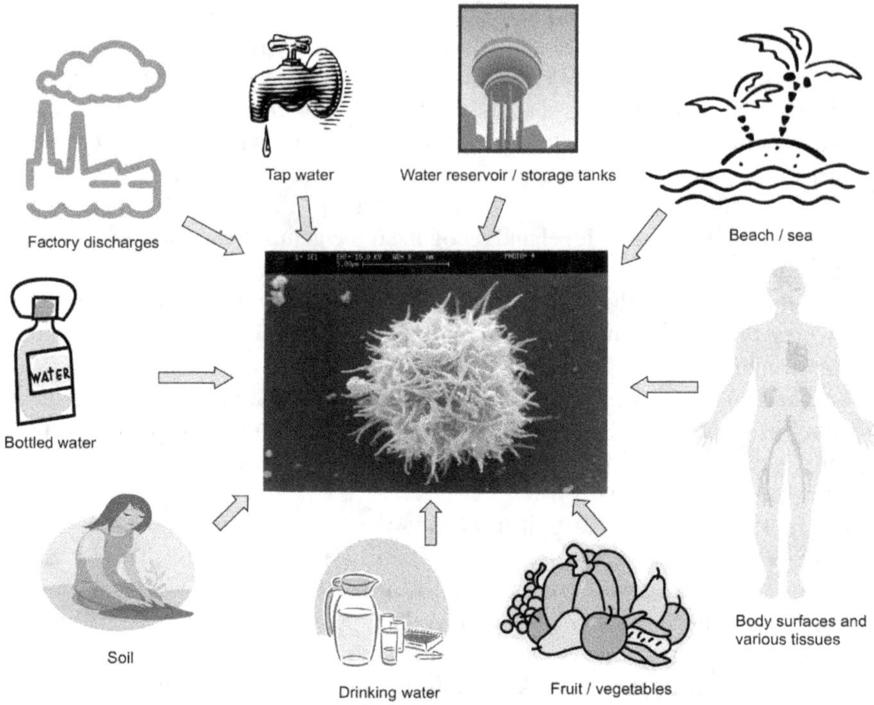

Figure 5. *Acanthamoeba* has been isolated from natural environments, man-made environments and clinical settings.

a diverse natural environments, as well as man-made environments, and human tissues as described above (Fig. 5).

3.1. Distribution in the environment

Acanthamoeba has been isolated from diverse environments, including sea water, ocean sediments, beaches, pond water, stagnant water, soil, mud, shoreline soils, fresh water lakes, streams, frozen swimming areas, salt water lakes, river water, McMurdo Sound-Dry Valley region of Antarctica, water-air interface, and even from the air, indicating the ubiquitous nature of this organism. The isolation of *Acanthamoeba* from Antarctica is significant, in that Antarctica has a very dry atmosphere, i.e., the lowest mean temperature on earth, with persistent strong cold dry winds, and around 5 months of darkness. The entire continent, except for a few coastal regions is covered with permanent ice sheets, over which the precipitation is very low. In addition, *Acanthamoeba* has been isolated from pigeon droppings, fresh water fish, as well as other healthy, diseased, and dead animals. For example, in a survey, gut contents of reptiles captured at the Canary Islands (Tenerife, Spain) were tested for the presence of *Acanthamoeba*. It was interesting to note that *Acanthamoeba* was isolated from the gut contents of 157, out of 508 reptiles captured. Based on the above, it is now widely accepted that *Acanthamoeba* is ubiquitously present in the environment and that we commonly encounter this organism in our routine lives.

3.2. Distribution in the man-made environment

In man-made environments, *Acanthamoeba* has been isolated from bottled mineral water, distilled water bottles, thermally-polluted factory discharges, cooling towers of the electric and nuclear power plants, Jacuzzi tubs, hydrotherapy, ventilation ducts, humidifiers, air-conditioning units, sewage, compost, soil fertilized with municipal sewage wastes, farm soil not receiving sewage wastes, a freshwater stream receiving discharged effluent from sewage lagoons, deep ravine receiving surface water runoff from a cattle farm, vegetables, surgical instruments, contact lenses and their cases, suggesting our common exposure to this organism.

Of note, *Acanthamoeba* isolated from thermally-polluted discharges are found to be more virulent compared with amoeba isolated from surface waters, as determined by axenic growth, cytopathic effects on mammalian cell cultures and virulence in mice.

3.3. Distribution in clinical settings

Acanthamoeba has been routinely recovered from hospitals (both moist areas as well as dust), dialysis units, portable and stationary eye wash stations, human nasal cavities, throat, pharyngeal swabs, lung tissues, skin lesions, human faeces, bone graft of the mandible of patients suffering from osteomyelitis, corneal biopsies, maxillary sinus, mandibular autografts, stool samples, cerebrospinal fluids and the brain necropsies.

To demonstrate our common exposure to this ubiquitous protozoan, by determining anti-*Acanthamoeba* antibody titres in healthy populations, only a handful surveys have been conducted. In one survey of asymptomatic individuals in the New Zealand, it was found that normal human sera from 100% of individuals tested, contained antibodies with titres ranging from 1: 20 to 1: 80 for *Acanthamoeba* spp. (Cursons *et al.*, 1980). The immunoglobulins (Igs) mainly belonged to IgG and IgM classes. However, in other surveys that tested the presence of anti-*Acanthamoeba* antibodies in 1,054 inhabitants of Bohemia, Czechoslovakia, it was found that out of 282 soldiers in active service, only 9 (3.2%) were positive. The titres of positive sera were never higher than 1: 80. In addition, 448 patients of a hospital in Prague were tested, and 41 (9.1%) were found to be positive. Again, the titres of positive sera were never higher than 1: 80. This study was further expanded to include patients who had central nervous system disorders. Out of 274 patients of psychiatric hospital, 9 (3.3%) were positive and the highest titres of positive sera were 1: 160. Interestingly, out of 50 hepatitis A patients, 26 (52%) were found to be positive for *Acanthamoeba* with the highest titres of positive sera being 1: 640. Overall, these findings showed conflicting results of our exposure to *Acanthamoeba* in the asymptomatic population. Given the free-living nature and wide distribution of this organism in the environment, it is difficult to understand these discrepancies. However, this may be due to different assay conditions in different laboratories without comparator sera.

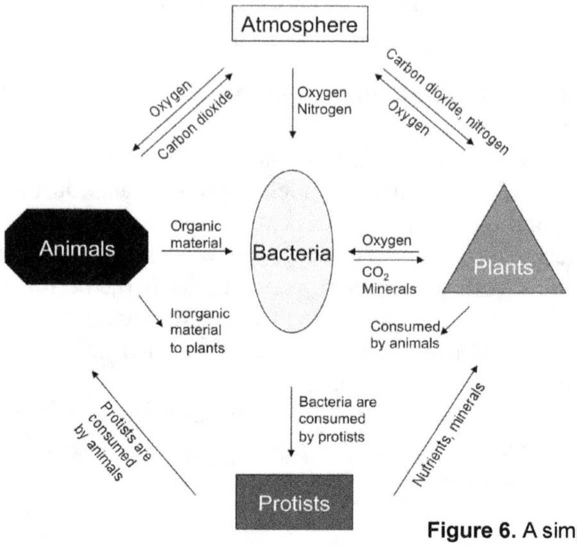

Figure 6. A simplified view of food chain.

In addition, studies on the distribution of *Acanthamoeba* in the environments, Czechoslovakia and Czech Republic *versus* New Zealand, as well as an understanding of the wider global geographical distribution of this organism may explain these findings. Alternatively, the life style of the inhabitants of these geographical locations or a lack of population diversity in these surveys may explain their varied exposure to *Acanthamoeba* in the environment. Of note, a recent study, in 2008, investigated the presence of anti-*Acanthamoeba* antibodies in asymptomatic individuals in London, England from diverse ethnic backgrounds. It was observed that up to 80% of individuals possess anti-*Acanthamoeba* antibodies, further confirming that *Acanthamoeba* is indeed widely distributed in the environment and we encounter this organism in our routine lives.

4. Ecosystem

An ecosystem is a natural unit consisting of all living organisms, i.e., plants, animals and micro-organisms (biotic factors) in an area functioning together with all of the non-living physical (abiotic) factors of the environment. A key feature of the ecosystem is that each living organism has an ongoing and continual relationship with every other elements (biotic or abiotic) that makes up the environment, in which they exist. Almost all ecosystems run on the energy captured from the sun by the primary producers (plants) via photosynthesis. This energy then flows through the food chains to the primary consumers (herbivores who eat and digest the plants), and on to the secondary and the tertiary consumers. In return, the consumers (e.g., animals) produce organic and inorganic matter. Inorganic matter is incorporated into plants, while organic matter is incorporated into living organisms via the primary decomposers (bacteria) (Fig. 6). In addition, the photosynthetic plants fix carbon from carbon dioxide, and nitrogen from the atmospheric nitrogen or nitrates present in the soil to produce amino acids. Much of the carbon and nitrogen contained in ecosystems is created by such plants, and is then consumed by the secondary and the tertiary consumers and incorporated into themselves. Nutrients are usually returned to the ecosystem via decomposition by bacteria and protists. The entire movement of chemicals in an ecosystem is termed "a biogeochemical cycle", and includes the carbon cycle and the nitrogen cycle.

Protists play a significant role in the microbial loop, a term used to describe a trophic pathway in the environment, where dissolved organic matter is re-introduced to the food web through the incorporation into bacteria (Fig. 6). Bacteria are consumed mostly by protists such as amoebae, flagellates and ciliates. The breakdown and dissolution of organic matter and/or bacteria results in the production of dissolved organic carbon and nutrients. Thus, the reclamation of organic carbon and nutrients into the food web results in additional energy available to the higher trophic levels. Because microbes are the base of the food web in most environments, the trophic efficiency of the microbial loop has a profound impact on the ecosystem.

Overall, in soil, protists have two major ecological roles: (1) influencing the structure of the microbial community, and (2) enhancing nutrient cycling. Both of these activities are associated with soil protists feeding on bacteria. It is well-established that these microscopic predators (i.e., protists) play a key role in regulating bacterial populations in the soil. Among protozoa, free-living amoebae are the dominant bacterial consumers in the soil and are responsible for up to 60% of the total decrease of bacterial population. Amoebae consume bacteria and excrete excess nutrients, thus enhancing mineralization of nutrients immobilized in the microbial mass. The primary decomposers (i.e., bacteria) directly decompose organic materials but are not efficient in quickly releasing minerals from their own mass. The secondary decomposers, such as protozoa, consume the primary decomposers and release mineral nutrients as waste products that are tied up in the primary decomposer's biomass. In this way, protozoa such as *Acanthamoeba* (as well as other grazers) are able to make nutrients available that would otherwise remain inaccessible for much longer. This was demonstrated with findings that the soil containing *Acanthamoeba* and bacteria showed significantly greater mineralization of carbon, nitrogen, and phosphorous, compared with the soil containing bacteria, but without *Acanthamoeba*. These activities of *Acanthamoeba* occur during their active trophozoite stage. *Acanthamoeba* can be found at high densities in various soil ecosystems. In addition to bacteria, *Acanthamoeba* preys on fungi and other protists. The harsh conditions, such as desiccation inhibit the feeding activity and induce encystation in *Acanthamoeba*, thus repressing mineralization.

In addition to consuming bacteria, amoebae promote bacterial populations in the soil. The mineral regeneration by the secondary decomposers (protists such as amoebae), relieves nutrient limitation for the primary decomposers. In the presence of carbon but a mineral limitation, the regeneration of minerals by grazers permits continuing growth by the primary decomposers. However

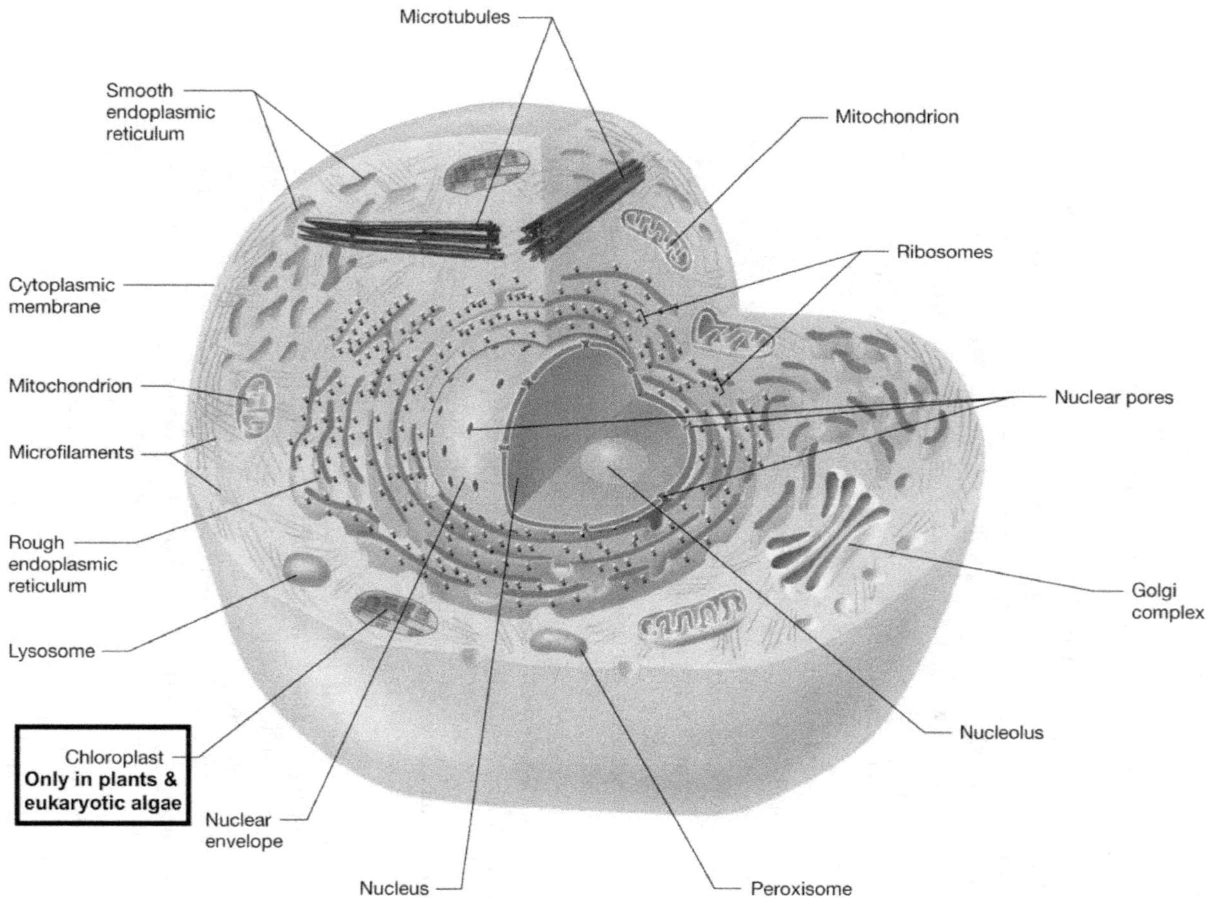

Figure 7. A eukaryotic cell.

with no available carbon source, there is a built up of mineralized nutrients. The role of *Acanthamoeba* in mineral regeneration leading to bacterial growth was clearly demonstrated with the findings that when nitrogen was limiting (but carbon present), nitrogen mineralization by *Acanthamoeba* permitted continued growth of bacteria (*Pseudomonas paucimobilis*) resulting in a greater bacterial biomass. It was further shown that when carbon was limiting, *Acanthamoeba* was almost entirely responsible for nitrogen mineralization, with bacteria (*Pseudomonas paucimobilis*) contributing little. Overall, *Acanthamoeba* seems to play an important role in the regulation of bacterial populations in the environment and the nutrient cycling, thus contributing to the functioning of the ecosystems.

5. Cell biology

Acanthamoeba (plural *Acanthamoebae*) is a genus of protozoa that moves by means of pseudopods. The word amoeba (plural amoebae) is often used to refer to it and its close relatives, now grouped as the Amoebozoa, or to all protozoa that move using pseudopods, otherwise termed amoeboids. In addition, the trophozoites of *Acanthamoeba* exhibit spine-like structures on their surface, known as acanthopodia. The acanthopodia are most likely important in adhesion to surfaces (biological or inert), movements and capturing prey. The acanthopodia or pseudopodia are cytoskeletal structures that protrude from the main body of the cell and possess hyaline (transparent) cytoplasm, which excludes various vacuoles and particles that are normally present in the interior of cell (Bowers and Korn, 1968).

Like in other eukaryotic cells, *Acanthamoeba* contains membrane systems including the rough endoplasmic reticulum, a small amount of smooth endoplasmic reticulum, free ribosomes, Golgi complex, food vacuoles (phagocytic or digestive vacuoles), contractile vacuoles, mitochondria, microtubules, and a nucleus (Fig. 7) (Chlapowski and Band, 1971). For simplicity, here the cellular structures in *Acanthamoeba* are divided into 3 categories, nuclear, cytoplasmic, and mitochondrial as described below.

5.1. Nucleus

Acanthamoeba trophozoite normally possesses a single nucleus (plural nuclei) that is approximately one sixth the size of the trophozoite (Fig. 8), but multinucleate amoebae have been observed. Like other eukaryotes, the nucleus is the largest organelle in *Acanthamoeba*. A nuclear envelope (consisting of two membranes, an inner and an outer membrane, both of which are made up of lipid bilayer) encloses the nucleus, and separates the cells' genetic material from the cytoplasm. The membranes of the nucleus are separated by a distance of about 350 Å (Å is called angstrom; one angstrom is equal to 0.1 nanometre or 1×10^{-10} metres) and the space between the two membranes is called "perinuclear space" (Bowers and Korn, 1968). This provides a barrier to prevent macromolecules from diffusing freely between the nucleoplasm and the cytoplasm. To allow the flow of molecules such as RNA and proteins (for transcription and translation), the envelope contains nuclear pores. As for other eukaryotes, there are numerous nuclear pores in

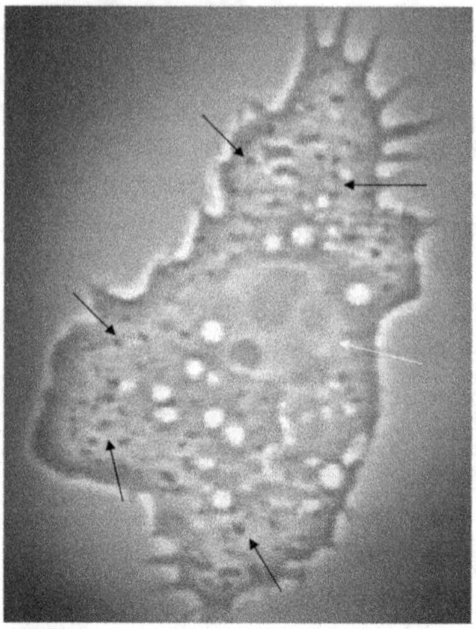

Figure 8. An *Acanthamoeba* trophozoite. Note a nucleus (shown by a white arrow) and large number of mitochondria (some mitochondria are shown by black arrows). Bar = 5 μm.

the nuclear envelope of *Acanthamoeba* nucleus and these are about 1,040 Å in diameter. The nuclear pores are large protein complexes that cross the nuclear envelope. The proteins that make up the nuclear pore complex are known as nucleoporins. The small water soluble molecules can pass through freely, but large molecules (nucleic acids and proteins) that are destined for the nucleus from the cytoplasm possess a nuclear localization signal (sequence of amino acids that acts like a tag). The signal sequence binds to proteins called, importins, which subsequently help docking to the nuclear pore complex. However, for export into the cytoplasm, large molecules possess a nuclear export signal. This signal sequence binds to proteins known as exportins, which mediate transport via the nuclear pore complex. The nuclear transport provides movement of molecules between the nucleus and the cytoplasm that is required for both gene expression and chromosomal maintenance.

The nucleus does not contain any membrane-bound sub-compartments, but the most prominent structures are the nucleolus (plural nucleoli; approximately 2.4 μm in diameter, occupying one eighth of the nuclear volume, while two nucleoli per nucleus are not uncommon), the nuclear lamina and the chromosomes. The nucleolus is a membraneless organelle. It forms around tandem repeats of the ribosomal DNA (i.e., DNA coding for rRNA). These regions are called nucleolar organizer regions. It consists of densely packed chromosomes, proteins and precursor RNA strands from which the subunits of ribosomes are formed. The nucleolus is mainly involved in the biogenesis and assembly of ribosomes (ribosome components, rRNA, ribosomal proteins), which are exported to the cytoplasm where they translate messenger RNA (mRNA). The first step in ribosomal assembly is transcription of the rDNA, by a protein called "RNA polymerase I", forming a large pre-rRNA precursor. This is cleaved into the subunits 5.8S, 18S, and 28S rRNA. The transcription, post-transcriptional processing, and the assembly of rRNA occurs in the nucleolus, aided by small nucleolar RNA molecules. Finally, the assembled ribosomal subunits pass through the nuclear pores into the cytoplasm, where they translate mRNA.

The nuclear lamina is a major constituent of the nucleus. It is a fibrillar network composed of intermediate filaments and membrane-associated proteins that add mechanical support, much like the cytoskeleton supports the cell as a whole. The nuclear lamina is associated with the inner face of the nuclear envelope whereas the outer face stays continuous with the endoplasmic reticulum. Besides providing mechanical support, the nuclear lamina regulates important cellular events such as the DNA replication and the cell division as well as participating in the chromatin organization and in anchoring of the nuclear pore complexes in the nuclear envelope.

The primary function of the nucleus in a cell is that it contains the genetic material, organized as multiple long linear DNA molecules in a complex with a large variety of proteins, such as histones, to form chromosomes. The nucleus maintains the integrity of these genes and controls the activities of the cell by regulating gene expression. The nucleus is crucial for the RNA synthesis. Only nucleated *Acanthamoeba* (and not enucleated cells) incorporate RNA precursors, confirming that RNA synthesis in *Acanthamoeba* is a nuclear function (Prescott, 1959). A small fraction of genes are also located in the mitochondria.

In addition to RNA polymerase I, RNA polymerase II and III have been identified from the trophozoites of *Acanthamoeba* and their kinetic properties and subunit composition indicate their similarities with those of the higher organisms. The class I RNA polymerases are involved in the synthesis of the rRNA precursors; type II, heterogeneous RNA; and type III, transfer RNA (tRNA) precursors. The molecular weight of the RNA polymerase I is approximately 473 kDa (10 putative subunits composition of approximate molecular weights of 185, 133, 41.5, 37, 35, 22.5, 17.5, 15.5, 13.3, and less than 10 kDa) and it is responsible for 25% of the total polymerase activity. The molecular weight of RNA polymerase II is approximately 530 kDa and it is responsible for 62% of the total polymerase activity (12 putative subunits composition of approximate molecular weights of 193, 178, 152, 38.5, 40, 22.5, 18, 15.5, 14, 13.3, 12.5, 12, and a less than 10 kDa). The molecular weight of RNA polymerase III is approximate 550 kDa (15 to 17 putative subunits composition of approximate molecular weights of 169, 138, 82, 52, 37, 34, 30, 28.5, 22.5, 17.5, 15.5, 13.3, and less than 12 kDa) (Spindler *et al.*, 1978; D'Alessio *et al.*, 1979; Detke and Paule, 1978).

5.2. Cytoplasm

Most of the cellular activities are carried out in the cytoplasm, including cellular respiration, and growth-related activities. The cytoplasm also contains the cytoskeleton which gives the cell its shape and produces the movement of the cell. The cytoplasm has three major elements as follows:

- **Cytosol** – translucent fluid in which the other cytoplasmic elements are suspended. The cytosol makes up more than 65% of the cell and is composed of water, salts and organic molecules. It holds various

A

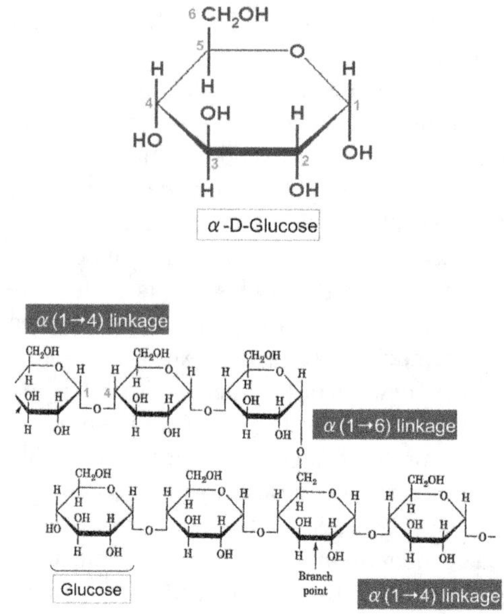

B

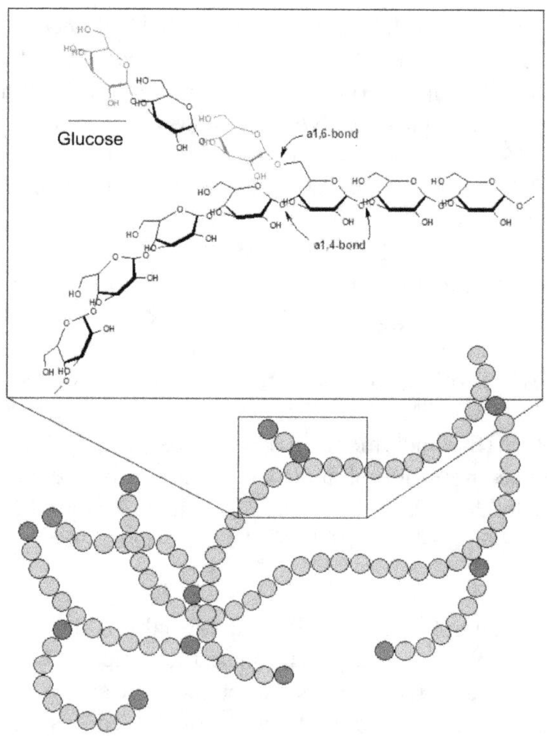

Figure 9. Glycogen structure. Glycogen is a polysaccharide of glucose (**A**). It is a highly branched polymer (**B**). Most of the glucose units are linked by α-1,4 glycosidic bonds, approximately 1 in 12 glucose residues also make 1,6 glycosidic bond with a second glucose, which results in the creation of a branch.

organelles such as the vacuoles, endoplasmic reticulum. In addition, it contains the cytoskeleton, various proteins, enzymes and large numbers of fibrils (approximately 50 Å in diameter that appear as bundles), and free ribosomes. The inner, granular and more fluid portion of the cytoplasm is referred to as endoplasm.

- **Organelles** (organs of the cell) – these are confined functional units within a cell (like organs of the body), and are suspended in the cytosol. Among other functions they provide the metabolic machinery for the cell. Some major organelles in *Acanthamoeba* are mitochondria, the endoplasmic reticulum, the Golgi complex, and the lysosomes.
- **Inclusion bodies** – these store nutrients, secretory products and pigment granules. The polysaccharide reserve (i.e., glycogen; Fig. 9) is present as granules or beta-particles, rather than the more commonly found rosettes and can be as large as 400 Å in diameter. In addition, lipid droplets of an average diameter of 0.5 μm are also found. The cytoplasmic reserve materials include both lipids and glycogen, each of which amounts to about 10% of the dry weight of the *Acanthamoeba* cell (Bowers and Korn, 1968).

5.2.1. Endoplasmic reticulum

Of the nuclear membrane, various sheet-like membranes extend into the cytoplasm, collectively called as the endoplasmic reticulum. *Acanthamoeba* possesses an extensive membrane network of cisternae (sac-like structures) of the endoplasmic reticulum, which could be either smooth (absence of ribosomes) or rough (rough appearance due to the presence of ribosomes, i.e., responsible for protein synthesis) (Bowers and Korn, 1968). Here, endoplasmic means "within the plasm" and reticulum means "network". Notably, ribosomes are not resident of the endoplasmic reticulum but bind to it, once protein synthesis initiates. The membrane of the rough endoplasmic reticulum is continuous with the outer membrane of the nuclear envelope. However, there is no continuous membrane between the rough endoplasmic membrane and the Golgi complex. The endoplasmic reticulum is primarily concerned with the protein synthesis and transport.

In contrast, the smooth endoplasmic reticulum synthesizes and transports lipids and steroids. In addition, the smooth endoplasmic reticulum has functions in the metabolism of carbohydrates and Ca^{2+} concentration, drug detoxification, and the attachment of receptors on the cell membrane proteins, while others break down energy rich glycogen and fats. They also are connected to the nuclear envelope. The smooth endoplasmic reticulum is free of ribosomes and hence has a smooth appearance. The channels formed are similar to those formed by the rough endoplasmic reticulum but are tubular and are concerned with making and transporting lipids and steroids.

Both rough and smooth endoplasmic reticulum possess substantial levels of phosphatases, glucose-6-phosphatase (2 to 3-fold higher activities in the rough than in the smooth endoplasmic reticulum), and 5′-nucleotidase. In contrast, the reduced nicotinamide adenine dinucleotide phosphate (i.e., NADPH)-cytochrome *c* reductase activities are primarily associated with

Table 1. Ribosome is composed of two subunits (S represents Svedberg units, based on their mass). In prokaryotes, 30S contains 16S, while in eukaryotes, 40S contains 18S.

Type	Large subunit	Small subunit
Prokaryotic	50S (5S, 23S)	30S (16S)
Eukaryotic	60S (5S, 5.8S, 28S)	40S (18S)

the smooth endoplasmic reticulum (Thompson and Schultz, 1971). The rough endoplasmic reticulum appears rough under the electron microscope because they are stubbed with ribosomes on the cytoplasmic surface of the endoplasmic reticulum.

5.2.2. Ribosomal RNA (rRNA)

The rRNA is the key component of the ribosome, which synthesizes proteins by decoding mRNA into amino acids. During this process, the rRNA interacts with the tRNA, which provides peptidyl transferase activity during translation. Like other eukaryotes, the *Acanthamoeba* ribosome contains a large subunit rRNA (5S, 5.8S, 28S) and a small subunit rRNA (18S). In addition, the *Acanthamoeba* mitochondrial rRNA molecule consists of a large subunit (5S, 23S) and a small subunit (16S) (Table 1).

The regions of the eukaryotic small subunit rRNA genes are typically 1,850 – 1,900 base pairs (bp) in length, however they can range in size. For example, the small subunit rRNA gene is 1,753 bp in *Tetrahymena thermophila*, 2,206 in *Crithidia fasciculate*, 2,255 bp in *Trypanosoma brucei*, and 2,305 bp in *Euglena gracilis*. The *Acanthamoeba castellanii* gene is 2,303 bp, making it one of the longest. Many of the extra base pairs in the small subunit rRNA genes of *Acanthamoeba* can be accounted for by elements of up to 156 bp in length which are located within non-conserved regions (Gunderson and Sogin, 1986). The repeat unit coding for the precursors to 5.8S, 18S and 26S rRNA has been cloned from *Acanthamoeba* (D'Alessio *et al.,* 1981). The cloned DNA demonstrated that the entire repeat unit is 12,000 bp in length and contains sites for *Eco*RI, *Sma*I, *Bgl*II, *Sst*I, *Bam*HI, *Pst*I, *Kpn*I, *Hind*III, and *Xba*I but not for *Xho*I or *Sal*I (D'Alessio *et al.,* 1981). All of the repeat units in the nuclear DNA appear to be identical and no introns are found. However, the regions which code for the two RNAs which comprise the 26S RNA are separated by a gap of approximately 200 bp (D'Alessio *et al.,* 1981). Unlike some other lower eukaryotes, the 5S RNA gene is not linked to this repeat unit (D'Alessio *et al.,* 1981). The pattern of ribosomal proteins analyzed by two-dimensional (2-D) gel electrophoresis revealed that the small subunit of cytoplasmic ribosome contains 25 proteins, while the large subunit contains 40 proteins. Furthermore, exponentially growing *Acanthamoeba* contains a 45 kDa phosphorylated small subunit ribosomal protein that is absent in the cyst ribosomes (Jantzen, 1981).

5.2.3. Proteins

Proteins are the most important macromolecules in the cells that play crucial roles in almost all functions of the cell. The word protein comes from a Greek word "proteios" meaning "first place". Proteins are made of amino acids (20 different amino acids) that are arranged in a linear chain and joined together by peptide bonds between the carboxyl and amino groups of the adjacent amino acid residues. The sequence of amino acids in a protein is defined by a gene and encoded in the genetic code.

Proteins are actively synthesized on the rough ribosomes that translate mRNA. However, not all proteins are synthesized on the endoplasmic reticulum-bound ribosomes. The mRNA is translated into a chain of amino acids, called "primary structure", which forms secondary structure such as alpha (α)-helix and beta (β)-sheet. This results in the formation of a single protein molecule achieving a three-dimensional tertiary structure. The interaction of more than one protein molecule, usually called "protein subunits" assemble to form quaternary structure.

Protein synthesis also occurs on the unbound ribosomes that are free in the cytosol. Generally speaking, secretory proteins and membrane proteins are made by the endoplasmic reticulum-bound ribosomes, while proteins intended for use within the cytosol are made on the free ribosomes. Proteins are generally divided into three main classes, which correlate with their typical tertiary structures: globular proteins, fibrous proteins, and membrane proteins. Almost all globular proteins are soluble and many are enzymes. Fibrous proteins are often structural; membrane proteins often serve as receptors or provide channels for polar or charged molecules to pass through the cell membrane. Like polysaccharides and nucleic acids, proteins participate in every process within *Acanthamoeba*. Many proteins are enzymes that catalyze biochemical reactions and are vital to metabolism. Proteins have structural or mechanical functions, such as actin and myosin in the cytoskeleton, providing shape to the cell. Other proteins are important in cell signaling, immune responses, cell adhesion, and the cell cycle. However, it's the motility proteins that have gained particular attention in *Acanthamoeba*.

Proteins synthesized from ribosomes enter the interior space or lumen of the endoplasmic reticulum. Subsequently, proteins bud off the endoplasmic reticulum in vesicles and arrive at the Golgi complex (named after its discoverer, Camillo Golgi) that plays important roles in the synthesis of complex polysaccharides. Here, proteins (and other substances) undergo post-translational modifications (most notably glycosylation, i.e., the addition of carbohydrates). Generally, in animal cells, carbohydrates are shown to be either linked to the nitrogen atom of an amino group (*N*-linked glycosylation, that is shown to be limited to asparagine amino acid) or to the oxygen atom of a hydroxyl group (*O*-linked glycosylation, i.e., limited to serine or threonine amino acids but may also be attached to hydroxylysine or hydroxyproline, which are derivatives of the amino acids lysine and proline, respectively). The modified contents are then passed onto other compartments of the cell by means of vesicles that arise by budding off the

Table 2. Six major classes of lipids

Classes of Lipids	Properties
Fatty acids	Building blocks of all membrane lipids except the sterols. A fatty acid is a long unbranched hydrocarbon chain with polar carboxyl group on one end (often called the "head") and nonpolar hydrocarbon "tail" on the other end. They normally contain an even number of carbon atoms usually ranging from 12 to 20 carbon atoms per chain. If carbon atoms do not possess any double bonds (have maximum number of hydrogen atoms attached to it), it is referred to as saturated fatty acids. By contrast, unsaturated fatty acids contain one or more double bonds, which affect the shape of the chain. Saturated fatty acids have long straight tails that pack together, while the presence of double bond results in a bend in the chain preventing tight packing.
Triacylglycerols (or triacylglycerides)	Consist of a glycerol molecule with three fatty acids linked to it. Triacylglycerols are synthesized in a stepwise manner with each one fatty acid added at a time, thus monoacylglycerols contain a single fatty acid, diacylglycerols contain two and triacylglycerols contain three fatty acids. Their main function is to store energy. Triacylglycerols containing saturated fatty acids are solid (or semi-solid) at room temperature and are called fats such as in fats in our bodies. This is due to the fact that saturated fatty acids have long straight tails that pack together. In contrast, triacylglycerols containing unsaturated fatty acids are liquid at room temperature, for example, vegetable oils (due to the presence of double bonds that prevent orderly packing of these molecules).
Phospholipids	Can be either phosphoglycerides or sphingolipids. Phosphoglycerides are predominant in membranes (e.g., phosphatidylinositol, phosphatidylcholine) and as for triacylglycerols contain a glycerol molecule (polar head group) with two fatty acids attached to it (nonpolar). Other phospholipids are sphingolipids which are not based on glycerol but sphingosine. Sphingosine forms an amide bond with fatty acid.
Glycolipid	These are derivatives of sphingosine or glycerol but contain carbohydrate group instead of a phosphate group. The carbohydrate group may contain one to six sugar units which can be either D-glucose, D-galactose or *N*-acetyl-D-galactosamine.
Sterols	These are derivatives of four-ringed hydrocarbons and are structurally distinct from other lipids. Sterols differ from one another in the number and position of double bonds and are limited to eukaryotic cells. The most common sterols in animal cells is cholesterol that is primarily found in plasma membranes. Cholesterol is the basic components for the synthesis of steroid hormones, which include male and female sex hormones, testosterone and estrogen respectively.
Terpenes	These are five carbon compounds, called isoprene that are basis of substances such as vitamin A and comprise of essential oils, flavours and fragrances.

Golgi complex. The process of glycosylation initiates within the lumen of the endoplasmic reticulum and is completed within the Golgi complex and finally destined for the cell membrane or for export. The Golgi complex comprises two or more polarized stacks of 3 to 5 cisternae. The cisternae are separated by a space of about 200 Å. The space between the membranes of a cisterna is also approximately 200 Å, except at the margins where the sacs are slightly dilated (Bowers and Korn, 1968). Finally, proteins are modified, concentrated and packaged into small spherical membrane-bound structures called "vesicles". The membrane-bound vesicles shuttle proteins between these two compartments. The rough endoplasmic reticulum works in concert with the Golgi complex to target new proteins to their destinations including lysosomal enzymes, secreted proteins, or integral membrane proteins as well as initial glycosylation.

5.2.4. Lipids

In contrast to the rough endoplasmic reticulum, the smooth endoplasmic reticulum does not possess ribosomes and thus no protein synthesis occurs on them. The smooth endoplasmic reticulum is involved in the synthesis of lipids and may play a role in the inactivation or detoxification of drugs that might otherwise be toxic or harmful to the cell. Furthermore, *Acanthamoeba* contains soluble glutathione transferase enzyme that may play an important role in the detoxification of drugs (Dierickx *et al.*, 1990).

A primary function in the biology and the pathogenesis of *Acanthamoeba* is attributed to pinocytosis and phagocytosis. During this process, *Acanthamoeba* converts a significant portion of their plasma membrane into phagocytic vesicles. Lipids are major constituents of membranes and a major group of the cellular components. Lipids play at least three important roles: energy storage, membrane structure, and transmission of chemical signals into and within the cell. Based on the chemical structure, there are six major classes of lipids including, fatty acids, triacylglycerols (make fats and oils), phospholipids (found in the membranes), glycolipids (involved in recognition phenomena), sterols and terpenes (Table 2).

The major phospholipids in *Acanthamoeba* are phosphatidylcholine (45%), phosphatidylethanolamine (33%), phosphatidylserine (10%), a phosphoinositide (6%), and diphosphatidylglycerol (4%) (Table 3). The phosphoinositide contains fatty acids, aldehyde, inositol, and phosphate in the ratio of 1.4:0.5:1.1. The main fatty acids chains in *Acanthamoeba* are oleic acids

Table 3. Major lipids in *Acanthamoeba* spp.

Lipids	% composition
Phospholipids	Phosphatidylcholine (45%) and Phosphatidylethanolamine (33%)
Fatty acids	Oleic acids (40 – 50%) and polyunsaturated fatty acids (20 – 30%)

(40 – 50%), and longer polyunsaturated fatty acids (20 – 30%), as modeled by linoleic acid (Ulsamer *et al.,* 1971).

In *Acanthamoeba*, approximately 80% of the fatty acids are unsaturated. The fatty acids of *Acanthamoeba* possess properties of plants, yeast and higher animals. Like plants and yeast, *Acanthamoeba* can synthesize linoleic acid *de novo*, which can act as a precursor for other (mainly C20) *n*-6-unsaturated fatty acids. And like higher animals, *Acanthamoeba* possesses large quantities of polyunsaturated fatty acids such as 5,8,11,14-eicosatetraenoate. The fatty acids of *Acanthamoeba* resemble qualitatively the fatty acids of the higher animals (Korn, 1963). The predominant fatty acids in *Acanthamoeba* are myristate, palmitate, 9-hexadecenoate, stearate, 9-octadecenoate, 9,12-octadecadienoate, 11-eicosenoate, 11,14-eicosadienoate, 8,11,14-eicosatrienoate, and 5,8,11,14-eicosatetraenoate, that are synthesized *de novo*. The pathway of the 5,8,11,14-eicosatetraenoate is shown as follows (Korn, 1964);

Acetate → stearate → oleate → linoleate → 11,14-eicosadienoate →

8,11,14-eicosatrienoate → 5,8,11,14-eicosatetraenoate

The variable temperature induces membrane-lipid adaptation in *Acanthamoeba*. This was demonstrated with observations of increased unsaturation of phospholipids during growth at low temperatures. In comparison to *Acanthamoeba* grown at 30°C, amoeba grown at 15°C showed decreased levels of oleate, and palmitate but increased levels of linoleate, icosatrienoate and arachidonate. Furthermore, *Acanthamoeba* grown at 15°C converts stearate via oleate into further desaturation and chain elongation products, and also that *n*-6 desaturation to produce linoleate is of key importance in the low-temperature adaptation of *Acanthamoeba* (Jones *et al.,* 1991; 1993).

In addition, *Acanthamoeba* is shown to contain diacylglyceryltrimethylhomoserine, an unusual non-phosphorous-containing polar lipid. Gas-liquid chromatography-mass spectrometric analysis of the acyl chain composition of this lipid has shown that 87.5% consists of *cis*-9-octadecenoic acid (Furlong *et al.,* 1986).

Acanthamoeba contains very low levels of glycolipids. Glucose accounts for about 60% of the sugars of the glycolipids of the whole cells and of the plasma membranes. Notable, sphingomyelin, cerebrosides, psychosine and glycoglycerides have not been detected in *Acanthamoeba*. Also, sugars such as sialic acid are not detected (Ulsamer *et al.,* 1971).

Interestingly, *Acanthamoeba* delta 12, delta 15-desaturase generated a highly unusual fatty acid, hexadecatrienoic acid [16:3 delta (9,12,15) (*n*-1)]. *Acanthamoeba* desaturase can catalyze the insertion of a double bond between the terminal two carbons of a fatty acid and represents a new addition to desaturase functionality and plasticity. *Acanthamoeba* bifunctional delta 12, delta 15-desaturase with a microsomal delta 6-desaturase is involved in the synthesis of the highly unsaturated C(16) fatty acid hexadecatetraenoic acid [16:4 delta (6,9,12,15) (*n*-1)] (Sayanova *et al.,* 2006).

Among sterols, the non-saponifiable fraction of the total lipids extracted from the trophozoites of pathogenic *Acanthamoeba* is shown to possess the following sterols:

- Stigmasta-5,7,22,25-tetraene-3 beta -ol (1)
- Stigmasta-5,7,14,22,25-pentaene-3 beta -ol (2),
- Ergosterol and
- 7-dehydrostigmasterol

Interestingly, these sterols were found to be absent in the non-pathogenic isolates of *Acanthamoeba* (Mehdi *et al.*, 1988).

5.2.5. Prostaglandins

Prostaglandins are lipid compounds that are derived enzymatically from the fatty acids. These hormone-like substances have been found to be continuously synthesised from the essential fatty acids by most cells throughout the body. They are generally quickly degraded, so they are short-acting substances. Prostaglandins have various functions including, their action as powerful vasodilators and so reducing blood pressure, inducing uterine muscle contraction, enhancing and inhibiting the inflammatory response of the body, enhance or inhibit the contraction of the smooth muscles of the intestinal wall, dilate blood vessels and increase their permeability to fluid and proteins etc. A prostaglandin contains 20 carbon atoms, including a 5-carbon ring. The prostaglandins belong to the prostanoid class of fatty acid derivatives; which is a subclass of eicosanoids. *Acanthamoeba* has been shown to contain prostaglandins. Interestingly, the synthesis of endoperoxide prostaglandins is shown to be similar in both pathogenic and non-pathogenic *Acanthamoeba*. The course of biosynthesis *in vitro* can be activated by various compounds such as glutathione, albumin, and *p*-chloromercuribenzoic acid, which are either activators or inhibitors of the enzymes. It was suggested that the course of the biosynthesis of prostaglandins *in vivo* is most probably activated by tissues or constitutional liquids surrounding the parasites (Hadas and Mazur, 1997).

5.2.6. Contractile vacuole and osmoregulation

Under the microscope, an actively feeding trophozoite exhibits one or more prominent contractile vacuoles (contractility meaning shortening), whose function is to expel water (and possibly excrete waste), and it is involved in the osmotic regulation. The contractile vacuole absorbs water by osmosis from the cytoplasm, and moves to the surface of the cell when full, and undergoes exocytosis. The contractile vacuole stores extra water. If the cell has a need for water, the contractile vacuole can release more water into the cell. But if water is in excess, the contractile vacuole will remove it, thus the contractile vacuole maintains the homeostatis of the cells' water content. The contraction of the vacuole is produced by myosin 1 in *Acanthamoeba* (discussed in section A11.1). Usually, contractile vacuoles do not contain any precipitates but contain alkaline phosphatase activity in the membrane and this activity may also be present in other vesicles (Bowers and Korn,

1973). During discharge, the contractile vacuole becomes associated with the plasmalemma and may account for the alkaline phosphatase activity in the plasma membrane (Wilkins and Thompson, 1974). In addition, it is shown that alkaline phosphatase activity in the plasma membrane of the stationary phase cells (high density populations) is remarkably reduced, as compared to exponentially growing cells (Wilkins and Thompson, 1974). The content of the contractile vacuole is hypotonic, compared with the cytoplasm. The ionic content of the contractile vacuole in other amoeba (e.g., *Pelomyxa carolinensis*), is high in Na^+ and low in K^+, with these ions accounting essentially all of the osmotically active content of the vacuole (Riddick, 1968). *Acanthamoeba* appears to lack a "sodium pump" similar to that found in many cells since no Na^+-K^+-adenosine triphosphatase (ATPase) activity is observed in the whole homogenates or the isolated membranes (Ulsamer *et al.*, 1971).

5.2.7. Lysosome, peroxisome, and digestive vacuoles

In addition to the contractile vacuole, other types of vacuoles present in the cytoplasm are lysosomes and digestive vacuoles. The digestive vacuoles contain precipitates and amorphous substances that are bounded by membranes similar to the plasma membrane, both in size and staining characteristics (Bowers and Korn, 1968). The size of digestive vacuoles may range from 0.1 μm to a size larger than the nucleus. While the lysosome (Greek "lysis" means "destruction", while "soma" means "body") digests and disposes of foreign particles, malfunctioning structures, food particles, engulfed viruses and bacteria, and worn out organelles. The contents of the lysosome include various digestive enzymes (acid hydrolases; discussed in section B6 and D6.4) that are capable of breaking down large biological molecule, including DNA, RNA, proteins, and certain lipids. Lysosomes fuse with vacuoles and dispense their enzymes into the vacuoles, digesting their contents. Generally, lysosomes are created by the addition of hydrolytic enzymes to the early endosomes from the Golgi complex. These enzymes include lipases, which digests lipids; glycosidases, which digest carbohydrates; proteases, which digest proteins; and nucleases, which digest nucleic acids. These enzymes are synthesized in the cytosol and the endoplasmic reticulum, where they receive a mannose-6-phosphate tag that targets them for the lysosome. The overall function of the lysosome is digestion of macromolecules from phagocytosis (ingestion of food particles or other extracellular material), and endocytosis (where receptor proteins are recycled from the cell surface). In addition, *Acanthamoeba* contains peroxisomes (prokaryotes lack peroxisomes) that are used to break down peroxide, which is toxic for the cell. Generally, peroxisomes participate in the metabolism of the fatty acids and other metabolites and contain enzymes (e.g., catalase) that rid the cell of toxic peroxides.

5.2.8. Plasma membrane

The plasma membrane surrounds the cytoplasm of a cell and, physically separates the intracellular components from the extracellular environment.

Table 4. Plasma membrane make up in *Acanthamoeba* spp.

Constituent	% of plasma membrane
Proteins	33
Phospholipids	25
Sterols	13
Lipophosphonoglycan	29

The plasma membrane plays a role in anchoring the cytoskeleton to provide shape to the cell. In addition, proteins embedded in the cell membrane can act as molecular signals that allow cells to communicate with each other. Protein receptors are found ubiquitously and function to receive signals from both the environment and other cells. These signals are transduced into a form that the cell can use to directly affect a response. Furthermore, protein ligands provide adhesins that allow adhesion of *Acanthamoeba* to the human cells, tissues and inert surfaces.

The plasma membrane provides a selective barrier that regulates what enters and leaves the cell, thus it regulates the transport of substances. The transport can be either passive, i.e., without the input of cellular energy, or active transport that requires energy. Lipid soluble substances, like alcohol, easily pass across the membrane by dissolving in it. Lipid insoluble substances cannot pass, and water passes through protein lined pores. In addition, the plasma membrane gives the cell shape and protection. The plasma membrane is made up of two layers of phospholipids with proteins embedded randomly in the layers. The plasma membrane is involved in the nutrient uptake, structural support and secreting proteins. To this end, all of the enzymes that are needed to carry out these tasks are present in the plasma membrane of *Acanthamoeba* as described below. Structurally, the plasma membrane is approximately 100 Å in thickness and consists at least of proteins, phospholipids, sterols, and lipophosphonoglycan (Table 4) (Dearborn and Korn, 1974) as follows:

- **Proteins (33%)**: as the plasma membrane is involved in the nutrient uptake, and recycling, all of the enzymes that are needed for such a cycle are present in the plasma membrane of *Acanthamoeba* including, phospholipase A2, acyl coenzyme A: lysolecithin acyltransferase, and acyl coenzyme A synthetase (involved in the metabolism of fatty acids (Victoria and Korn, 1975). In addition, phospholipase A1 and lysophospholipase are present in the plasma membrane of *Acanthamoeba* (Victoria and Korn, 1975). Phospholipases are enzymes that convert phospholipids into fatty acids and other lipophilic substances. In addition, the plasma membranes possess enriched 5′-nucleotidase activities. A Mg^{2+}-ATPase is enriched six-fold in the plasma membrane, relative to the homogenate of the whole cell. However, the presence of Na^{+}-ATPase, K^{+}-ATPase, adenyl cyclase, or glucosidase, each of which have been associated with the plasma membranes of the mammalian cells are not detected in the plasma membrane of *Acanthamoeba* (Ulsamer *et al.*, 1971). Interestingly,

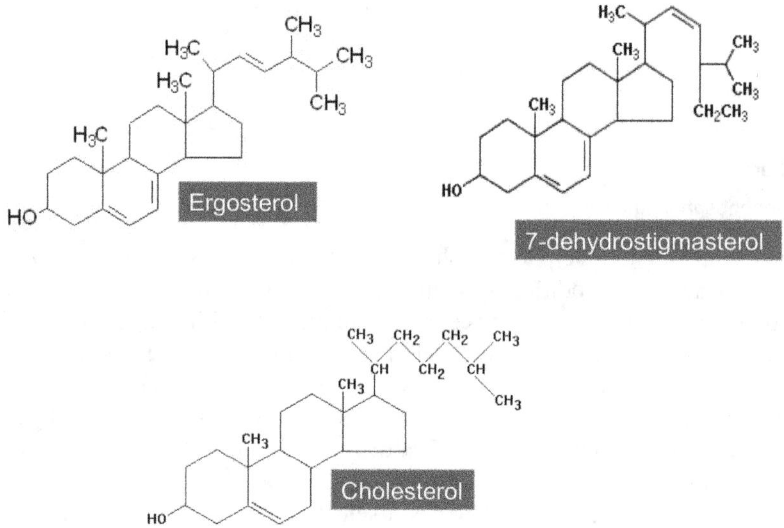

Figure 10. Structure of ergosterol, 7-dehydrostigmasterol, and cholesterol.

some of the proteins in the *Acanthamoeba* plasma membranes are distinct compared with their closely related genus. For example, the antigens of *Acanthamoeba* did not react with the anti-*Naegleria fowleri* sera indicating the lack of antigenic relationships between these organisms (Visvesvara and Healy, 1975), suggesting their potential use in the clinical diagnosis and in the development of therapeutic measures.

- **Phospholipids (25%)**: among phospholipids, 38.6% is phosphatidylethanolamine and 30% is phosphatidylcholine, 10% is phosphatidylserine (major fatty acid is 11,14-eicosadienoic acid), 6% is phosphoinositide (major fatty acid is linoleic acid), and 4% diphosphatidylglycerol (major fatty acid is arachidonic acid). The phosphoinositide is unique in that it contains fatty acids, aldehyde, inositol, and phosphate (Ulsamer *et al.*, 1969).
- **Sterols (13%)**: sterols consist almost entirely of 7-dehydrostigmasterol and ergosterol (Fig. 10) (Smith and Korn, 1968). In contrast, the most common sterol in animal cells is cholesterol.
- **Lipophosphonoglycan (29%)**: the presence of lipophosphonoglycan (a glycosphingolipid) is unusual as it is absent from the mammalian cells (Dearborn and Korn, 1974). Lipophosphonoglycan is a macromolecule that accounts for about 29% of the mass of the plasma membrane of *Acanthamoeba* (Korn *et al.*, 1974). Lipophosphonoglycan has same uniform distribution on the cell surface. Lipophosphonoglycan was observed to be absent from the nuclear membranes of *Acanthamoeba*, but vacuolar membranes (including contractile vacuole) possess lipophosphonoglycan. Approximately, 80% of its mass has been accounted for as follows: 26% neutral sugars (glucose, mannose, galactose, xylose, ratio, 5:4:1:1), 3.3% amino sugars (galactosamine, glucosamine, ratio,

3:1), 10% aminophosphonate (2-aminoethylphosphonate, 1-hydroxy-2-aminoethylphosphonate, ratio, 1:1), 8% inositol, 13% phytosphingosines, 3.2% acid hydrolyzable phosphate, and 14% fatty acids [long chain saturated fatty acids (C-16 to C-28) and 2-hydroxy fatty acids]. The presence of 2-hydroxy fatty acids is unusual as they are minor components of the eukaryotic microorganisms. However, 3-hydroxy fatty acids do occur in lipopolysaccharides of the Gram-negative bacteria, which may bear some resemblance to 2-hydroxy fatty acids of *Acanthamoeba* (Dearborn and Korn, 1974). Although both contain hydroxyl fatty acids as well as neutral and amino sugars, the functional similarities between these two molecules remain unknown. Interestingly, the sugars and acidic groups of lipophosphonoglycan are exposed on both sides of the plasma membrane of *Acanthamoeba* (Bowers and Korn, 1974). This is interesting as the presence of sugars of glycoproteins and glycolipids on the external surface of the mammalian cells is well established, but there is limited evidence of carbohydrate moieties on the inside surface of mammalian plasma membranes. There are two forms of lipophosphonoglycan in *Acanthamoeba*, distinguishable by their electrophoretic mobility (shown by sodium dodecyl sulfate-polyacrylamide gel electrophoresis; SDS-PAGE), a fast lipophosphonoglycan and a slow lipophosphonoglycan. They also are different in their sugar contents. The fast lipophosphonoglycan contains glucose, mannose, and xylose, whereas, the slow lipophosphonoglycan contains glucose, mannose and galactose (Dearborn *et al.*, 1976; Baily and Bowers 1981).

In addition, the plasma membranes of *Acanthamoeba* are involved in the storage of excess cellular Ca^{2+}, in addition to Ca^{2+} containing vesicles like in other eukaryotes. The plasma membrane of *Acanthamoeba* shows the presence of a physiologically active system involved in capturing excess of inflowing Ca^{2+} (Sobota and Przelecka, 1981). Furthermore, electron-dense deposits were observed at the protoplasmic side of the plasma membrane, when log-phase cells were fixed either with glutaraldehyde or formaldehyde supplemented with Ca^{2+} (Sobota and Przelecka, 1981). The appearance of the deposits depends on the age of the *Acanthamoeba* culture. In the 24-hour old culture, the deposits are very small with a diameter of 26 nm, while at logarithmic and stationary-phase cultures, the diameter is larger, 70 – 80 nm and the deposits are localized only on the plasma membrane (Sobota and Przelecka, 1981a). Interestingly, during differentiation, the deposits (170 nm) appear in the mitochondria, while no deposits are observed on the plasma membrane suggesting that depending on the stage of the life cycle, either the plasma membrane or the mitochondria may be involved in storage of excess cellular Ca^{2+}, in addition to Ca^{2+} containing vesicles like in other eukaryotes (Sobota and Przelecka, 1981a).

5.2.8.1. Surface properties

The plasma membranes of the exponentially growing trophozoites possess a negatively charged surface coat bearing sugar residues. Also,

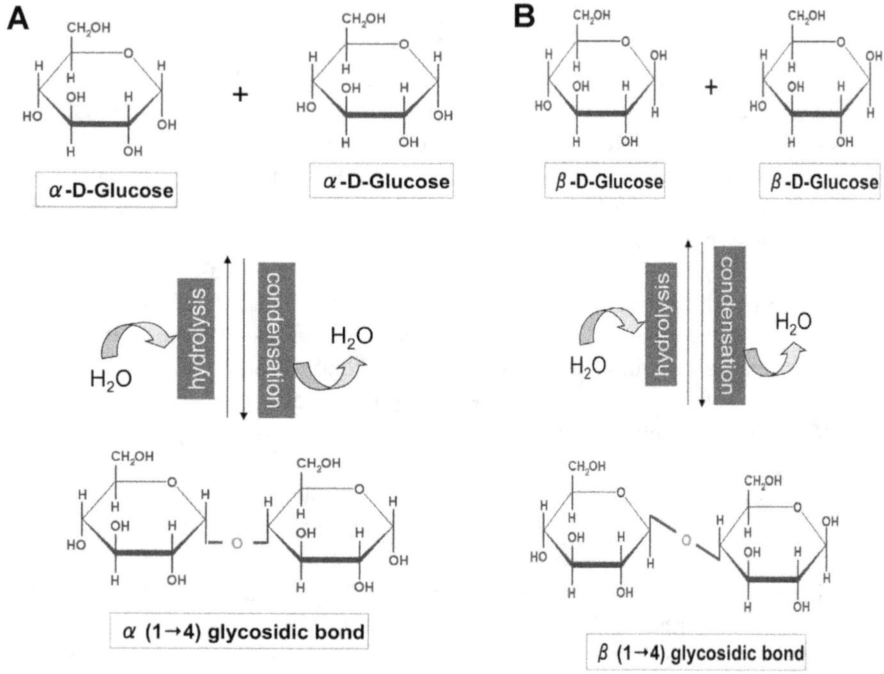

Figure 11. Glycosidic bond is formed when two sugars are joined together via alpha-linkage (**A**) or beta-linkage (**B**) and a molecule of water is removed. The two monosaccharides are bonded via a dehydration reaction (also called a condensation reaction) that leads to the loss of a molecule of water and formation of a glycosidic bond.

in the mature cysts, the external layer of their wall is negatively charged (Przelecka and Sobota 1982). Fluorescein-conjugated lectins revealed the expression of *N*-acetyl-glucosamine, *N*-acetyl-galactosamine, mannose and glucose on trophozoite and cyst surfaces. The trophozoites and cysts of a given species or genotype of *Acanthamoeba* may expose similar surface carbohydrate residues, foremost amongst which is *N*-acetyl-glucosamine (for *Acanthamoeba castellanii* belonging to the T4 genotype), in varying orientation and availability (Elloway *et al.*, 2004). However, this may differ between different genotypes of *Acanthamoeba*. Concanavalin A agglutinated *Acanthamoeba*. The pre-treatment of *Acanthamoeba* with surface-active alkyltrimethylammonium salts (surfactant) enhanced concanavalin A-mediated agglutination of the amoebae. Electron microscopic studies revealed that surfactant did not alter cellular shape or surface morphology. These findings suggested that concanavalin A-mediated agglutination in surfactant-treated amoebae may be due to a fluidizing effect of the surfactants on the plasma membrane of the *Acanthamoeba* (Paatero *et al.*, 1981; Bose *et al.*, 1989).

5.2.9. Carbohydrates: glycogen storage and cell wall synthesis

Carbohydrates are needed to fulfil nutritional requirements as well as being utilized to synthesize cell walls of encysting *Acanthamoeba*. The carbohydrates are stored in the cell as polysaccharide molecules, such as glycogen. Polysaccharides are chains of monosaccharides linked together by either alpha (α) or beta (β) glycosidic bonds (Fig. 11). Saccharides that have alpha-linkages are readily digested by animals and thus suitable for storage such as starch (most common in plant cells) or glycogen (most common in animal cells). In contrast, saccharides that have beta-linkages are not digestible and play important roles in structural functions such as cellulose and chitin. *Acanthamoeba* contains large number of glycogen (α-linked monosaccharides)-containing vacuoles. *Acanthamoeba* trophozoites possess α,α-trehalose (α-D-glucopyranosyl α-D-glucopyranoside) (10 – 15% of the total). Similarly, *Acanthamoeba* cysts predominantly possess α,α-trehalose (α-D-glucopyranosyl α-D-glucopyranoside), however glucose and to a lesser extent maltose (4-*O*-α-D-glucopyranosyl D-glucopyranose) also is present (Deslauriers *et al.*, 1980). In addition, *Acanthamoeba* cysts contain cellulose. Cellulose is a polymer of glucose, the repeating polymer is β-D-glucose and the linkage is $\beta(1\rightarrow 4)$, that link carbon atom 1 to carbon atom 4 of the adjacent glucose unit, forming long, unbranched chains. Several cellulose chains (about 2,000) associate laterally to form microfibrils that are about 25 nm in diameter. Because mammals do not possess an enzyme that can hydrolyze $\beta(1\rightarrow 4)$, thus they can not utilize cellulose. Cows and sheep depend on bacteria and protists to digest cellulose for them. In fungal cell walls, cellulose may be present either as $\beta(1\rightarrow 4)$ or $\beta(1\rightarrow 3)$ linkages. Of interest, chitin consists of *N*-acetyl-glucosamine units only, linked by $\beta(1\rightarrow 4)$ bonds. Thus an understanding of the mechanisms associated with glycogen storage and cellulose biosynthesis during encystation will be of value in the understanding of the biology and the pathogenesis of *Acanthamoeba*.

5.2.10. Ionic distribution

The key biologically important elements required for normal physiological functions are K^+, Na^+, Mg^{2+}, Ca^{2+}, chlorine, phosphorous, and sulphur. In *Acanthamoeba*, higher K^+ content, as compared with the other cations was detected. Total content of the cation-forming elements: K^+, Na^+, Mg^{2+}, and Ca^{2+} was *ca.* 360 mM per kg dry weight of the cells. The phosphorus content was estimated as 492 mM per kg of dry weight. The content of chlorine, a basic cellular anion, was 173 mM per kg of dry weight. The low level of chlorine appears not be sufficient to balance all the cations in *Acanthamoeba*. Interestingly, sulphur is distributed uniformly, whereas the distribution of K^+ is non-uniform; the high level of the element is observed in the cortex region of these cells. The distribution of phosphorous was similar to that of K^+. Quantitative correlation between phosphorus and K^+ as well as similar distribution of these elements suggests that in *Acanthamoeba* phosphorus is an essential anion which, being non-uniformly distributed in the cell, determines also a non-uniform distribution of K^+ (Sobota *et al.*, 1981b; 1984).

5.2.11. Polyamine

Polyamines are organic compounds that have two or more primary amino groups, such as putrescine, spermidine, and spermine. Generally, polyamines are considered as growth factors in both eukaryotic organisms as well as prokaryotic cells. The inhibition of polyamines results in amoebistatic effects, i.e., inhibition of growth of *Acanthamoeba*.

The presence of five polyamines is described in *Acanthamoeba*, including 1,3-diaminopropane, spermidine, spermine, norspermidine, and putrescine. The trophozoites of *Acanthamoeba* exhibited high concentration of 1,3-diaminopropane as the major polyamine (Poulin *et al.,* 1984; Kishore *et al.,* 1993).

Putrescine is synthesized at low levels in *Acanthamoeba* by decarboxylation of ornithine (ornithine is synthesized from arginine or citrulline) with the action of ornithine decarboxylase. Next, putrescine attacks S-adenosyl-methionine (involved in the polyamine biosynthesis) and converts it to spermidine. Spermidine in turn attacks another S-adenosyl-methionine and converts it to spermine and also leads to the synthesis of diaminopropane. To this end, *Acanthamoeba* contains ornithine decarboxylase, arginase, and urease, however arginine decarboxylase is absent (Kim *et al.,* 1987).

As indicated above, the polyamine found in the greatest concentration in the growing *Acanthamoeba* is 1,3-diaminopropane, followed by spermidine and a low level of putrescine. Later studies determined the presence of polyamine metabolites in *Acanthamoeba*. To this end, the presence of N8-acetyl-spermidine, acetyl-spermine and acetyl-cadaverine is described (Zhu *et al.,* 1989). Furthermore, it is shown that *Acanthamoeba* contains polyamine oxidase that plays a key role in the catabolism of polyamines and regeneration of spermidine and putrescine. The *Acanthamoeba* polyamine oxidase acts preferentially on N8-acetyl-spermidine as the substrate while N1-acetyl-spermidine was a poor substrate. The free polyamines did not serve as a good substrate for this enzyme. The *Acanthamoeba* polyamine oxidase yielded diaminopropane as the end product of catabolism, which is likely to be involved in the biosynthesis of this unusual polyamine present in large amounts in this organism (Shukla *et al.,* 1992).

In addition, the intracellular levels of S-adenosyl-methionine metabolites in *Acanthamoeba* demonstrated the presence of S-adenosyl-methionine and S-adenosyl-homocysteine (Zhu *et al.,* 1989). The enzyme, S-adenosyl-methionine decarboxylase (approximate molecular weight of 88.8 kDa) has been isolated from *Acanthamoeba* (Hugo and Byers, 1993). This enzyme is involved in the methyl group transfers and also known to catalyze the conversion of S-adenosyl-methionine (involved in polyamine biosynthesis) to S-adenosyl-5'-3-methylpropylamine. This suggests that S-adenosyl-methionine decarboxylase may be involved in a variety of cell methylation reactions in *Acanthamoeba* by causing accumulation of its substrate, S-adenosyl-methionine and could arrest the polyamine synthesis by blocking production of decarboxylated S-adnosyl-methionine needed for spermidine synthesis. Notably, the *Acanthamoeba* S-adenosyl-methionine decarboxylase is strongly activated by putrescine. The enzymatic properties of this enzyme suggest that it has characteristics intermediate between the bacterial and the eukaryotic S-adenosyl-methionine decarboxylase (Gupta *et al.*, 1987).

5.2.12. Adenylate cyclase

Acanthamoeba contains the endoplasmic reticulum localized enzyme, adenylate cyclase. The adenylate cyclase catalyzes the conversion of the adenosine-5'-triphosphate (ATP) to 3',5'-cyclic adenosine monophosphate (cyclic AMP) and pyrophosphate. The cyclic AMP acts as a second messenger and plays a key role in the intracellular signal transduction pathway. In contrast to the mammalian form, the activity of *Acanthamoeba* adenylate cyclase is inhibited by sodium fluoride and is unaffected by guanosine-5'-triphosphate (GTP), adenosine, epinephrine, prostaglandin E1, propranolol, and meclofenamic acid. These findings suggest that the adenylate cyclase of *Acanthamoeba* is structurally different from that of most mammalian cells (Chlapowshi and Butcher, 1986).

5.2.13. Purine and pyrimidine

Purines and pyrimidines make up the two groups of nitrogenous bases. These bases make up a crucial part of both deoxyribonucleotides and ribonucleotides, and the basis for the universal genetic code, i.e., DNA and RNA.

Acanthamoeba is capable of interconverting purine nucleotides. In addition, it has been shown to contain deaminase, phosphorylase, kinase, phosphoribosyltransferase and 5'-nucleotidase activities towards a number of purine compounds. The results of both the incorporation studies and the enzyme analysis suggested that hypoxanthine is of central importance in the parasite's purine metabolism (Hassan and Coombs, 1987).

Acanthamoeba utilizes formate and glycine as precursors for the purine biosynthesis and to grow in a medium devoid of any source of pre-formed purine compounds. *Acanthamoeba* also incorporates into its nucleic acids radiolabel derived from adenosine and guanosine indicating the presence of purine salvage pathway. Thus, the presence in *Acanthamoeba* of active *de novo* purine biosynthesis in addition to a purine salvage pathway may be a general characteristic of free-living protozoa, since parasitic protozoa are devoid of *de novo* purine biosynthesis (Ogbunude and Baer 1993).

5.3. Mitochondria

The chemical reactions which produce energy occur in the mitochondrion (plural mitochondria) and the energy is stored as ATP. Glucose and oxygen are used to produce ATP, carbon dioxide and water. Collectively these reactions are called aerobic respiration. The ATP consists of adenosine (an adenine ring and a ribose sugar), and three phosphate groups (triphosphate). The structure of the mitochondrion is characterized by a double membrane which creates two areas within the organelle. The area between the membranes houses the enzymes of the Kreb's cycle and is called the matrix. The internal membranes of the mitochondria form internal compartments called the cristae, and these contain the enzymes of the electron transport system.

The *Acanthamoeba* trophozoite possesses large numbers of mitochondria, generating the energy required for the metabolic activities involved in feeding, as well as movement, reproduction and other cellular functions. The mitochondria are in either of two forms: elongate or dumbbell-shaped and spherical (Bowers and Korn, 1968). The cristae are formed by branching tubular extensions of the inner mitochondrial membrane and are about 600 Å in diameter (Bowers and Korn, 1968). *Acanthamoeba* contains *a*-, *b*- and *c*-type cytochromes (discussed in section B8), which are not dissimilar to those reported for other mammalian systems (Edwards *et al.,* 1977). These cytochromes appear to be mainly due to the contribution of mitochondrial components; the only possible contribution from microsomal cytochromes is in the *b*-region (Edwards *et al.,* 1977). The *a*-type cytochromes appear unusual with two discrete maxima at 441 nm and at 449 nm (Edwards *et al.,* 1977). An ATP-sensitive potassium channel is present in the mitochondrial inner membrane of *Acanthamoeba*. This is similar to mammalian mitochondria containing K^+ ion transport system (Kicinska *et al.,* 2007).

6. Nuclear genome

Using pulsed field gel electrophoresis to resolve chromosome-sized DNA molecules, an *Acanthamoeba* isolate belonging to the T4 genotype was shown to contain about 20 small chromosomes ranging in size from 2.2 x 10^5 bp to greater than 2 x 10^6 bp (Rimm *et al.,* 1988). However, the actual number of nuclear chromosomes remains uncertain. Later studies described that the *Acanthamoeba* nuclear genome is polyploidy. The trophozoites contain cellular, nuclear and mitochondrial DNA with the nuclear DNA comprising 80 – 85% of the total DNA. In addition, the cytoplasmic non-mitochondrial DNA has been reported (Ito *et al.,* 1969), but its origin is not known. The total cellular DNA ranges between 1 and 2 pg for a single-cell uninucleate *Acanthamoeba* during the exponential phase (Byers *et al.,* 1990). The measurements of the nuclear DNA content (*Acanthamoeba castellanii* Neff strain, belonging to the T4 genotype) showed a total DNA content of 10^9 bp. The measurement of kinetic complexity suggested a haploid genome size of approximately around 4 x 10^7 bp (Byers *et al.,* 1990). Pulse field gel electrophoresis suggests a genome of approximately 3.3 x 10^7 bp, which expresses more than 5,000 transcripts. For comparison, the haploid genome size of *Saccharomyces* is approximately 2 x 10^7 bp, *Plasmodium yoelii* 2.5 x 10^7 bp, *Aspergillus fumigatus* 3.0 x 10^7 bp, *Coccidoides immitis* 2.8 x 10^7 bp, *Cryptococcus neoformans* 1.9 x 10^7 bp, *Tetrahymena thermophila* 1.8 x 10^7 bp, *Trichomonas vaginalis* ~7.5 x 10^7 bp, *Entamoeba histolytica* 2.0 x 10^7 bp, and *Dictyostelium discoideum* is approximately 5 x 10^7 bp (reviewed in Byers *et al.,* 1990).

A sequence survey of approximately 0.5X coverage of the genome of *Acanthamoeba* was recently determined. The data predicted that *Acanthamoeba castellanii* isolate belonging to the T4 genotype exhibits a greater biosynthetic capacity than other amoeba including, *Dictyostelium discoideum* or *Entamoeba histolytica*, providing an explanation for the ability of *Acanthamoeba* to inhabit a diversity of environments (Anderson *et al.*, 2005). In an analysis of 200 genes, it was calculated that there are on average 3 exons per gene (for comparison, *Entamoeba histolytica* possess 1.3 exons per gene, and *Dictyostelium discoideum* possess 2.3 exons per gene). The data obtained is summarized below:

- **Small RNAs**: 118 potential tRNA genes were identified. These genes encode 45 distinct species of the cytoplasmic tRNA (including separate initiator and elongator $tRNA^{MET}$) that account for all codons except CGG (Arg), UUA/UUG (Leu) and UGU/UGC (Cys). In addition, 5S rRNA functional genes (well below 100 genes) are reported.
- **Core metabolism**: the genome sequence analysis predicted a broad biosynthetic capacity thus increasing the diversity of its food source. Unlike, *Entamoeba histolytica* that can not synthesize most amino acids, *Acanthamoeba* exhibits impressive ability to synthesize essential nutrients. Genes were identified that are involved in the synthesis of threonine, tryptophan, methionine, serine, cysteine, histidine, arginine, chorismate, heme, riboflavin, pantothenate, molybdopterin, purines, pyrimidines, thymidylate, deoxynucleotides and folate is also present, suggesting that *Acanthamoeba* can obtain all of its required nutrients from a variety of sources. In addition, the *Acanthamoeba* genome is shown to possess polyhydroxybutyrate depolymerase. Polyhydroxybutyrate is produced and stored in bacteria in significant amounts, where it is degraded to acetyl-CoA, which enters the Krebs cycle. The fact that *Acanthamoeba* possesses polyhydroxybutyrate dpolymerase suggests that it can utilize polyhydroxybutyrate as a food source during bacterial grazing. The *Acanthamoeba* genome also possesses alginate lyase. Alginate is produced by some bacteria (e.g., *Pseudomonas* spp.) and algae as a biofilm matrix. The presence of alginate lyase in *Acanthamoeba* suggests that it can breakdown biofilms to release bacteria to support its growth and/or amoeba can utilize alginate directly as a food source.
- **Phagocytosis**: *Acanthamoeba* uses phagocytosis to obtain food and if the target is human cell, this may also result in infection. To this end, *Acanthamoeba* genome is shown to possess superoxide-generating NADPH oxidase and lysozyme, which are utilized in the lysosomal pathway, and play an active role in *Acanthamoeba* phagocytosis.
- **Proteases**: In the available genome, genes for serine, cystine and metalloproteases are identified.
- **Signal transduction**: *Acanthamoeba* genome possesses many serine/threonine protein kinases, tyrosine protein kinases, tyrosine protein phosphatases, as well as receptor kinases. Notably, 21 histidine kinases and 11 response regulator receiver domains are identified in the genome. These findings suggested the capacity of *Acanthamoeba* to sense and interact with the environment and respond to the external signals. Generally in eukaryotes, histidine kinases are involved in the development or detection of stressful conditions, thus histidine kinases may play an important in the differentiation of *Acanthamoeba*.

Table 5. Genes identified in *Acanthamoeba* mitochondrial DNA (taken from Burger *et al.*, (1995) J Mol Biol 245, 522–537. The number of genes are indicated in bracket.

Ribosomal RNA (2)*
- large subunit (rnl) and small subunit (rns)

Transfer RNA (16)*

Electron transport / oxidative phosphorylation (16)*

Respiratory chain (13)*
- NADH dehydrogenase (*nad1, nad2, nad3, nad4, nad4L, nad5, nad6, nad7, nad9, nad11*)
- Apocytochrome *b* (*cob*)
- Cytochrome oxidase (*cox1/2, cox3*)
- ATP synthase complex (3)*
  - F_o-ATPase (*atp6, atp9*)
  - F_1-ATPase (*atp1*)

Ribosomal protein (16)*
- Small subunit (10)*: *rps2, rps3, rps4, rps7, rps8, rps12, rps13, rps14, rps19*
- Large subunit (6)*: *rpl2, rpl5, rpl6, rpl11, rpl14, rpl16*

Conserved ORFs of unknown function (2)*
- *orf142* (= *ymf19*;Aca), homologous to *orfB* of liverwort and angiosperm mtDNA
- *orf124* (= *ymf39*;Aca), homologous to *orf25* of liverwort and angiosperm mtDNA

ORFs unique to *Acanthamoeba* mtDNA (3)*
- *orf83* (= *ymf49*), *orf115* (= *ymf50*), *orf349* (= *ymf51*)

Group I intronic ORFs (3)*
- i1-orf (= *ymf46*), i2-orf (= *ymf47*), i3-orf (= *ymf48*) (in the first, second, and third introns, respectively, of *rnl*)

*Number of genes

- **Cellular differentiation**: In the available *Acanthamoeba* genome, at least the cellulose synthase genes were identified, which are necessary in the formation of cysts as cellulose is incorporated in the inner wall of cyst. In addition, genes for cellulose (belong to glycosyl hydrolase family 5) and cellobiosidase are identified, which can degrade cellulose to glucose, and are necessary for the emergence of the trophozoite from the cyst stage.
- Stress tolerance: *Acanthamoeba* possesses trehalose-6-phosphate synthase. Trehalose is known to be involved in protection from harsh conditions such as desiccation, osmotic stress, extremes of temperatures and pH, confirming the ability of *Acanthamoeba* to survive harsh conditions.

7. Mitochondrial genome

Acanthamoeba mitochondrial DNA was originally calculated to be about 3.4 x 10^7 daltons, from the measurements of renaturation kinetics, sedimentation coefficient and electron micrographs (that constitute 10 to 20% of the total cellular DNA (Hettiarachchy and Jones, 1974). In comparison, it was shown that the mitochondrial DNA in other organisms is 3.0 – 3.5 x 10^7 Daltons in *Tetrahymena pyriformis*, 5 x 10^7 Daltons for yeast, and 7 x 10^7 Daltons for higher plants.

Later, the complete primary sequence of the *Acanthamoeba* mitochondrial genome was determined (Burger *et al.,* 1995). The genome size of mitochondrial DNA from *Acanthamoeba castellanii* belonging to the T4 genotype is now known to be a single circular linkage group of 41,591 bp. The A + T content calculated from the primary sequence was 70.6%. Genes identified in *Acanthamoeba* isolate belonging to the T4 genotype mitochondrial DNA are shown in the Table 5.

The *Acanthamoeba* mitochondrial DNA contains genes for two rRNAs (a large subunit and a small subunit), 16 tRNAs and 33 proteins, as well as eight open reading frames (ORFs) of undefined function. Two of the ORFs have homologs in other organisms, three are unique to the *Acanthamoeba* mitochondrial genome, and three are free-standing intronic ORFs (i.e. not contiguous with the upstream exons) that may encode site-specific endonucleases. In addition, the *Acanthamoeba* mitochondrial genome specifies 12 respiratory proteins whose genes are found in animal mitochondrial DNA. The *Acanthamoeba* mitochondrial DNA contains ATP synthase subunit 9 gene, i.e., *atp9* (found in fungal and plants but not animal mitochondrial DNA) but not ATP synthase subunit 8 gene, *atp8* (present in the mitochondrial DNA of the vertebrate animals, fungi, and plants). Furthermore, the *Acanthamoeba* mitochondrial genome contains a single continuous ORF (*cox1/2*) encoding subunits 1 and 2 of cytochrome oxidase (COX1 and COX2) of the respiratory chain.

The *Acanthamoeba* mitochondrial DNA encodes 16 tRNA genes, ten of which flank the large ribosomal subunit gene, *rnl* in two clusters of five genes each (cluster 1 upstream and cluster 2 downstream). The tRNA genes specific for 13 amino acids have been identified in the *Acanthamoeba* mitochondrial DNA, including two each for isoleucine (Ile) and leucine (Leu). In contrast, *Prototheca wickerhamii* mitochondrial DNA encodes 26 tRNAs, a set sufficient to support mitochondrial protein synthesis in this organism. Presumably the missing tRNAs are imported from the cytosol, as is also the case in certain other mitochondrial systems, including those of plants, ciliate, other protozoa, and trypanosomatid flagellates.

The *Acanthamoeba* mitochondrial large subunit ribosomal gene, *rnl* and small subunit ribosomal gene, *rns* encode conventional, continuous long subunit and small subunit rRNAs, with a strong structural resemblance to eubacterial 23S and 16S rRNAs. Notably, neither 5S rRNA gene, nor 5S rRNA itself has been identified in the *Acanthamoeba* mitochondria. The organization of the rRNA genes is unusual, with the small subunit ribosomal gene, *rns* downstream of large subunit ribosomal gene, *rnl* in the direction of transcription and the two separated by a cluster of five tRNA genes, an arrangement highly reminiscent of that in the mitochondrial DNA of the ascomycete fungus, *Schizosaccharomyces pombe*.

The NADH dehydrogenase subunit 2 gene, *nad2* in *Acanthamoeba* mitochondrial DNA encodes a protein that is approximately 130 to 150 amino acid residues longer at its *N*-terminus than the homologous protein specified by the animal mitochondrial DNA. Furthermore, compared to both the animal and the fungal mitochondrial DNA, the *Acanthamoeba* mitochondrial DNA encodes extra NADH dehydrogenase and ATP synthase genes. In most other organisms, these additional genes are localized in the nuclear DNA. The *Acanthamoeba* mitochondrial DNA encodes a number of ribosomal protein genes.

During eukaryotic evolution, many of the bacterial-type genes, including those for ribosomal proteins, have been transferred from the mitochondrion to the nucleus. Indeed in animals, all ribosomal protein genes are nuclear-located, while in plants and certain protists (like *Acanthamoeba*), some are still in the mitochondrion. For example, the ribosome of *Escherichia coli* has about 22 proteins in the small subunit (labelled S1 to S22) and 34 proteins in the large subunit (L1 to L36). However, the *Acanthamoeba* mitochondrial ribosomal proteins are predicted to be substantially longer than their counterparts in other systems, having either *N*-terminal (S4, S7) or *C*-terminal (S2) extensions or internal inserts (S3). In these cases, it is possible that the post-translational processing reduced the size of the final mature protein. The *Acanthamoeba* mitochondrial DNA encodes one ribosomal protein (L11) that has not previously been reported in other mitochondrial DNA.

The content of identified genes in the *Acanthamoeba* mitochondrial DNA closely matches that of the 55,300 bp mitochondrial DNA of the heterotrophic chlorophycean alga, *Prototheca wickerhamii*, and also that of the 186,600 bp liverwort (*Marchantia polymorpha*) mitochondrial genome. A notable feature of these three mitochondrial DNAs is the high content and similar array of ribosomal protein genes, as well as the presence of two distinctive ORFs, *orf25* and *orfB*. In contrast, the gene content is considerably reduced in the much smaller (15,800 bp) mitochondrial genome of another chlorophycean alga, *Chlamydomonas reinhardtii*, which like those of characterized animal and fungal mitochondrial DNA, lacks several electron transport / oxidative phosphorylation genes and does not encode any ribosomal proteins.

8. Motility

Generally, the cell shape is maintained due to their surface elasticity, which arises mainly from an actin-rich cytoplasmic cortex. However, during locomotion, phagocytosis or fission, cells become partially fluid-like. Like in other eukaryotic cells, the motility in protozoa occurs both at the cellular-level (depending on cilia, flagella, cellular appendages adapted for propulsion, or amoeboid movement) and at the sub-cellular level (e.g., highly ordered microtubules of the mitotic spindle play a key role in the separation of chromosomes during the cell division).

Acanthamoeba moves relatively fast compared to other cells, with a locomotory rate of approximately 0.8 μm per second and the movement involves the formation of a hyaline pseudopodium. The manner of

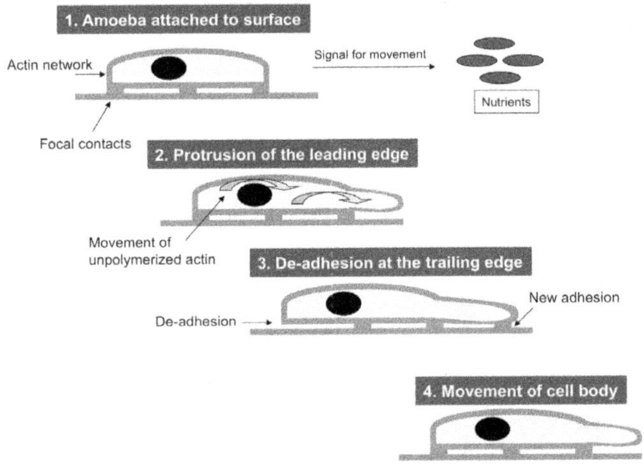

Figure 12. Steps involved in amoeboid movement including: extension of the cell membrane, formation of focal adhesions, de-adhesion and movement of the cell body.

Acanthamoeba movement is similar both at solid substrata and water-air interface. Interestingly, adhesion forces developed between *Acanthamoeba* and the water-air interface are greater than gravity, and thus the amoebae are also transported passively without detachment (Preston *et al.,* 2001). For movement, the ventral surface of *Acanthamoeba* forms a protean platform, called the "associated contact", establishing stable, focal adhesions on the substratum beneath. The gap between the cell and the substratum in the "associated contact" can be modulated by the concentration of ambient electrolyte. For example, dilution of the electrolyte from 50 mM KCl to 2 mM results in an increase in the cell-substratum gap. This results in an increased cell adhesion, and a marked decrease in the *Acanthamoeba* motility. From the "associated contact" the cytoplasm is protruded using pseudopodia-like structures (Fig. 12). *Acanthamoeba* has an outer thick, gelatinous cytoplasm called the "ectoplasm" and an inner layer of more fluid cytoplasm called the "endoplasm". As the pseudopodia-like structures extend from the cell, fluid endoplasm streams forward in the direction of extension and coagulate into the ectoplasm at the tip. At the same time in the rear of the cell, the ectoplasm changes into more fluid endoplasm and streams towards the front (direction of movement) of the cell. These changes in the actin cytoskeleton are referred to as gelation-solation (Preston and King, 1984). The gelation and solation serve two functions; (1) cell motility (i.e., cytoplasmic structures generate the force for cellular movement, and (2) serve as cytoskeletal elements by forming a solid gel.

9. Molecular basis of motility

The basis of intracellular and cellular movements is the characteristic movements of molecular structures that convert chemical energy into mechanical energy. This energy comes from the hydrolysis of ATP, which is coupled with the changes in the shape of specific proteins that mediate

Table 6. Selected motor proteins of eukaryotic cells

Molecules	Function
Microtubule-associated proteins	
• Cytoplasmic dynein	• Motion towards minus end of microtubule
• Axonemal dynein	• Activation of sliding in flagellar microtubule
• Kinesins	• Motion towards plus end of microtubule
Microfilament-associated (actin-binding) proteins	
• Myosin I	• Motion along actin filaments
• Myosin II	• Slides actin filaments

movement. Overall, these structures are part of the cytoskeleton. The cytoskeleton is a cellular "scaffolding" that is present in all cells. The cytoskelton provides the cell with structure and shape, enables cellular motion (using structures such as flagella, cilia and amoeboid motion), and plays important roles in both the intracellular transport (the movement of vesicles and organelles within the cell) and the cellular division. In general, eukaryotic cells contain three main kinds of cytoskeletal filaments, intermediate filaments, microtubules and microfilaments. The microfilaments and microtubules of the cytoskeleton provide basic scaffolding for specialized motor proteins or mechano-enzymes, which interact with the cytoskeleton to produce motion at the molecular-level (Table 6). Microtubules are abundant in cells and are normally used for a variety of intracellular movements (as well as for cilia and flagella), based on interactions between specialized motor molecules and microtubules, while microfilaments interact with the members of the myosin family of motor molecules. The interaction of these cytoskeletal systems provides cellular shape and motility.

9.1. Intermediate filaments

There is no evidence of intermediate filaments in *Acanthamoeba* as such. *Acanthamoeba* has intermediate filaments-like proteins around the nucleus but no cytoplasmic evidence. Thus the information presented below is for a general understanding for readers, as intermediate filaments are important cytoskeletal structures in human cells as well as other eukaryotes. These structures are stable filaments that are heterogeneous constituents of the cytoskeleton. They mainly function in the maintenance of cellular shape by bearing tension. It may be useful to think of intermediate filaments and microfilaments as cables, and of microtubules as cellular support beams. The intermediate filaments organize the internal tri-dimensional structure of the cell, anchoring organelles and serving as structural components of the nuclear lamina. They also participate in some cell-cell and cell-matrix junctions.

9.2. Microtubules

Microtubules-based movements are used for the intracellular transport of organelles (the mitotic spindle), as well as for cilia- and flagella-based

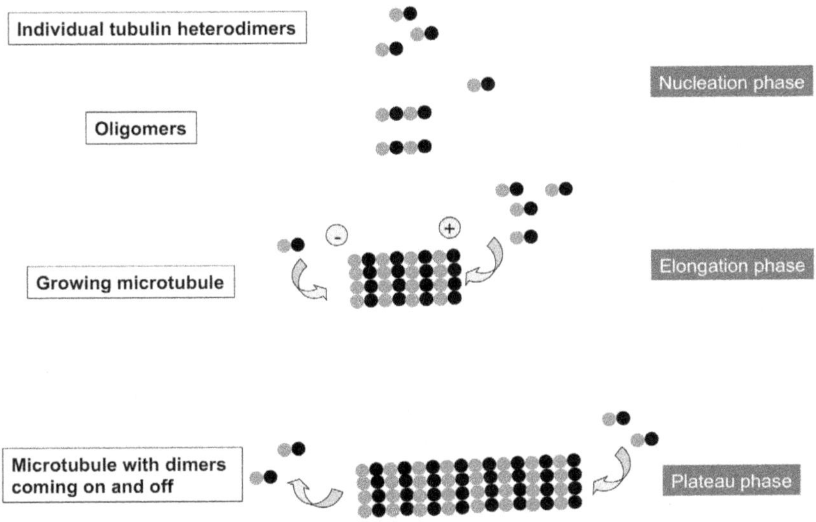

Figure 13. Microtubule assembly, *in vivo*.

motility. Microtubules provide a rigid set of tracks for the transport of a variety of membrane-enclosed organelles. Microtubules are quite straight, found throughout the cytoplasm and are about 250 Å in diameter (Bowers and Korn, 1968). They branch out from the Golgi complex where they originate. Although *Acanthamoeba* is shown to exhibit microtubule-associated movement of mitochondria and small particle, however there is little evidence that *Acanthamoeba* has organized Golgi, thus information below is for general understanding. In general, microtubules are classified into two general groups, which differ in organization and structural stability.

1. The first group is axonemal microtubules: these are highly organized, stable microtubules found in specific sub-cellular structures associated with cellular movement, including cilia, flagella and the basal bodies to which these appendages are attached. The central shaft (or axoneme) of a cilium or flagellum consists of highly ordered bundle of axonemal microtubules and associated proteins. These are only briefly described here for general understanding as *Acanthamoeba* do not possess cilia or flagella at any stage of their life cycle.
2. The second group is cytoplasmic microtubules: these are loosely organized microtubules found in the cytosol of most eukaryotic cells, including *Acanthamoeba*. They perform a variety of functions including directional movements of vesicles/organelles within the cell as well as the formation of mitotic and meiotic spindles that are essential for movement of chromosomes during reproduction. The basic subunit of the microtubule is the heterodimer protein, tubulin. It is composed of one molecule of α (alpha)-tubulin and one molecule of β (beta)-tubulin. As soon as individual α-tubulin and β-tubulin are synthesized, they bind non-covalently to each other to produce αβ-heterodimer that does not

dissociate under normal conditions. The tubulin dimers aggregate into clusters called oligomers. These oligomers serve as nuclei from which new microtubules can grow, thus this process is called "nucleation" (Fig. 13). The assembly of tubulin heterodimers occurs on both ends of the microtubule but one end grows much faster and is called the plus end, while the other is the minus end and requires the presence of guanosine-5′-triphosphate (GTP). The α-tubulin binds one GTP and β-tubulin also binds to one GTP, which is hydrolyzed to guanosine-5′-diphosphate (GDP), sometime after the heterodimer is added to a microtubule (Fig. 14).

Microtubules originate from a structure called the microtubule-organizing centre, that serves as a site at which microtubule assembly is initiated and acts as an anchor for one end of these microtubules. During interphase *Acanthamoeba* has a microtubule-organizing centre called the centrosome, centered in the Golgi complex. The centrosome is associated with two centrioles that are surrounded by a diffuse granular material known as pericentriolar material. The pericentriolar material is suggested to be the origin site for the microtubules. Notably, in ciliated or flagellar cells, centrioles are known to be involved in the formation of basal bodies (important for the formation of cilia and flagella). In non-ciliated or non-flagellar cells such as *Acanthamoeba*, centrioles may serve to recruit pericentriolar material to the centrosome. A number of drugs affect microtubule assembly, including colchicines and nocodazole. These drugs are known as anti-mitotic drugs because they disrupt the mitotic spindle of the dividing cell, blocking further progress of mitosis.

Although microtubules provide an organized set of tracks, along which organelles can move, they do not directly generate the force necessary for movement. The mechanical work needed for movement depends on the microtubule-associated motor proteins. The microtubule-associated motor proteins attach to organelles and then walk along the microtubule tracks using ATP to provide the needed energy. Furthermore, the microtubule-associated motor proteins recognize the polarity of the microtubule with each microtubule-associated motor protein having a preferred direction of movement. Although, it has not been shown in *Acanthamoeba*, at present there are two

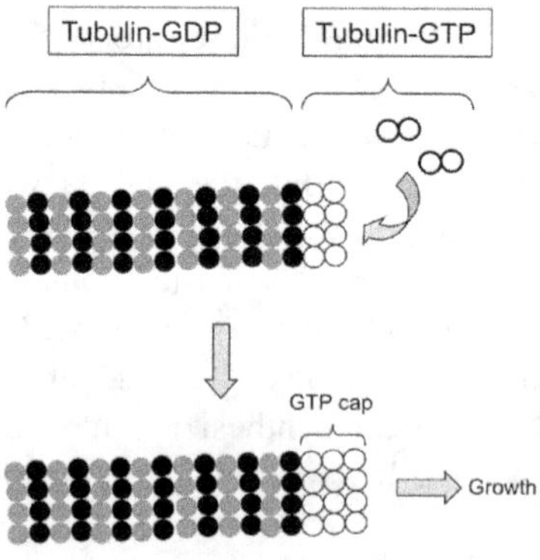

Figure 14. Role of GTP cap in microtubule assembly. If tubulin concentration is high, tubulin-GTP is added to microtubule tip faster than the incorporated GTP can be hydrolyzed. The resulting GTP cap stabilizes the microtubule tip and promotes further growth. At low concentration, the rate of growth decreases, thereby allowing GTP hydrolysis but creating unstable tip that favours depolymerization.

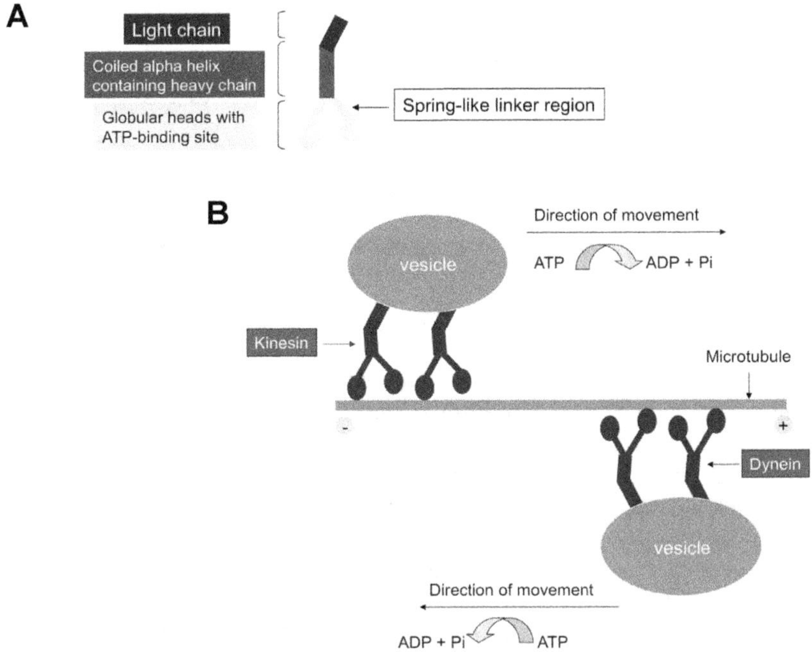

Figure 15. Structure of kinesin (**A**) and microtubule-based motility (**B**).

major families of the microtubule-associated motor proteins in eukaryotic cells, kinesins and dyneins.

1. Kinesins: consists of three parts, a globular head that attaches to microtubules and is involved in the ATP hydrolysis, a coiled-helical region and a light-chain region that is involved in attaching the kinesins to other proteins or organelles. Kinesins move along microtubules in 8 nm steps, i.e., one globular heads moves forward to make an attachment to a new β-tubulin, followed by detachment of the trailing globular head, which is then able to make a new attachment. This movement is coupled with the ATP hydrolysis to ADP at specific sites in the globular head (Fig. 15A). Since their discovery, many members of kinesins have been identified with varying structures and functions.
2. Dyneins: consists of two basic types, cytoplasmic dyneins and axonemal dyneins (Fig. 15B). Cytoplasmic dyneins contain two heavy chains that interact with microtubule, three intermediate chains, and four light chains. However unlike kinesins they can not bind to vesicles on their own and require additional proteins, a complex called "dynactin". Axonemal dyneins play an important role in the function of cilia and flagella. Both cilia and flagella have a common structure consisting of an axoneme, which is connected to a basal body and surrounded by cell membranes. Notably, *Acanthamoeba* does not possess cilia or flagella at any stage of their life cycle.

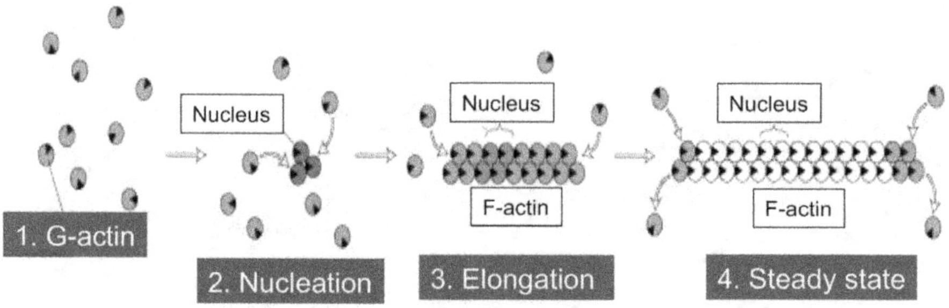

Figure 16. Steps involved in actin polymerization.

The vesicles moving from the endoplasmic reticulum to the Golgi complex are shown to be carried by dynein-like motors that move towards the minus end of the microtubule (as indicated above, the minus ends are anchored in the centrosome). These vesicles fuse together and become part of the flattened membrane stacks that make up one side of the Golgi complex. Once in the Golgi complex, proteins move through the stacks of the Golgi membranes. The finished proteins emerge, packed in the vesicles, from the other side of the Golgi complex. Then, the plus end directed microtubule-associated motor proteins (kinesins) carry the finished vesicles away from the Golgi complex towards the periphery of the cell. Thus microtubule-associated motor proteins and microtubules allow the two-way flow of the vesicle traffic to and from the Golgi complex.

9.3. Microfilaments

Acanthamoeba has long been studied as a model eukaryotic cell with special emphasis on the actin cytoskeleton. Much of what is known about actin comes from the study of *Acanthamoeba*. The movement of other molecules and cellular components also occur along another major filament system, the microfilaments. This filament is composed of two inter-wound "actin" chains, also referred to as "actin microfilament". Actin microfilaments are most concentrated just beneath the cell membrane, and are responsible for resisting tension and maintaining cellular shape, forming cytoplasmatic protrutions such as pseudopodia observed in the amoeboid motion. In addition, actin microfilaments also participate in the cell-to-cell or cell-to-matrix junctions, cytokinesis, and cytoplasmic streaming in most cells.Thus, this system is the basis of cellular movements in *Acanthamoeba* and is discussed in detail here. Similar to microtubules, ATP-dependent motors exert force on actin microfilaments. Here, actin microfilaments are described first, and then a discussion of mechano-enzymes (motor proteins) required for generating force for *Acanthamoeba* movements.

9.3.1. Actin polymerization to form actin microfilaments

Actin filaments are involved in various types of movements including well-known muscle contractions, amoeboid movements, i.e., locomotion of cells over a surface and cytoplasmic streaming. The basic unit of microfilaments is actin, which is a single polypeptide consisting of approximately

42 kDa. Once synthesized, it folds into a roughly U-shaped molecule with a central cavity that binds to ATP or ADP. Individual actin molecules are referred to as G-actin (globular actin). Like tubulin heterodimers in microtubules, G-actin molecules polymerize to form microfilaments and in this form, actin is referred to as F-actin (filamentous actin) (Fig. 16). The role of actin-based movement in *Acanthamoeba* has been shown with the finding that cytoplasm dissociated from *Acanthamoeba* retains the ability to stream and contract in the presence of ATP indicating that the force for cellular movement is generated in the cytoplasm. The soluble proteins extracted from *Acanthamoeba* under cold conditions, when warmed up to room temperature, form a gel under physiological conditions, a process known as "gelation reaction". The gelation in *Acanthamoeba* has two steps, a lag period followed by a rapid rise in viscosity and it requires 0.5 mM ATP and 1.5 mM $MgCl_2$, and is inhibited by Ca^{2+}. The optimum pH is 7.0 and temperature 25 – 30°C (MacLean-Fletcher and Pollard, 1980).

In addition to the actin, there are actin-related proteins that are broadly defined as proteins sharing significant (30 – 60%) amino acid identity with the conventional actin isoforms, which are among the most highly conserved of known proteins. These were discovered first in *Acanthamoeba*. There are three classes of actin-related proteins including, actin-related protein-1, actin-related protein-2, and actin-related protein-3. At least two actin-related proteins from the crude extracts of *Acanthamoeba* have been identified, including a 47 kDa and 44 kDa that interact with proteins of the actin cytoskeleton, in particular with profilin and conventional actin and are present in the cortex of *Acanthamoeba* (Machesky *et al.,* 1994). Actin-related protein-2 is localized in the spikes, lamellae, amoebastomes, i.e., cup-like projections and the cytoplasm of *Acanthamoeba*, while the actin-related protein-3 is localized in the amoebastomes and the cytoplasm of *Acanthamoeba* (Kelleher *et al.,* 1994).

9.3.2. Regulators of actin polymerization

Like most motile cells, *Acanthamoeba* contains a large pool of unpolymerized actin despite intracellular ionic conditions that favour actin polymerization. This is because of the interaction of actin with one or more of the regulatory proteins. Cells control the polymerizations at various steps including nucleation, elongation and depolymerization. This requires the function of a variety of molecules including, various actin-binding proteins as well as inositol phospholipids, which are found on the inner leaflet of the plasma membrane, and the small regulatory GTP-binding proteins known as Rho, Rac and Cdc42, that regulate and modify the function of the actin, thus called "regulators" of polymerization and depolymerization and are described below.

9.3.2.1. Actin-binding proteins

If the concentration of ATP-bound G-actin is high, microfilaments will assemble until G-actin is limiting. However, free G-actin is not always available and may be bound by actin-binding proteins that block nucleation.

Some proteins bind to the end of the growing tip and block further incorporation of the actin monomers, thus inhibiting elongation and are called "capping proteins", while others induce depolymerization of the filamentous F-actin, thus severing existing actin filaments. Overall, these proteins are known as "actin-depolymerizing factors". In addition, other proteins promote the assembly of, or cross-link actin filaments and can regulate the polymerization as well as support the gel (Pollard, 1976). Several cytoplasmic proteins have been identified in *Acanthamoeba* including myosin (discussed later), myosin cofactor, profilin-I, profilin-II, actophorin, and actobindin, which bind to monomeric G-actin.

Of these, profilin is one of the best studied proteins. Profilin primarily binds to the actin monomers and inhibits their assembly, but also inhibits the rate of filament elongation and reduces the concentration of F-actin at steady state (Tobacman and Korn, 1982). Thus primarily it regulates actin polymerization by inhibiting nucleation, a first event in polymerization and prolongs the lag phase by inhibiting the actin polymerization. In addition, the elongation rates at both the barbed and the pointed ends of growing filaments are inhibited by profilin. These effects of profilin can be explained by the formation of a 1:1 complex with actin with a dissociation constant of 1 to 4 μM. The $MgCl_2$ strongly inhibits these effects of profilin most likely by binding to the high affinity divalent cation site on the actin (Tseng and Pollard, 1982). An actin monomer bound to profilin can not participate in nucleation or add to the pointed end of an actin filament. The molar concentration of profilin in *Acanthamoeba* extracts is about equal to the molar concentration of actin that remains non-polymerized under polymerizing conditions. Therefore, profilin is one of several proteins that regulate the organizational state of actin in the cytoplasm of *Acanthamoeba* (Reichstein and Korn, 1979). A seven-fold molar excess of *Acanthamoeba* profilin increases the rate of exchange of ATP bound to G-actin with ATP in solution about 17-fold. Profilin binding to actin is rapid and reversible and profilin does not interact with ATP (Mockrin and Korn, 1980). *Acanthamoeba* profilin has a significant lower molecular weight, i.e., 12 kDa compared with 15.22 kDa of vertebrate profilin and a different amino acid composition (Reichstein and Korn, 1979; Mockrin and Korn, 1980). For example, *Acanthamoeba* profilin does not contain cysteine and only one methionine (mammalian profilin contains 3 amino acids of cysteine and 6 amino acids of methionine). In contrast, *Acanthamoeba* profilin contains 1 residue of trimethyllysine (at position 103), a modification which has not yet been detected in mammalian profilin. *Acanthamoeba* profilin contains more alanine and less leucine and lysine than has been reported for the mammalian profilin (Reichstein and Korn, 1979). *Acanthamoeba* profilin has a pI of about 6.4 under denaturing conditions and contains about 28% α-helix and 29% β structure. *Acanthamoeba* has three isoforms of profilin, i.e., profilin-IA, profilin-IB, and profilin-II) that are present throughout the cytoplasm. *Acanthamoeba* profilin-I and profilin-II consists of 125 amino acids, have a blocked *N*-terminus (Ampe *et al.,* 1988).

Another protein is α-actinin, which is a homodimer composed of two polypeptides of approximate molecular weight of 90 kDa, each appearing as a rod of 30 – 40 nm long with globular regions at each end. Each polypeptide has a highly conserved actin-binding site located near the *N*-terminus of the polypeptide chain. Each isoform is able to cross-link F-actin into networks at low concentrations, while at high concentrations, α-actinin and actin filaments form bundles (Pollard *et al.,* 1986; Wachsstock *et al.,* 1993). The physical properties of isotropic networks of cross-linked actin filaments are different from the gel of bundles: the network behaves like a solid because each actin filament is part of a single structure that encompasses all the filaments. Bundles of filaments behave more like a very viscous fluid because each bundle, while very long and stiff, can slip past other bundles.

Additional actin-binding proteins are actobindin (i.e., a monomer of 9.8 kDa) and actophorin (i.e., a monomer of 15 kDa, an actin-depolymerizing factor that severs F-actin filaments, a property not possessed by actobindin) (Lambooy and Korn, 1986; Cooper *et al.,* 1986). Actobindin (comprising of 88 amino acids), with bivalent affinity (binds two monomeric G-actin) and it is a potent inhibitor of early phase of actin polymerization by binding to *Acanthamoeba* G-actin, thus reduces the concentration of F-actin at steady state and inhibits the rate of filament elongation (Vandekerckhove *et al.,* 1990). Actobindin inhibits the nucleation but not the elongation phase of the actin polymerization. Actophorin is present in the amoeba at a concentration of 20 μM. Actophorin differs from all of the other actin-binding proteins in *Acanthamoeba*, because it appears to have 2 distinct effects on actin. First, actophorin forms a complex with actin, thus acting like profilin in blocking spontaneous polymerization and secondly, severing actin filaments at steady state by shortening and fragmenting pre-formed actin filaments (Cooper *et al.,* 1986; Maciver *et al.,* 1991). Actophorin decreases the viscosity of actin filaments, but increases the stiffness and viscosity of mixtures of actin filaments and the cross-linking protein, α-actinin (Maciver *et al.,* 1991a). It was thought that the shortened actin filaments rotate into positions favourable for lateral interactions with each other via cross-linking protein α-actinin. This may result in the formation of actin filament aggregate into an interlocking meshwork of bundles (Maciver *et al.,* 1991a). Later studies revealed that actophorin binds to ADP-bound monomers with higher affinity, than ATP-bound monomers (Maciver and Weeds, 1994). This indicated that actophorin is less efficient at severing actin filaments in the presence of ADP compared to the ATP (Maciver and Weeds, 1994).

In addition, four small proteins are identified with approximate molecular weights of 23, 28, 32, and 38 kDa, each of which caused pure actin filaments to form gel (Maruta and Korn, 1977a). By gel filtration, it was estimated that the 23 kDa gelation factor is a monomer, while 28, 32, and 38 kDa factors are dimers in their native states. It was described that these proteins account for more than 95% of the gelation factor activity in the gel. Later the existence of another gelation protein, called GP-85 was described (Pollard, 1981). The

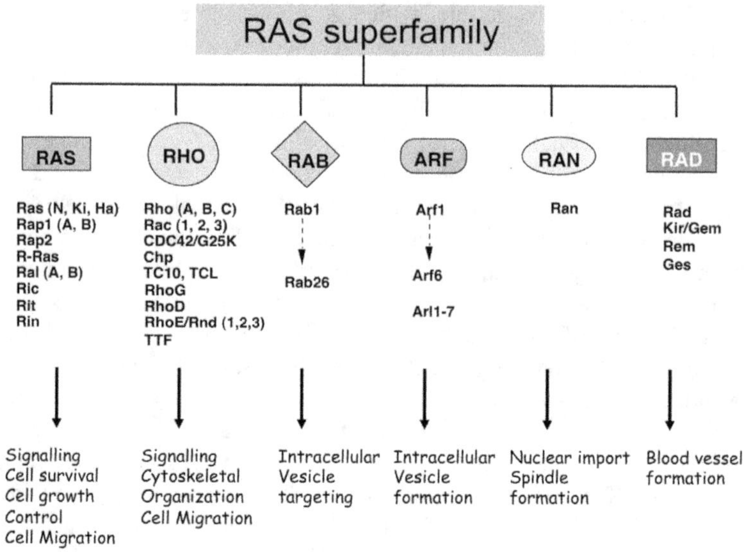

Figure 17. RAS superfamily of GTP-binding protein.

molecular weight of GP-85 was shown to vary depending on the concentration of reducing conditions. The major band is 60 kDa in 10% β-mercaptoethanol, 85 kDa in 1% β-mercaptoethanol, and 90 kDa without β-mercaptoethanol. Its actin cross-linking activity is inhibited by micromolar concentration of Ca^{2+} and promoted by ATP, KCl and $MgCl_2$. In addition, two gelation factors were isolated, a GF-46 that has both subunit and native molecular weight of approximately 46 kDa; and a gelation factor, GF-210, which is composed of 6 subunits, each with an approximate molecular weight of 35 kDa, which cross-links actin filaments in an ATP-sensitive manner (Albanasi *et al.*, 1986; 1987).

A number of drugs affect microfilament formation and assembly. For example, cytochalasins prevent the addition of new actin monomers and inhibit actin polymerization. In contrast, latrunculin A acts by sequestering actin monomers, preventing their addition to microfilament resulting in the loss of microfilaments.

In addition to the aforementioned actin-binding proteins, a coronin-like protein has been identified in *Acanthamoeba* and named "Ahcoronin". The protein is composed of 454 amino acids which contain the characteristic WD repeats of coronin and coronin-like proteins. The *C*-terminal region of the gene was also predicted to have a high tendency of forming a coiled-coil, another structural characteristic of coronin. The cDNA showed a 50% homology to mammalian coronins. Ahcoronin occurs as a single copy and expressed as a transcript of approximately 1,400 bp in *Acanthamoeba*. Ahcoronin was localized in the cells' periphery and in the leading edge consistent to that of actin, as well as observed to localize around phagocytic cups but was disassembled later during phagocytosis (Baldo *et al.*, 2005).

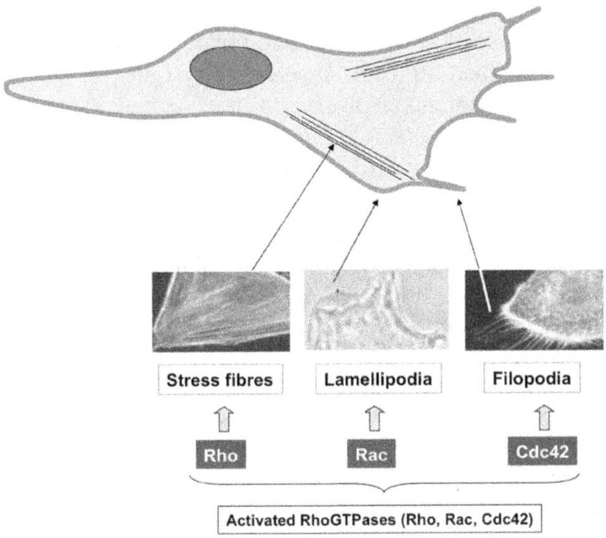

Figure 18. Activation of RhoGTPases (RhoA, Rac1, Cdc42) leads to distinct phenotypes such as stress fibre formation, lamellipodia, or filopodia formation.

9.3.2.2. Inositol phospholipids-based regulation of actin microfilaments

A derivative of inositol is phosphatidylinositol that can be phosphorylated by specific kinases in the cytoplasm to yield a variety of products including polyphosphoinositides, which then bind to actin-binding proteins. For example, phosphatidylinositol 4-monophosphate and phosphatidylinositol 4,5-biphosphate bind to profilin. This prevents hydrolysis of these lipids (i.e., phosphatidylinositol 4-monophosphate and phosphatidylinositol 4,5-biphosphate) by phospholipase C-gamma, thus inhibiting the production of second messengers, inositol triphosphate and diacylglycerol, suggesting that profilins are involved in transmembrane signaling. Furthermore, phosphatidylinositol 4,5-biphosphate regulate the activity of profilin to interact with actin by inhibiting the binding of profilins to actin, thus profilins link transmembrane signaling and regulation of the cytoskeleton. Of note, both *Acanthamoeba* profilins, profilin-I and profilin-II can bind to the phosphatidylinositol-4,5-biphosphate.

9.3.2.3. GTP-binding proteins regulate actin microfilaments

The Rho family of GTP-binding proteins (belonging to RAS superfamily; Fig. 17) are the regulators of actin polymerization and include Rho, Rac and Cdc42. Depending on the specific pathways involved, actin polymerization results in three kinds of phenotypes as follows (Fig. 18):

1. Stress fibres formation: these occur in cells that adhere very tightly to the underlying substratum and do not move well. The organized bundles of microfilaments stretch from the tail or trailing edge of the cell to the front of the cell. The GTP-binding protein involved in stress fibres formation is Rho activation. However, these are not found in *Acanthamoeba*.

2. Lamellipodia: these occur in cells with relative rapid movements. In these cells, the cell cortex, which lies immediately beneath the plasma membrane, enriched in actin is cross-linked into a gel or loosely organized microfilaments. However, at the leading edge microfilaments form highly oriented, polarized bundles with their barbed ends oriented towards the tip of the protrusion. The GTP-binding protein involved in lamellipodia formation is Rac activation.
3. Filopodia: these are similar to lamellipodia except actin is more organized. Cells control actin polymerization to regulate their movements. The GTP-binding protein involved in stress fibres formation is Cdc42 activation.

9.3.2.4. Other key regulators

Adenosine-5′-triphosphate (ATP), $MgCl_2$, and temperature are important regulators of actin polymerization (gel forms at 25°C and can be liquefied by cooling). It is important to note that these gels do not creep in response to applied stress on a time scale compatible with the most cell-surface movements. These findings can explain why cells do not collapse under the constant shear forces that often exist in tissues.

10. *Acanthamoeba* actin

Although actin is very conserved among various organisms, however, based on their sequence similarity, actins can be divided into two major groups; the muscle-specific actins (α-actins) and non-muscle actins (β-actins and γ-actins). In the case of β- and γ-actins, it is shown that these actins localize to different region of the cell and have different interactions with the actin-binding proteins. In *Acanthamoeba*, actin represents 20% of the total protein.

10.1. Properties of *Acanthamoeba* actin and its filaments

The structure of purified F-actin is show to be an elongated globular molecule about 60 Å in diameter with a pronounced asymmetric shape in perpendicular to the plane of the sheet (Aebi *et al.*, 1980; Pollard *et al.*, 1970). *Acanthamoeba* actin is similar to muscle actin in the amino acid composition, and sequence, molecular weight, viscosity of the F-actin, stoichiometry of bound nucleotide, stoichiometry of binding to heavy meromyosin and the ability to undergo polymerization to F-actin filaments. However, *Acanthamoeba* actin contain *N*-methyllysines, which is absent in muscle actin).

At low ionic strength, purified actin is monomeric (i.e., G-actin); polymerization into filaments (F-actin) occurs when the ionic strength is raised to physiological levels. Actin is a major component of the gel that forms filaments, and these filaments must bind to each other (i.e., cross-link) to form solid gel. The polymerization of *Acanthamoeba* actin and muscle actin is qualitatively similar with a rate limiting nucleation step followed by rapid polymer extension to give long helical filaments (Weihing and Korn, 1971; Gordon *et al.*, 1976). Actin filaments have been found to have discrete helical structure and appear to be constructed from two strands of actin molecules wound around one another. Later studies revealed that crystalline sheets of *Acanthamoeba* actin exist in 3 different polymorphic forms, rectangular sheets, square sheets, and cylinders and each of the 3 forms is constructed from common basic lattices associated in different ways (Aebi *et al.,* 1981). Actin filaments are present in *Acanthamoeba*, some of which are associated with the plasma membrane.

Acanthamoeba actin is only one third as effective as an activator of the muscle heavy meromyosin ATPase as compared with muscle actin. Also, other studies showed that muscle tropomyosin binds to *Acanthamoeba* actin much more weakly than muscle actin. Moreover, tropomyosin stimulates the *Acanthamoeba* actin-activated heavy meromyosin ATPase, while it inhibits muscle actin heavy meromyosin ATPase (Yang *et al.,* 1977). Although, the tropomyosin binds to seven actin monomers (both in the case of muscle actin or *Acanthamoeba* actin), however meromyosin binding to muscle actin in the presence of 2 mM ATP begins at 4 mM Mg^{2+} and is complete at about 4.75 mM Mg^{2+}, while meromyosin binding to *Acanthamoeba* actin initiates at 6 mM Mg^{2+} and reaches saturation at 8.5 mM Mg^{2+} (Yang *et al.,* 1979).

11. *Acanthamoeba* myosin

Myosins are a family of motor proteins that are responsible for the actin-based motility. Myosin molecules are composed of a head and a tail domain. The head domain binds the filamentous actin, and uses ATP hydrolysis to generate force and to "walk" along the filament towards the plus (+) end. The tail domain generally mediates interaction with cargo molecules and/or other myosin subunits.

The term "myosin" was originally used to describe a group of similar but non-identical ATPases found in striated and smooth muscle cells. They play crucial roles in force-generating processes using ATPase activities, i.e., a property that generates chemical energy, which is transmitted into the mechanical force for *Acanthamoeba* cellular functions such as cell movements, endocytosis, cell division or even chromosome segregation. Generally in the absence of F-actin, their ATPase activities can be expressed in the presence of K^+ or Ca^{2+}, but not Mg^{2+}, which is inhibitory. On the other hand, actin-activated myosin ATPase activity requires Mg^{2+} for which no other cation can substitute. There are two major types of myosins identified in *Acanthamoeba*, myosin-I and myosin-II and both are products of separate genes (Maruta *et al.,* 1979; Pollard and Korn, 1973; Pollard *et al.,* 1978). Myosin-I has one head, while myosin-II has two heads.

11.1. Structure of *Acanthamoeba* myosins

11.1.1. Myosin-I

Acanthamoeba myosin-I is a single-headed molecule, with an approximate molecular weight of 180 kDa. Myosin-I was discovered first in *Acanthamoeba*. There are three form of myosin-I (myosin-IA, myosin-IB, myosin-IB), each consisting of one heavy and one light chain. By SDS-PAGE, it was shown that *Acanthamoeba* myosin-IA consists of one 140 kDa heavy chain and a

17 kDa light chain; myosin-IB consists of one 125 kDa heavy chain and a 27 kDa light chain; and myosin-IC consists of one 130 kDa heavy chain and a 14 kDa light chain. Of note, myosin-IB is composed of a myosin globular head fused not to the typical rod-like myosin tail structure but rather to unusual glycine-, praline-, and alanine-rich *C*-terminal domain (Jung *et al.*, 1987). Furthermore, the heavy chain (125 kDa) of myosin-IB contains two F-actin-binding sites, among which one is associated with the catalytic site and second allows myosin-IB to cross-link actin filaments but has no direct effect on the catalytic activity (Brzeska *et al.*, (1988).

In comparison, the molecular weight of rabbit myosin is approximately 470 kDa and consists of two heavy chains (each chain consisting of approximate molecular weight of 200 kDa) and four light chains (two light chains of approximate molecular weight of 18 kDa and two light chains of approximate molecular weight of 20 kDa or 16 kDa). While the molecular weight of muscle myosin is 460 kDa; slime mold myosin is 460 kDa and platelet myosin is 500 kDa. Of note, generally actin is 10-fold higher than myosin I.

11.1.2. Myosin-II

Acanthamoeba myosin-II is a two-headed globular molecule with an approximate molecular weight of 400 kDa. The globular heads are joined to a coiled-coil rod-like tail. Myosin-II consists of two heavy chains (each approximately 170 kDa) and four light chains (two light chains of approximate molecular weight of 17 kDa and two light chains of approximate molecular weight of 17.5 kDa). Myosin-II has two globular heads attached to a 90 nm long tail and the heads are 15 nm apart. Myosin-II is rich in acidic residues and contains about 32 residues of cysteine, while cysteine is absent from myosin-IA.

11.2. Activation of myosin-I and myosin-II (also see section 14.6)

Actin-activated Mg^{2+}-ATPase activity of *Acanthamoeba* myosin-IA, myosin-IB, and myosin-IC are only observed when their heavy chains are phosphorylated. In contrast, actin-activated Mg^{2+}-ATPase activity of *Acanthamoeba* myosin-II is only observed when their heavy chains are dephosphorylated.

11.3. Localization of *Acanthamoeba* myosin

11.3.1. Myosin-I

Myosin-I is a class of monomeric myosins that function at the membrane to mediate motility. Myosin-I has been found in association with the plasma membranes and at the surface of organelles by the immunofluorescence assays. The plasma membrane treatment with saponin disrupted staining suggesting that myosin-I (in particular myosin-IC) was associated with the membrane phospholipids (Baines and Korn, 1990). This is shown by the binding of myosin-I to acidic phospholipids, i.e., phosphatidylserine, phosphatidylinositol 4,5-biphosphate and phosphatidylglycerol.

Later, it was shown that myosin-IA occurs almost exclusively in the cytoplasm (where it accounts for about 50% of the total myosin-I), in the actin rich cortex beneath the phagocytic cups, and in association with small cytoplasmic vesicles. Most of the cytoplasmic myosin-IA was phosphorylated (i.e., activated) (Baines *et al.*, 1994).

Myosin-IB is the predominant isoform associated with the plasma membrane, large vacuole membranes and phagocytic membranes, but only a fraction of membrane-associated myosin-IB is phosphorylated (10 – 20%). Like myosin-IA, the majority of the cytoplasmic myosin-IB is phosphorylated (60 – 100%) (Baines *et al.*, 1994). The membrane-associated phosphorylated myosin-IB was concentrated in the actively motile regions of the plasma membrane.

Myosin-IC accounts for a significant fraction of the total myosin-I associated with the plasma membrane and large vacuole membranes and is the only myosin-I isoform associated with the contractile vacuole membranes. Like myosin-IB, only a fraction of the membrane-associated myosin-IC is phosphorylated (10 – 20%). However, most of the cytoplasmic myosin-IC is phosphorylated (60 – 100%) (Baines *et al.*, 1994). Interestingly, the membrane-associated phosphorylated myosin-IC was concentrated in the membranes of contracting contractile vacuole than of the filling contractile vacuole. As the total amount of the contractile vacuole-associated myosin-IC remained constant, it must be phosphorylated at the start of the contractile vacuole contraction. This was shown with the findings that antibodies that inhibit the activity of *Acanthamoeba* myosin-IC *in vitro* interfere with the expulsion of excess water by the contractile vacuole *in vivo*, leading to overfilling of this organelle and cell lysis. Myosin-IC may be responsible for generating the force required to contract the vacuole and may also be involved in transfer of water to the contractile vacuole during refilling (Doberstein *et al.*, 1993).

Overall it can be concluded that myosin-IA may function in the cytoplasmic vesicle transport, and myosin-I-mediated cortical contraction associated with phagocytosis, myosin-IB is required in the pseudopod extension and phagocytosis / pinocytosis, and myosin-IC in contractile vacuole function (Baines and Korn, 1990; Baines *et al.*, 1992).

11.3.2. Myosin-II

Myosin-II is present throughout the cytoplasm in the form of filaments. It is most concentrated in the cortex, but is absent from thin membrane extensions such as microspikes and broad pseudopodia (Baines and Korn, 1990; Yonemura and Pollard, 1992). Myosin-II is not detected in the purified plasma membranes of *Acanthamoeba* (Miyata *et al.*, 1989).

11.4. Myosins and motility

Acanthamoeba myosin-I that was bound to substrates of the planar lipid membranes on glass, moved actin filaments at an average velocity of 0.2 μm per sec, suggesting their role in the motility (Zot *et al.*, 1992). To demonstrate the role of myosin-II in *Acanthamoeba* motility, motile cells were micro-injected with monoclonal antibodies specific for myosin-II, which led to decreased motility from 7 μm per min to approximately 0.8 μm per min (Xu *et al.,* 1997). However, levels of antibody far in excess of endogenous myosin-II concentrations could not completely block the amoeboid movement. The morphology of moving anti-myosin-II antibody-injected cells was unusual, suggesting a greater defect in the ability to retract the trailing edge of the cell rather than to extend the leading edge. The endosomes frequently disappeared from the injected cells. The inability of antibodies to myosin-II to complete inhibit *Acanthamoeba* motility suggests that although myosin-II probably plays a role in these motilities, the cell either routinely uses or can draw upon another cytoplasmic motor to maintain locomotion, organelle movement, contractile vacuole activity, and endocytosis.

11.5. Polymerization of *Acanthamoeba* myosin

Myosin-I are non-filamentous, single headed, globular molecules. Unlike myosin-II, myosin-I is soluble at low (physiological) ionic strength and does not form filaments. In contrast, monomers of myosin-II polymerize to form filaments (Pollard and Korn, 1973; Pollard *et al.,* 1978). It is characteristic of myosin-II monomers that they spontaneously self-assemble into bipolar filaments when transferred from a high ionic strength environment (e.g., more than 400 mM KCl) to low ionic strength (less than 150 mM KCl). The polymerization of myosin into filamentous structures is a multistep process and can be grouped into three steps. First, *Acanthamoeba* myosin-II monomers form into minifilaments (containing 8 to 16 molecules each). Minifilaments are short enough that each molecule overlaps in an anti-parallel fashion with at least one other molecule in the filament, thus forming anti-parallel dimer. In the second elongation step, addition of myosin molecules to an existing anti-parallel structure occurs such that the filament length increases and involves only parallel interactions between the incoming molecules and the molecules already in the filament, and finally completed to octameric minifilaments. Minifilaments are the smallest, highly homogenous bipolar structures formed by myosins. Unlike vertebrate minifilaments, which elongate into synthetic filaments, *Acanthamoeba* minifilaments do not elongate into larger structures (Sinard *et al.,* 1989). Myosin-II forms filaments of two different sizes. Thin bipolar filaments 7 nm wide and 200 nm long, consist of 16 myosin-II molecules and thick bipolar filaments of variable width (14 – 19 nm). Thick filaments are formed rapidly and reversibly in the presence of 6 – 10 mM $MgCl_2$. Acid pH also favours thick filament formation but they are much more complex than the thin ones (Pollard, 1982).

11.6. ATPase activity of myosin-I and myosin-II

11.6.1. Myosin-I

Myosin-I contain ATPases that are capable of supporting contractile and motile-like activities. Myosin-IA, myosin-IB, and myosin-IC have K^+-EDTA-ATPase and Ca^{2+}-ATPase activities but high Mg^{2+}-ATPase activity only in the presence of actin. The actin-activated Mg^{2+}-ATPase activities of *Acanthamoeba* myosin-IA, IB, IC are expressed only when a single site in their heavy chains is phosphorylated by a myosin-I heavy chain-specific kinase (*Acanthamoeba* myosin-I heavy chain kinase is a monomeric protein consisting of a single 97 kDa polypeptide chain). Notably, there only is one phosphorylation site on the heavy chain of myosin-IA, IB, IC, and that phosphorylation is required for actin activation of the Mg^{2+}-ATPase activity of myosin-I (Maruta and Korn, 1981). Phosphorylation occurs at threonine ~300 residue in myosin-IA, at Ser-315 in the myosin-IB, and at Ser-311 in myosin-IC (Brezeska *et al.*, 1989; Matura and Korn, 1977; Maruta *et al.*, 1979; Pollard and Korn, 1973; Pollard *et al.*, 1978). Myosin-IA, IB, IC are capable of cross-linking F-actin and possess two independent actin-binding sites. The catalytic and phosphorylation sites of *Acanthamoeba* myosin-IA and myosin-IB are on the heavy chain, while phosphorylation of the light chain is not observed. Actin activated Mg^{2+}-ATPase activity is lost when phosphorylated *Acanthamoeba* myosin-I is dephosphorylated by phosphatases. Actin-activated Mg^{2+}-ATPase activity is lost when phosphorylated *Acanthamoeba* myosin-I is dephosphorylated by phosphatases. However, phosphorylation and dephosphorylation have no effect on K^+-EDTA-ATPase and Ca^{2+}-ATPase activities of *Acanthamoeba* myosin-I (Maruta and Korn, 1977).

11.6.2. Myosin-II

Myosin-II is found to have both the Ca^{2+}-ATPase and K^+-EDTA-ATPase activities and exhibits low Mg^{2+}-ATPase activity, however Mg^{2+}-ATPase activity is stimulated about 25-fold in the presence of actin and cofactor protein. In myosin-II, the globular heads contain actin-activated Mg^{2+}-ATPase activity. *Acanthamoeba* myosin-II Mg^{2+}-ATPase activity is more activated by *Acanthamoeba* actin than muscle actin. The catalytic and phosphorylation sites of *Acanthamoeba* myosin-II is on the heavy chain while phosphorylation of the light chain is not observed. Unlike myosin-I, it's the dephosphorylation rather than phosphorylation that is required for actin activation of the Mg^{2+}-ATPase activity of myosin-II. Each heavy chain of myosin-II is phosphorylated at three or possibly four sites (on serine residues at positions 11, 16, 21, and possibly 6 from the carboxy-terminus of each heavy chain) by myosin-II heavy chain kinase, and this completely inhibits Mg^{2+}-ATPase activity. Three phosphorylation sites are within close proximity to each other within a peptide segment no larger than about 9 kDa and these 3 sites are located at the tip of the tail of the heavy chain. While the heavy chain phosphorylation inhibits the actin activated Mg^{2+}-ATPase activity, the dephosphorylation by phosphates restores full activity. However, phosphorylation has no effect on the Ca^{2+}-ATPase activity.

Despite several differences, both *Acanthamoeba* myosin-I and myosin-II do cross-link F-actin, reversibly bind to actin filaments and demonstrate the activation of Mg^{2+}-ATPase by actin, which is related to the generation of force required for movement (Pollard and Korn, 1973). Overall, cyclic interactions of *Acanthamoeba* actin, myosin and ATP in concert with other proteins generate force for the cell motility by a mechanism that is closely related to that utilized for muscle contraction.

From: *Acanthamoeba*: Biology and Pathogenesis. By: Naveed Khan.

Section B

Life Cycle and Genotyping

1. Life cycle

Acanthamoeba undergoes two stages during its life cycle; an active trophozoite stage during which *Acanthamoeba* reproduces, and a dormant cyst stage (Fig. 1).

1.1. Trophozoite stage

The trophozoite stage dominates when the growth conditions are optimal, i.e., abundant food supply, neutral pH, appropriate temperature (i.e., 30°C) and osmolarity between 50 – 80 mOsmol. However, the generation time differs between isolates belonging to different species/genotypes from 8 – 24 hours, during which the cell passes through a series of discreet stages, collectively called as the growth cycle. The growth cycle in *Acanthamoeba* is asexual and occurs by binary fission, resulting in two genetically identical daughter cells. Most of the cellular contents are synthesized continuously so the cell mass gradually increases as the cell approaches division. This is followed by cell division and involves two overlapping events in which the nucleus divides first, followed by the division of the cytoplasm termed "cytokinesis".

Binary fission in eukaryotes begins with DNA replication. DNA replication starts from multiple origins of replication, which opens up into "replication bubbles" (prokaryotic DNA replication usually has only 1 origin

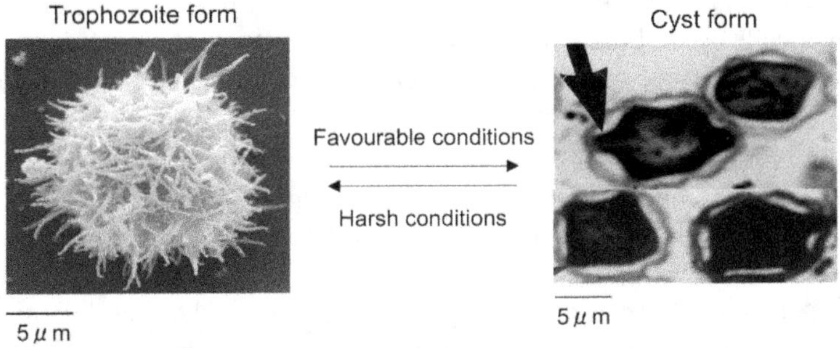

Figure 1. The life cycle of *Acanthamoeba* spp. The infective form of *Acanthamoeba* is known as trophozoites stage, as observed under light microscope. Under harsh conditions such as lack of food, extremes in pH, temperature and osmolarity, or desiccation, trophozoites differentiate into double-walled cysts, as indicated by arrows. Bar = 5 μm.

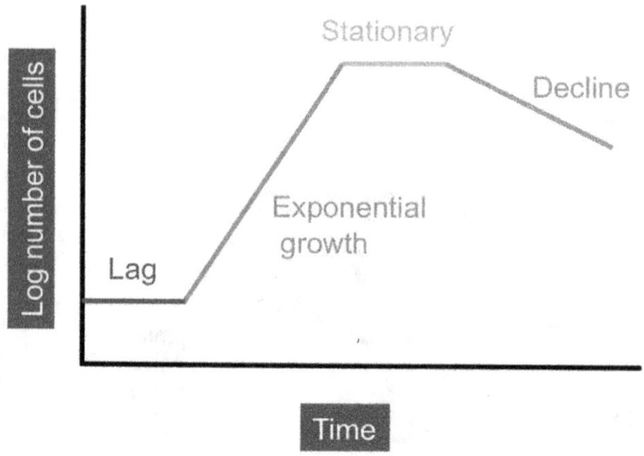

Figure 2. Growth phases in asynchronous cultures of *Acanthamoeba* spp.

of replication). The replication bubble separates the DNA double strand, each strand acts as a template for the synthesis of a daughter strand, until the entire DNA is duplicated. The replicated chromosomes are separated and the nucleus divides. The cytoplasm then gradually constricts between the two separating nuclei, ultimately forming two equally-sized daughter individuals, each with a nucleus. The offspring grows to the size of the parent cell before dividing again.

Growth in *Acanthamoeba* is a highly regulated process but it varies significantly depending on the synchrony of amoebae population. Here, *Acanthamoeba* growth cycle in the asynchronous population, under optimal conditions is described, followed by synchronous cultures. In axenic cultures, *Acanthamoeba* shows a typical exponential growth phase, followed by a period of reduced growth rate, called "growth decelartion phase", and finally the stationary phase during which no further increase in cell density occurs (Band and Mohrlok, 1973; Steven and Pachler, 1973).

1.2. Growth cycle

Under optimal growth conditions, one cell of *Acanthamoeba* divides into two daughter cells by the binary fission. This process halts or continues depending on the availability of nutrients and/or environmental conditions and can be subdivided into the following (Fig. 2):

- Lag phase: lag phase is due to the time taken for the amoebae to adapt or otherwise alter the growth media. During this period, the individual *Acanthamoeba* mature, involving the synthesis of RNA, enzymes and other molecules.
- Exponential phase (also called the log phase): During this period, the number of amoebae continues to double. Under optimal growth conditions, doubling continues at a constant rate. Because at some point, nutrients become depleted, and toxins build up, the exponential growth cannot continue indefinitely.

Table 1. Growth phases in asynchronous cultures of *Acanthamoeba*.

Growth phases in *Acanthamoeba*	% growth phase
Interphase	
G1 phase (pre-synthetic gap phase)	<1%
S phase (synthesis phase)	2 – 3%
G2 phase (post-synthetic phase)	85 – 90%
M phase (mitotic phase)	8 – 10%

- Stationary phase: During this phase, the growth rate slows as a result of nutrient depletion and accumulation of toxic products.
- Death or encystation phase: During this phase, amoebae run out of nutrients and depending on the stage of their growth cycle, they will either die or transform into cysts.

However, the aforementioned phases are not well defined as amoebae cells do not reproduce in synchrony and their exponential phase may not achieve a constant rate, and involve a deceleration phase, as well as a constant response to pressures such as nutrient concentrations and increasing waste concentrations, to either reproduce or to transform into a dormant cyst form.

Below, I have described growth cycle in *Acanthamoeba* in asynchronous cultures, which represent the natural mode of replication in these organisms. The cell cycle during growth phase is the series of events that take place in a eukaryotic cell leading to its replication. In general, these events can be divided in two periods as follows (Table 1);

i) Interphase (can be further sub-divided into 3 phases, G1, S, G2), during which the cell grows, accumulating nutrients needed for mitosis and duplicating its DNA, and
ii) Mitotic phase (M phase), during which the cell splits itself into two daughter cells.

1.2.1. G1 phase

Following the cell division (after the M phase), each daughter cell begins interphase of a new cell cycle (Fig. 2). The first phase within interphase is called the G1 phase, hence also called the pre-synthetic gap. During this phase the synthesis of various enzymes that are required in S phase occurs. During the G1 phase, a major decision is made whether and when a cell divides again. Cells that are arrested in the G1 phase awaiting for a signal that will trigger re-entry into the cell cycle are said to be in the G0 (G zero) phase. However, *Acanthamoeba* trophozoites show a lack of the G1 phase in asynchronous cultures.

1.2.2. S phase

During the S phase (synthesis phase), DNA synthesis or replication takes place and the new DNA is synthesized. DNA is replicated by a semi conservative mechanism in which the two strands of the double helix unwind and each strand serves as a template for the synthesis of a complementary strand. In

Acanthamoeba, there is a short S phase of approximately 2 – 3% of the total growth cycle.

1.2.3. G2 phase

Next, the cell then enters the G2 phase, which lasts until the cell enters mitosis. It follows successful completion of DNA synthesis and chromosomal replication during the S phase, hence also called post-synthetic gap. During the G2 phase, significant protein synthesis occurs, mainly involving the production of microtubules, which are required during the process of mitosis. Inhibition of protein synthesis during G2 phase prevents the cell from undergoing mitosis. Cell division in exponentially growing *Acanthamoeba* trophozoites is largely occupied with the G2 phase (85 – 90%) (post-synthetic gap).

1.2.4. M phase

The M phase (mitosis) consists of the nuclear division (karyokinesis) and the cytoplasmic division (cytokinesis). During the nuclear division, a cell separates the chromosomes into two identical sets into two daughter nuclei. Immediately after chromosomal segregation, the cytokinesis occurs, which divides the nuclei, cytoplasm, organelles and cell membrane into two daughter cells containing roughly equal shares of these cellular components. Mitosis and cytokinesis together define the mitotic (M) phase of the cell cycle, i.e., division of the mother cell into two daughter cells, which are genetically identical to each other and to their parent cell. Overall, mitosis can encompass the separation of chromosomes (i.e., nuclear division) and cytoplasmic division (i.e., cytokinesis). Although visibly striking, the events of the mitotic phase account for a relatively small portion of the total cell division of *Acanthamoeba* (8 – 10%) in asynchronous cultures. The beginning of mitosis is marked by condensation (coiling and folding) of the cell's chromatin. The chromosome condensation is an important step otherwise the long chromatin fibers would become tangled during the distribution of chromosomes. As the chromatids become visible, the nuclear envelope breaks into fragments. Guided by microtubules of the mitotic spindle, the sister chromatids separate and each, now a fully-fledged chromosome, move to the opposite ends of the cell. Of note, *Acanthamoeba* trophozoites' microtubules are found inserted in a dense material similar to centriolar satellites and are centered in the Golgi complex, which is often the location of centrioles in other cell types (Bowers and Korn, 1968).

As described above, mitosis is immediately followed by cytokinesis, which divides the cell into two daughter cells. In this phase, new nuclear membranes envelope the two groups of daughter chromosome as cell division is completed. Notably in *Acanthamoeba*, there could be multiple rounds of chromosomes replication and nuclear division in the absence of cytokinesis, thereby producing multinucleated cells. The cytoplasmic division begins with slight indentation of the cell surface which deepens into a cleavage furrow that encircles the cell. Ultimately the opposite surfaces make contact. The cleavage depends on a bundle of actin microfilaments called the contractile ring, which forms just beneath the plasma membrane. Tightening of the contractile ring involves interaction between actin microfilaments and the protein myosin II. The end of cytokinesis marks the end of the M phase. In culture, *Acanthamoeba* are often observed having difficulty in the act of final separation, in the related *Entamoeba* and *Dictyostelium* (both fellow members of the Amoebozoa) so called "midwife amoeba" are known to undergo this final session although mechanistic details are unknown.

1.3. Cell division in asynchronous cultures

Overall, asynchronous cultures of *Acanthamoeba* show a lack of the G1 phase (pre-synthetic gap), 2 – 3% S phase (synthesis), 85 – 90% G2 phase (post-synthetic gap) and 8 – 10% M phase (mitosis) (Fig. 2).

1.4. Cell division in synchronous cultures

In trophozoite populations, in which synchrony is achieved, the cell cycle appears to consist of 37 – 47% G1 phase (pre-synthetic gap), 13% S phase (synthesis), 40 – 50% G2 phase (post-synthetic gap) plus the M phase (mitosis) (Band and Mohrlok, 1973; Byers *et al.*, 1990; 1991).

1.5. Stationary phase

During the stationary phase, significant changes occur in *Acanthamoeba*. The total dry mass increases during the stationary phase, although the total protein remains constant. DNA decreases about 50% and RNA increases about 75%. Glycogen decreases 50% during the RNA buildup but later increases again (reviewed in Byers *et al.*, 1991).

2. What is the best stage in *Acanthamoeba* cell division to induce encystation?

Initially, the formation of cysts can be reversed by transfer of *Acanthamoeba* back to the growth medium. However, at some point during the immature cyst stage, the cells become committed to the differentiation process and continue to form mature cysts even after transfer to the growth medium (Weisman and Moore, 1968). Later studies revealed that when *Acanthamoeba* cells of the growth stage are transferred into the non-nutrient medium, only 7% of cells encyst. In contrast, 50 – 70% of *Acanthamoeba* cells of the stationary phase culture develop into cysts (Stohr *et al.*, 1987). It was argued that this difference could be due to some processes essential to the encystation that occurred only during the stationary phase. Synchronization of *Acanthamoeba* cells by release from the stationary stage reveals that encystation is initiated at a particular time in the late G2 phase, suggesting that in contrast to growing cells, the high encystation competence of the stationary stage cells is due to arrest of *Acanthamoeba* cells at a particular point in the G2 phase (Jantzen and Schulze, 1988). Overall, it can be concluded that trophozoites are generated in the mitotic division cycle, whereas cells committed at late G2 phase of the cell cycle develop into cysts in response to starvation or other adverse environmental conditions.

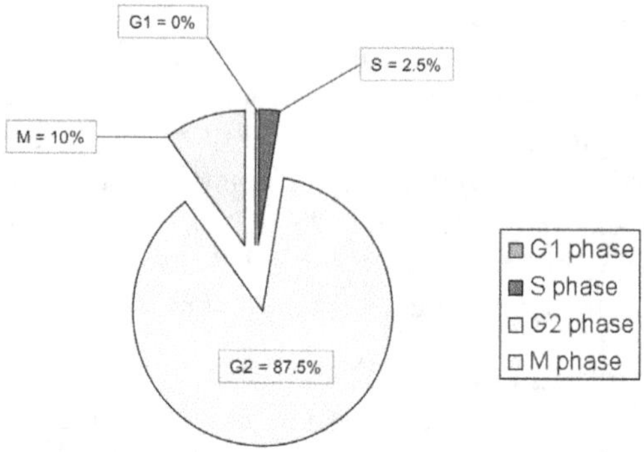

Figure 3. The various phases of cell division in asynchronous cultures of *Acanthamoeba* spp.

3. Encystation

Acanthamoeba can be permanently maintained in the trophozoite stage with an abundant food supply, neutral pH, appropriate temperature (i.e., 30°C) and osmolarity between 50 – 80 mOsmol. However, adverse environmental and nutritional conditions (i.e., lack of food, hyper- or hypo-osmolarity, extremes in temperatures and pH, high cell densities, and certain chemicals used in medical treatment) induce the transformation of trophozoites into the cyst stage, a process known as encystation (Neff and Neff, 1969; Band and Mohrlok, 1973). Encystation can occur both under harsh conditions as well as after the growth phase (Fig. 1). Encystation is accompanied by morphological changes, termination of cell growth, and biochemical differences. In simple terms, the trophozoite becomes metabolically inactive (minimal metabolic activity) after an initial burst of metabolic activity, and encloses itself within a resistant shell (Fig. 3). The cellular levels of RNA, proteins, triacylglycerides and glycogen declines substantially during the encystation process resulting in decreased cellular volume and dry weight (Weisman, 1976). In addition, during the encystation, excess food, water and particulate matter is expelled with a decrease in cytoplasmic mass and the trophozoite condenses itself into a double-walled cyst with the wall serving only as a shell to help survive hostile conditions.

3.1. At the morphological-level

It is interesting to note that the mean diameter of rounded *Acanthamoeba* cells (Neff strain) was found to be 26.5 μm, while the mean diameter of the cyst stage was 16.2 μm (Bowers and Korn, 1969). The decrease in surface area during encystation was calculated to be more than 65% and decrease in the median cell volume was about 80%. The large reduction in cell surface implies a significant loss of the cell membrane, however the morphological analysis suggests possible long infoldings of the surface membrane that occur early in the encystation. While dense cytoplasm of encysted cells suggests that volume change occurs partially as a result of dehydration. Under the scanning electron microscope, the morphological changes include thickening or shortening of the acanthopodia until the surface is covered with short stubby processes, followed by the development of thick, interconnecting ridges that characterize the surface of the mature cyst (Pasternack *et al*., 1969).

3.2. At the organelle-level

Nuclear activity: During encystation, the nuclear volume decreases to at least 40%, while the nucleolar volume decreases to about 75%. A partial expulsion of the peripheral chromatin during encystation is observed (Volkonsky, 1931). In the mature cysts, the nucleolus develops as a large, dense droplet. The functional activity within the nucleus and of the nucleolar proteins is important for *Acanthamoeba* encystation. For example, when *Acanthamoeba* trophozoites are halved using microsurgical procedures, the nucleated halves encyst but the non-nucleate halves do not form cysts. In addition, there are clear differences between nucleolar proteins from the stationary phase cells and the exponential growth phase cells, both in the cell function, and at the molecular-level. Even after the induction of encystation, normal DNA synthesis is continued at a reduced rate, when differentiated products (cell wall components) are formed (Roti Roti and Stevens, 1975). However, there is no net accumulation of DNA occurring during the differentiation process, indicating that the DNA synthesis occurring at this time may be balanced by its breakdown (Roti Roti and Stevens, 1975). However, there is a decrease in the DNAse activity in *Acanthamoeba* cultures undergoing encystation (Martin and Byers, 1976). These findings suggested that the nuclear activity (i.e., DNA synthesis) is required during the encystation process. However, it remains unclear whether inhibition of the DNA synthesis prevents or induces encystation. This is due to studies with conflicting findings, which showed that the DNA synthesis inhibitor (e.g., hydroxyurea) prevents growth and induces encystation, while other studies revealed that hydroxyurea actually delayed encystation. On the other hand, some studies showed that 5-bromodeoxyuridine (that can be incorporated into DNA in the place of thymidine) has no effect on encystation (Roti Roti and Stevens, 1974). A likely explanation for these contradicting findings may be due to *Acanthamoeba* cells in different phases of their growth. For example, it is shown that in the presence of the DNA synthesis inhibitor, *Acanthamoeba* cells in the exponential growth phase divide once, before growth stops and encystation begins. However, cells in the growth deceleration phase do not divide, prior to the encystation. This suggests that the encystation in *Acanthamoeba* can occur in the presence of the DNA synthesis inhibitor. From the aforementioned studies, it can be concluded that factors that control the induction of cyst formation are associated with the growth of *Acanthamoeba* and are not directly related with termination of the DNA synthesis (Roti Roti and Stevens, 1974).

3.2.1. RNA and protein synthesis

Similar to changes in the DNA levels during encystation (as described above), studies have shown that the RNA levels preceding to encystation are dependent on the status of *Acanthamoeba* cultures. In the cultures approaching the stationary phase (high cell densities), onset of the growth deceleration phase is accompanied initially by an increase in the cellular RNA levels (Stevens and Pachler, 1973). The maximum RNA content occurs at the stationary phase, followed by its decline with the appearance of cysts.

In contrast, *Acanthamoeba* cells (in the exponential phase) whose growth is immediately terminated by transferring them into the non-nutrient medium do not exhibit an increase in the RNA levels; however, RNA levels decline with the appearance of cysts (Stevens and Pachler, 1973). The mature cysts from either the stationary phase or obtained from experimental induction contain an equal amount of RNA (approximately 50% of the exponential levels). Overall, it can be concluded that the RNA levels decrease 50% during encystation. Moreover, during encystation, there is a decline of 70 – 80% of the rRNA within 1 hour of transfer into the non-nutrient medium (Schulze and Jantzen 1982), while rRNA synthesis could not be detected after the cells had been incubated in the non-nutrient medium for 7 hours. However, the amount of RNA polymerase I, II, III activities from whole cells remains constant during encystation (Stevens and Pachler, 1973; Detke and Paule, 1978; 1979). The presence of actinomycin or cycloheximide, inhibitors of RNA and protein synthesis respectively, prevents the formation of mature cysts, suggesting that synthesis of biomolecules is crucial for the encystation process. The endoplasmic reticulum of the encysting cells is predominantly cisternal in form. The densely stained content of the endoplasmic reticulum that is often present in the trophozoite is not observed in the cyst. Encystation is associated with less synthesis of proteins soon after transfer of growing cells into the non-nutrient medium. For example, there is a decline of 70 – 80% of ribosomal proteins within 1 hour of transfer to the non-nutrient medium (Schulze and Jantzen, 1982). Nuclei from the exponential growth phase show 50 protein bands on the SDS-PAGE, but at the beginning of the encystation, 10 proteins disappeared.

3.2.2. Golgi complex

The changes in the Golgi complex are observed both at the level of form and distribution. For example, Golgi complex of the trophozoite occur as large aggregates of cisternal stacks but occur as smaller aggregates in the encysting cells, accompanied by the appearance of dense staining material within the Golgi complex that are also present on the cell surface. This dense staining content of the Golgi complex gradually diminishes at the time the endocyst is being formed and is usually absent in the mature cysts. The volume fraction of the Golgi complex in encysting cells increases about six-fold early in encystation and thereafter returns to its original volume fraction of about 0.5% (Bowers and Korn, 1969).

3.2.3. Phagocytic and pinocytic activity

Once *Acanthamoeba* cells enter the stationary phase, phagocytic activity ceases, while the pinocytic activity is halved (Chambers and Thompson, 1976). The reduced pinocytic activity remains sensitive to respiratory inhibitors. The water expulsion vesicles remain free of any dense content throughout encystation but are often partially collapsed. Early in encystation, large numbers of vacuoles are observed in the encysting cells. Some of these vacuoles possess acid phosphatase, a positive criteria for lysosomes. The volume fraction of lysosomes increases for 18 hours during induced encystation, and many lysosomes are present in the mature cysts. In addition to lysosomes, peroxisomes have been isolated from *Acanthamoeba* (Muller, 1969; Muller and Moller, 1969). A variety of lysosomal enzymes have been identified in *Acanthamoeba* including α-glycosidases, β-glycosidases, amylase, β-galactosidase, β-*N*-acetyl-glucosaminidase, β-glucuronidase, protease, phosphatase, acid hydrolase, RNAse, and DNAse (reviewed in Weisman, 176). However, the source of the hydrolytic enzymes is not well-understood. Overall, the digestive vacuoles disappear during encystation and their contents are discharged.

3.2.4. Respiration

The respiration rates of whole cells and isolated mitochondria as well as the intracellular ATP levels of cells diminish (after an initial burst in metabolism) throughout the course of encystation, which may be due to the development of various droplets and dense granules in mitochondria. Significant changes are observed within mitochondria. Within 1 – 2 hours of induced encystation, coiled lamellate structures appear within mitochondria that are formed by compaction and by coiling of tubular entities that lie in an enlarged space between cristae. The lamellar bodies are lost in the later stages of encystation. The mitochondria of dehydrated cysts are much more condensed and still contain the tubular cristae but are poorly defined relative to those in the trophozoites. The mitochondria are incorporated into autolysosomes but remain at a constant fraction of the cell cytoplasm (Bowers and Korn, 1969).

3.2.5. Cytoskeletal proteins

The cytoplasmic microtubules can be observed at all stages of the encystation. During outer wall (i.e., ectocyst) deposition, microtubules seem to be confined to the central region of the *Acanthamoeba* cell. Later, during inner wall (i.e., endocyst) deposition, microtubules are found in all regions of the cell. Actin synthesis is reduced about two-fold during early stages of the development. Subsequently, levels of the actin mRNA does not change. However the actin protein synthesis is completely shut down during late stages of the encystation, suggesting translational regulation of the actin synthesis (Jantzen, 1981; Byers *et al*., 1991). This is supported by findings of actin turnover during the encystation. For example, during the exponential growth phase, no actin degradation was observed, however as *Acanthamoeba* cells enter the stationary phase there is a dramatic increase in the amount of the actin, however the encystation process is accompanied by a rapid reduction in actin level (Robin and Maher, 1976).

3.2.6. Glycogen

Among macromolecules, glycogen is the most rapidly degraded molecules during encystation. In the initial phase of encystation, cellular levels of glycogen decrease by one half to two thirds within 4 hours and then fall to 10 – 20% of the original amount. Lysosomal hydrolases are probably most likely responsible for the rapid degradation of glycogen. Both lysosomes and peroxisomes have been isolated from *Acanthamoeba* (Muller, 1969;

Muller and Moller, 1969). To this end, *Acanthamoeba* is known to contain various enzymes including α-glycosidases (also β-glycosidases, although it is needed for cellulose degradation), amylase, β-galactosidase, β-*N*-acetyl-glucosaminidase, β-glucuronidase, protease, phosphatase, acid hydrolase, RNAse, and DNAse (reviewed in Weisman, 176). However, it is reasonable to suppose that some of the autolysed material is used as a source of precursors for the cell wall synthesis or as a source of energy for the cell that has no external food source (Bowers and Korn, 1969). Hydrolytic degradation of glycogen by lysosomal enzymes, would ultimately yield glucose, which must be phosphorylated by ATP for its conversion to the glucose-1-phosphate prior to UDP-glucose (uridine diphosphate glucose) and cellulose formation. Glycogen phosphorylase would yield glucose-1-phosphate directly (Verma and Raizada, 1975). Indeed, glycogen phosphorylase has been reported to be the main tool used for glycogen breakdown in order to maintain the free glucose levels during the encystation. The role of glycogen phosphorylase in *Acanthamoeba* encystation was shown by silencing the catalytic domain of the glycogen phosphorylase using the RNA interference method. It was shown that glycogen phosphorylase enzyme is required for cyst wall assembly, mainly for the formation of the endocyst (inner cell wall) of *Acanthamoeba* (Lorenzo-Morales *et al.*, 2008). The intracellular environment is maintained at pH between 6 and 7.5 (Deslauriers *et al.*, 1980).

3.2.7. Lipids

As indicated above, *Acanthamoeba* shows a significant decrease in the surface area during encystation (more than 65%). A large decrease in cell surface implies a significant loss of the cell membrane. In addition, smooth endoplasmic reticulum is also absent that is normally present in the trophozoite. It is shown that the levels of lysophosphatidylcholine, lysophosphatidylethanolamine, phosphatidylcholine, phosphatidylethanolamine and alkali-stable phospholipids I, and II are significantly lower in the mature cysts of *Acanthamoeba*, compared with the trophozoites. The major constituent sterols of trophozoites, ergosterol, 7-dehydrostigmasterol and 7,22,25-tridehydrostigmasterol are not detected in the cysts (Mehdi and Garg, 1987).

3.3. Other molecular changes

The autolysis and excretion of cytoplasmic constituents are consistent with the decrease in the dry weight of cysts relative to trophozoites. In support, soluble proteins, amino acids and nucleotides derivatives and carbohydrates appear in the medium of encysted cultures is observed.

During encystation, *Acanthamoeba* releases polyphosphate into the encystation medium (Deslauriers *et al.*, 1980). This was further supported by findings that immature cysts show detectable levels of nucleotide diphosphate and triphosphate similar to the vegetative cells, while mature cysts only excreted polyphosphate as well as a component which has a chemical shift of a phosphodiester (Deslauriers *et al.*, 1980). The role of cyclic AMP in the encystation is demonstrated with the findings that upon induction of encystation, the levels of cyclic AMP rise to 2 to 3-fold within

6 – 8 hours (due to changes in adenylate cyclase), and then reduce to levels similar to the exponential growth phase levels. The adenylate cyclase activity of *Acanthamoeba* during exponential growth phase is generally low but rises 2 to 4-fold during the stationary phase and reaches to peak at the time when cysts are detectable in the cultures. Fractionation of the homogenates of trophozoites indicated that adenylate cyclase activity is associated with the particulate microsomal fraction (Achar and Weisman, 1980). Regardless of the levels of cyclic AMP, the precise mechanism of their action is not known, although generally cyclic AMP acts via protein kinase-mediated modification of enzymes and proteins and may influence transcriptional-, translational-, and post-translational processes. The role of cyclic AMP in encystation is thought to be mediated via glycogen degradation by converting glycogen phosphorylase to the active form but this remains to be clearly defined. Interestingly, it is reported that cyclic AMP stimulates the phospholipase A activity of *Acanthamoeba* homogenates (Hax *et al.*, 1974).

In addition to the aforementioned, the activities of isocitrate lyase, glycolate and maleate are involved in the encystation. This is shown by findings that the inhibitors of isocitrate lyase, glycolate and maleate at 8.5 to 34 mM concentration inhibited encystation by 18 – 67% (Mehdi and Garg, 1987). Furthermore, it is shown that the activity of isocitrate lyase in the encysting cells steadily increase up to 24 hours of encystation, followed by a slight decrease during 24 – 32 hours, but the activity of isocitrate dehydrogenase steadily decreased up to 32 hours of encystation, beyond which time no enzyme activity was detected.

Acanthamoeba also possesses phenol oxidase activity. The localization of phenol oxidase is the cytoplasm of encysting *Acanthamoeba* and in the outer cyst wall, but whether this enzyme plays a role in encystation remains to be determined (Sykes and Band, 1985).

3.4. Cell wall synthesis

During encystation, the trophozoite condenses itself into a rounded structure (i.e., pre-cyst or immature cyst), accompanied by synthesis of a chemically and structurally complex cell wall, which matures into a double-walled cyst with the wall serving only as a shell to help amoebae survive hostile conditions. The mature cyst wall consists of a laminar, fibrous outer layer, (i.e., ectocyst), and an inner layer of fine fibrils, (i.e., endocyst). Periodic acid-Schiff's (PAS) staining of *Acanthamoeba* cysts revealed the cytoplasmic components and the inner wall (endocyst) and the outer wall (ectocyst) of cysts are stained, suggesting the presence of glycans (Figs. 4 and 5).

Both layers are normally separated by a space, except at certain points where they form opercula in the centre of ostioles (exit points for excysting *Acanthamoeba* trophozoite). The ectocyst is the first to appear in the encysting cells, even before the cell rounds up completely and consists of several layers. It appears as an amorphous, discontinuous layer about 10 μm just outside the plasma membrane (Bowers and Korn, 1969). By the time sufficient material accumulates to form a layer covering the entire surface, the cell has become

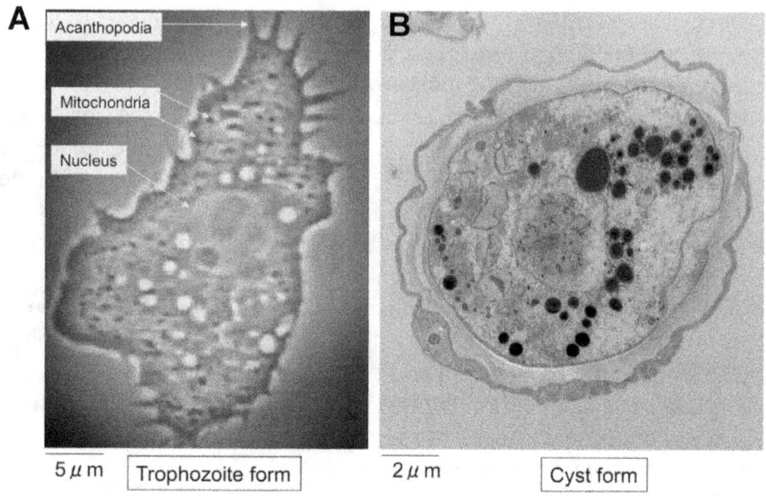

Figure 4. A) *Acanthamoeba* trophozoite under a phase-contrast microscope. **B)** Transmission electron micrograph of an *Acanthamoeba* cyst.

nearly spherical. In mature cysts, the ectocyst terminates in loose fibrous layer and the entire ectocyst is 0.3 – 0.5 μm in thickness and fibrils appear to be less than 50 Å in diameter (Bowers and Korn, 1969). The endocyst differs from the ectocyst in texture and appears finely granular, however fibrils appear to be less than 50 Å in diameter as for ectocyst. As indicated above, endocyst is consistently separated from the ectocyst except for operculum. The ectocyst is thinner in the operculum than elsewhere in the wall whereas the endocyst is generally thicker than elsewhere in the wall. During encystation, there is no consistent or well-defined space between the plasma membrane and the

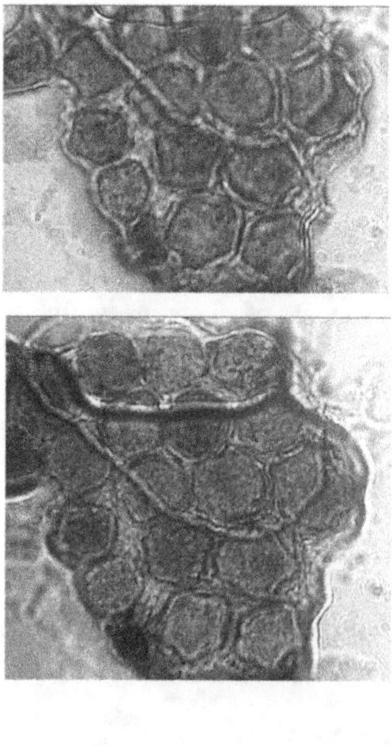

Figure 5. *Acanthamoeba* cyst walls possess glycans. *Acanthamoeba* cysts were stained with Periodic Acid-Schiff (PAS) and observed under an optical light microscope. The results revealed staining of the inner wall (endocyst) as well as the outer wall layer (ectocyst), suggesting the presence of glycol-containing components. X400.

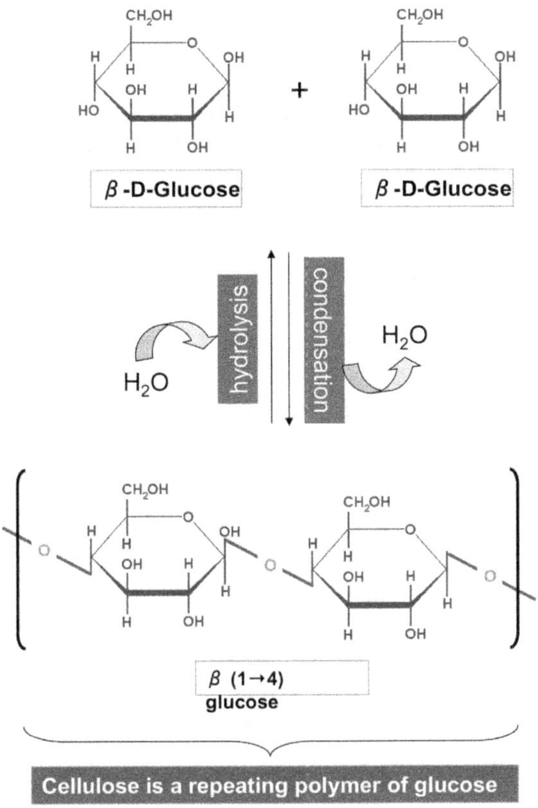

Figure 6. The synthesis and structure of cellulose.

endocyst; but in the mature cyst there is a space of about 0.1 μm between the endocyst and the plasma membrane. The endocyst is shown to possess cellulose (not present in the trophozoite stage) (Figs. 4 and 5) that accounts for nearly one third of the dry weight of the cyst wall (Neff and Benton, 1962; Tomlinson and Jones, 1962). The basic unit of cellulose is similar to starch or glycogen, i.e., glucose. However, starch is a polymer of α-glucose, while cellulose is a polymer of β-glucose (Fig. 6). The ineffectiveness of cellulase against intact cysts indicates that cellulose exists as an inner and inaccessible layer of the wall (Neff and Neff, 1969; Barrett and Alexander, 1977). In support of this assertion, extracts of growing cultures have been shown to contain cellulose-degrading system (Deichmann and Jantzen, 1977). At pH 4, the cellulose degradation products are cellobiose and glucose and at pH 6, higher molecular weight oligosaccharides are produced (Deichmann and Jantzen, 1977). During the encystation process the activities of both cellulases decline (Deichmann and Jantzen, 1977).

It is shown that the ectocyst consists of proteins and polysaccharides (Barrett and Alexander, 1977). A major component of the outermost cyst wall is a protein of molecular weight of 70 kDa (Rubin *et al.*, 1976). This protein, which is synthesized during the encystation process represents 15%

of the dry weight of the mature cysts (Neff and Neff, 1969). The presence of hydroxyproline in the ectocyst is reported (Bauer, 1967). On this basis, it is clear that the two walls are also different in their chemical composition. Overall, the cyst wall composition for *Acanthamoeba castellanii* belonging to the T4 genotype has been shown to contain 33% protein, 4 – 6% lipid, 35% carbohydrates (mostly cellulose), 8% ash, and 20% unidentified materials (Neff and Neff, 1969), although the cyst wall composition varies between isolates belonging to different species/genotypes.

The precursor of cellulose is glucose, source of which is glycogen. Glycogen is a storage polysaccharide that is highly branched and consists of glucose repeating subunits linked by $\alpha(1\rightarrow 4)$ glycosidic bonds. It is also shown that glucose of glycogen and/or other glucose are converted to cellulose without significant dilution (Stewart and Weisman, 1974). This is consistent with the findings that within the first four hours of encystation, the cellular levels of glycogen decrease to between one-half and one-third of that found in *Acanthamoeba*, growing in the normal growth medium (Weisman *et al.*, 1970). With analogy to higher eukaryotes, this is also consistent with an increase in glycogen phosphorylase activity. Glycogen degradation would ultimately yield glucose, which must be phosphorylated by ATP for its conversion to the glucose-1-phosphate that stimulates cellulose formation by UDP-glucose (uridine diphosphate glucose) (Potter and Weisman, 1972; 1976). The glucose subunits from UDP-glucose are incorporated into the cell wall as $\beta(1\rightarrow 4)$-glucans (i.e., cellulose) (Potter and Weisman, 1972; 1976). The levels of UDP-glucose increases 5-fold within 5 hours of the induction of encystation.

It is noteworthy that neither adenosine-5'-diphosphate glucose (ADP-glucose), thymidine-5'-diphosphate glucose (TDP-glucose), cytidine-5'-diphosphate glucose (CDP-glucose), nor guanosine-5'-diphosphate glucose (GDP-glucose) was able to replace UDP-glucose as the glucosyl donor, suggesting that the availability of UDP-glucose during encystation plays a key role in the cell wall biosynthesis. However, it is shown that the specific activity of uridine diphosphate glucose pyrophosphorylase (UDP-glucose pyrophosphorylase), which catalyzes the synthesis of UDP-glucose from UTP-glucose remained relatively constant during the 20 hours after the induction of encystation but rapidly decreased thereafter (Rudick and Weisman, 1974). This suggests that the increased concentration of UDP-glucose most likely reflects an increase in the availability of glucose 1-phosphate (and/or UTP-glucose) (Rudick and Weisman, 1974). In contrast, during the developmental programs of *Dictyostelium discoideum*, there is a 10-fold increase in the UDP-glucose pyrophosphorylase specific activity. However, the activity of β-glucan synthetase increases some 30-fold, whose function is to catalyze the formation of $\beta(1\rightarrow 4)$ glucan (cellulose), similar to the plant cells [in contrast, fungal cell wall cellulose can contain either $\beta(1\rightarrow 3)$ glucan or $\beta(1\rightarrow 4)$ glucan]. Both alkali-soluble and alkali-insoluble β-glucan occur in the *Acanthamoeba* cyst walls (Potter and Weisman, 1972). This results in the formation of cellulose, i.e., rigid, linear rods that aggregate laterally into microfibrils (25 nm in diameter and composed of 2,000 cellulose chains). Thus encystation is associated with the appearance of the β-glucan synthetase, cyst wall cellulose and the formation of the mature cyst (Potter and Weisman, 1972).

Overall, UDP-glucose is used as a glucosyl donor in the synthesis of various biomolecules including, glycogen, cellulose, glycolipids, and glycoproteins (reviewed in Weisman, 1976). Lactate dehydrogenase, amylase, and aminotransferase activities have been described, as have enzymes for phosphogluconate pathway (reviewed in Weisman, 1976). However, the pathway for the conversion of the acetyl-coenzyme-A (acetyl-CoA) to glucosyl residues is not understood. Acetyl CoA can be metabolized to malate and oxaloacetate by the action of glyoxylate and the Krebs cycle enzymes, but no enzymes favouring the conversion of either of these dicarboxylic acids to phosphoenolpyruvate or pyruvate have been detected in *Acanthamoeba*. Pyruvate carboxylase, malic enzymes, phosphoenolpyruvate carboxykinase or phosphoenolpyruvate carboxyphosphotransferase in extracts of *Acanthamoeba* trophozoites have not been detected (reviewed in Weisman, 1976). The phosphoenolpyruvate carboxylase and pyruvate phosphate di-kinase were detected; the reaction of the carboxylase is considered irreversible in the direction of oxaloacetate, while the di-kinase catalyzes the formation of phosphoenolpyruvate from pyruvate.

3.5. Cyst stage

Encystation results in the synthesis of at least two major products that are not detected in trophozoites; cellulose (Tomlinson and Jones, 1962) and an acid-insoluble protein-containing material (Neff and Neff, 1969). The former has been reported to be the major component of the inner wall of the cyst. Enzymes for the synthesis and breakdown of cellulose are identified in *Acanthamoeba*, which likely participate in encystation and excystation (Anderson *et al.*, 2005). Later studies identified a cyst-specific protein of 21 kDa that is absent in the trophozoite stage of *Acanthamoeba*. This cyst-specific protein-21 was purified from guanidine-HCl extract of the cyst wall and shown to be hydrophilic. Interestingly, the expression of cyst-specific protein-21 mRNA was restricted within early stages of encystation, suggesting that the biosynthesis of cyst-specific protein-21 is regulated at the mRNA level (Hirukawa *et al.*, 1998). The matrure cyst possesses two walls and it is 5 – 20 μm in diameter but again this varies between isolates belonging to different species/genotypes. The outer cyst wall, the ectocyst, is wrinkled with folds and ripples and contains proteins and lipids. The inner cyst wall, the endocyst, contains cellulose and hence stains positive using the Periodic acid-Schiff reagent. Both walls meet at pores or ostioles that are covered by convex-concave plugs or opercula. Cysts are viable airborne cells that can withstand prolonged starvation, desiccation, cold, heat, thus may help spread *Acanthamoeba* in the environment and/or carry these pathogens to the susceptible hosts. For example, previous reports have shown that cysts can remain viable for several years while maintaining their pathogenicity, thus presenting a role in the transmission of *Acanthamoeba* infections (Mazur *et al.*, 1995). The ability of *Acanthamoeba* trophozoites to switch into cysts provides an excellent system for the study of cellular differentiation/morphogenesis.

3.6. Excystation

Cysts possess pores known as ostioles, identified by the presence of an operculum surrounded by a circular ridge that is evident on the surface of the mature cysts. Ostioles are possibly used to monitor environmental changes. The trophozoites emerge from the cysts under the favourable conditions leaving behind the outer shell. In the pre-emergent stage, a contractile vacuole becomes evident and moves towards the wall. *Acanthamoeba* trophozoite then pulls away from the endocyst and moves freely within the cyst wall. The first indication of emergence of trophozoites is visible by scanning electron microscopy is the appearance of a cytoplasmic bud pushing through the ostioles from which the operculum had been removed (Chambers and Thompson, 1972). The emerging cytoplasmic bud does not possess the long acanthopodia, which are characteristics of the trophozoites. These acanthopodia do not appear until the trophozoites have completely excysted. The hole through which the trophozoites emerge is apparent on the surfaces of the empty cyst walls, however otherwise surfaces of the empty cyst walls are indistinguishable from those of mature cysts and exhibit sculptured interconnecting ridges surrounded by shallow craters (Chambers and Thompson, 1972). Thus *Acanthamoeba* trophozoites emerge without the digestion of the cyst wall (Deichmann and Jantzen, 1977). The emerged trophozoites actively reproduce as described above, thus completing the cycle. Interestingly, 5-bromodeoxyuridine (that can be incorporated into the DNA in place of thymidine) inhibits the excystation of *Acanthamoeba* (Roti Roti and Stevens, 1974).

4. Regulators of life cycle

Both the encystation and excystation processes require active macromolecule synthesis and can be blocked by cycloheximide (a protein synthesis inhibitor) and actinomycin D. In addition, recent studies have shown that *Acanthamoeba* serine proteases are crucial in the differentiation of *Acanthamoeba castellanii* belonging to the T4 genotype. It was observed that the inhibition of serine proteases attenuates *Acanthamoeba* metamorphosis, as demonstrated by the arrest of both encystation and excystation. This suggested that *Acanthamoeba* serine proteases are not unique to the trophozoite and are involved in the host invasion and cellular cytolysis, but are equally a requisite for trophozoite metamorphosis and its re-emergence during excystation and may prove potential target for future therapy. In addition, *Acanthamoeba* is shown to possess trehalose-6-phosphate synthase suggesting that trehalose plays a role in the stress adaptation and regulates encystation. Detection and response to a number of stress conditions is likely accomplished with a large set of signal transduction histidine kinases and a set of putative receptor serine/threonine kinases, similar to those found in *Entamoeba histolytica*, has been identified in *Acanthamoeba* (discussed in section A6) (Anderson *et al.*, 2005).

Several other molecular determinants have also been identified to be related with the encystation, including protein kinase C, proteasome, heat shock protein, cullin 4, autophage protein 8, cyst specific protein-21, and ubiquitin-conjugating enzymes (Moon *et al.*, 2008).

In addition, encystation may involve xylose isomerase, P-type ATPase, and subtilisin-like serine protease (Moon *et al.*, 2007). Inhibitors of isocitrate lyase, glycolate and maleate at 8.5 to 34 mM concentration inhibited encystation by 18 – 67% (Mehdi and Garg, 1987).

Changes in the polyamine levels were observed during the growth of amoeba suggesting their possible involvement in *Acanthamoeba* encystation. *Acanthamoeba* contains several polyamines including, 1,3-diaminopropane, spermidine, spermine, norspermidine, and putrescine. The polyamine found in the greatest concentration in the growing cells is 1,3-diaminopropane, followed by spermidine and a low level of putrescine. Generally, polyamines are considered as growth factors in both eukaryotic organisms as well as prokaryotic cells. The inhibition of polyamines results in amoebistatic effects. These polyamines are significantly decreased in concentration, as *Acanthamoeba* trophozoites differentiate into cysts. For metabolites of polyamine and S-adenosylmethionine, it was shown that N8-acetylspermidine and acetylspermine are present both in trophozoites and cycts, while acetylcadaverine is present only in growing *Acanthamoeba* and N1-acetylspermidine only in cysts. S-adenosylmethionine and S-adenosylhomocysteine are present in growing trophozoites but S-adenosylmethionine was undetectable or barely detectable in cysts. S-adenosylhomocysteine also decreased in concentration during encystation. The developmental transition from growing *Acanthamoeba* trophozoites to dormant cysts is characterized metabolically by a threshold adjustment in the concentration of S-adenosylmethionine, S-adenosylhomocysteine, and of the polyamines (especially 1,3-diaminopropane and spemidine (Zhu *et al.*, 1989). The enzyme, S-adenosyl-L-methionine decarboxylase (approximate molecular weight of 88.8 kDa) is present in *Acanthamoeba* (Hugo and Byers, 1993). Inhibition of this enzyme could affect a variety of cell methylation reactions by causing accumulation of its substrate S-adenosylmethionine and could arrest polyamine synthesis by blocking production of decarboxylated S-adnosylmethionine needed for spermidine synthesis. Interestingly, the addition of polyamines to growing cultures of *Acanthamoeba* blocked the induction of encystation by diamidines, suggesting the involvement of polyamine biosynthesis in *Acanthamoeba* encystation.

5. What stimulates encystation?

5.1. Food deprivation

The unavailability of nutrients is shown to induce encystation in *Acanthamoeba*. This is demonstrated by plating *Acanthamoeba* on non-nutrient agar plates seeded with *Escherichia coli* as a food source. *Acanthamoeba* clears the bacterial lawn by consuming *Escherichia coli*. Following bacterial consumption, *Acanthamoeba* trophozoite transforms into a cyst form (assisted by inevitable dehydration of the agar plate (reviewed in Byers *et al*., 1991).

5.2. High cell density

It is shown that *Acanthamoeba* trophozoites release growth inhibitor factors. Under favourable conditions (e.g., low cell density, abundant food supply), these factors are neutralized by other, as yet unknown, *Acanthamoeba* components. However, the growth inhibitor factors presumably inhibit *Acanthamoeba* growth at high cell densities, which would lead to encystation (Pigon, 1981).

5.3. Surface receptors (hyper- and hypo-osmolarity, extremes in pH)

Findings that the monoclonal antibodies induce encystation in *Acanthamoeba* suggested that surface receptors expressed on the surface of trophozoites may be involved in the differentiation of *Acanthamoeba*. Furthermore, the monoclonal antibody bound specifically to a 40 kDa trophozoite protein, which initiated encystation in *Acanthamoeba* trophozoite. Interestingly, the same monoclonal antibody, when bound to cysts, prevented *Acanthamoeba* excystation. Washing away the antibodies allowed both trophozoites and cysts to resume their normal cellular activity. Overall, these findings suggest that the cell surface receptors play a role in differentiation by allowing cells to react to the external signals.

The general process by which the receptors cause cellular change is similar in almost all systems. Receptors are either transmembrane macromolecules themselves or are closely associated with the transmembrane molecules. The conformational change in the receptor produced by the binding of an extracellular ligand is thought to be the actual extra- to intracellular signal transmission device. Consequently anything (hyper- and hypo-osmolarity, extremes in pH) that causes appropriate conformational change in the receptor should produce the biological effect of the natural ligand. These findings may explain the presence of cysts in the infected corneal tissues or the brain tissues. Generally, conditions in the tissues are highly favourable for the parasite, i.e., abundance of nutrients, neutral pH, oxygen levels, and physiological relevant temperatures. It is suggested that perhaps antibodies binding to the cell surface receptors may stimulate encystation in *Acanthamoeba*, hence the presence of cysts are observed frequently in the infected tissues, however this remains to be determined (Yang and Villemez, 1994).

5.4. Encystation-inducing factors

It is shown that the medium from the encysting mid-log phase cells contains an encystation enhancing activity, which when transferred to the low-density early-log phase cultures causes high levels of encystation (Akins and Byers, 1980). The encystation enhancing activity has a molecular weight of between 1 to 10 kDa and is resistant to nucleases, proteases, amylase, boiling, and

1N NaOH (sodium hydroxide). However, the encystation enhancing activity is inactivated by snake venom phosphodiesterase, autoclaving and acidic or harsher alkaline treatments and may be of mitochondrial origin (Akins and Byers, 1980).

5.5. Chemical factors

Chemical factors that induce encystation include inhibitors of the DNA synthesis, including fluorodeoxyuridine, mitomycin C, trenimon, and hydroxyurea; the mitochondrial inhibitors, diminazine (Berenil), ethidium bromide, erthyromycin, chloramphenicol and acetate (reviewed in Byers *et al.*, 1991). However, the inhibition is dependent on the cell density at the time of inhibition with the early log-phase amoebae yielding 0 – 20% cysts, mid-log phase *Acanthamoeba* yielding 55% (with chloramphenicol or erythromycin); 65% (with ethidium bromide); and 95% (with berenil) (Akins and Byers, 1980).

6. Feeding

Acanthamoeba feeds on microorganisms present on surfaces, in diverse environments (Brown and Barker, 1999) and even at the air-water interface (Preston *et al.*, 2001). Feeding in *Acanthamoeba* occurs during the trophozoite stage and many food vacuoles can be seen in the cytoplasm of the cell. *Acanthamoeba* trophozoites actively feed on small organic particles as well as other microbes. The spiny structures on the surface or acanthopodia that arise from the surface of *Acanthamoeba* trophozoites may be used to capture food particles, which usually are bacteria (Weekers *et al.,* 1993), but yeast (Allen and Dawidowicz, 1990; Bunting *et al.*, 1979; Nero *et al.*, 1964), and other protists are also "preyed upon".

Acanthamoeba depends upon pinocytosis and phagocytosis for the uptake of nutrients. Both processes are energy dependent and can be inhibited by metabolic inhibitors, sodium azide (NaN_3), sodium cyanide (NaCN), sodium fluoride (NaF), iodoacetate, and 2,4-dinitrophenol and by incubation at temperatures below 12°C but are not affected by inhibitors of the glycolysis (Bowers and Olszewski, 1972; Bowers, 1977). Phagocytosis is a specialized form of endocytosis (discussed later in section D6.4), while pinocytosis is a non-specific process (continuous in growing cells) that is used to take up large volumes of solutes/medium, through membrane invaginations and thus requires turn over of the cell surface (Bowers and Olszewski, 1972). *Acanthamoeba* uses both phagocytosis and pinocytsis for the uptake of food particles and large volumes of solutes with optimal uptake at 30 – 35°C (Allen and Dawidowicz, 1990; Alsam *et al.,* 2005a; Bowers and Olszewski, 1972; Weisman and Korn, 1967).

For phagocytosis, there is an optimal size for the formation of the phagocytic vesicle that can be satisfied by one bead with a diameter of 1.3, 1.9 or 2.7 μm (Korn and Weisman, 1967). In contrast, smaller beads are accumulated at the surface of the cell until they reach the critical size, at which point they are ingested collectively within one vesicle (Korn and Weisman, 1967). In ingesting particular matter greater than 1 μm, *Acanthamoeba* trophozoite closely surrounds each particle with its plasma membrane (Korn and Weisman, 1967; Weisman and Korn, 1967). Following particle uptake, the membrane-enclosed particles then bud-off into the cytoplasm giving rise to a newly formed phagosome. Subsequently, phagosome (within 30 min) fuses with other vacuoles (lysosomes) to acquire hydrolytic enzymes (acid hydrolases) such as acid phosphatase, acid β-glucosidase, *N*-acetyl-glucosaminidase, α-glucosidase, galactosidase, *N*-acetyl-β-hexosaminidase, (Baldwin and Bowers 1995), bacteriolytic *N*-acetyl-muramidase, which also is secreted (Drozanski, 1978; Muller, 1969; Wetzel and Korn, 1969). This results in the formation of phago-lysosomes. To this end, it is shown that the acid hydrolases exhibit distinct patterns from the mitochondrial and peroximal patterns, indicating their lysosomal origin. Notably, the metabolic inhibitors, i.e., potassium cyanide (KCN) and 2,4-dinitrophenol have no effect on the fusion of the phagosome with the lysosome, suggesting that fusion reaction can proceed in the absence of oxidative metabolism (Oates and Touster, 1978). Overall, these data indicate that the newly formed phagosomes rapidly enter into *Acanthamoeba*'s lysosomal system with many of phagosomes becoming phago-lysosomes soon after ingestion (Oates and Touster, 1976). For pinocytosis, the calculated volume of fluid taken in during pinocytosis in the culture medium is about 2 μl per hour per 10^6 amoebae at 30°C (Bowers and Olszewski, 1972). Solutes of varying molecular weights, including albumin (molecular weight of 65 kDa), insulin (molecular weight of 5 kDa), glucose (molecular weight of 180 Da) and leucine (molecular weight of 131 Da), enter amoebae at a similar rate. There is no uptake at 0°C indicating that the uptake process is not simply a physical binding of solutes to the amoebae surfaces.

The fate of vacuoles within *Acanthamoeba* is dependent on the nature of particles, latex beads *versus* food particle. *Acanthamoeba* exhibit the ability to distinguish vacuoles containing digestible and indigestible particles. Vacuoles containing food particles are retained and digested, while latex beads are exocytosed, upon presentation of new particles. It is estimated that during active uptake, the rate of turnover of the surface membrane by *Acanthamoeba* is 10 – 50% of its surface per hour, which is remarkably high, compared to the other cell types (Bowers and Olszewski, 1972; Bowers and Olszewski 1983). For example, the macrophages internalize 50% of its surface in 2 – 5 hours during active pinocytosis (Cohn, 1971). In *Amoeba proteus*, up to 50% of the surface may be ingested during one pinocytotic cycle, but then a rest period is required before a new cycle can be initiated (Chapman-Andersen 1965). The internalization and recycling of *Acanthameba* cell surface membranes involves controlled local degradation of phospholipids, which would then reform following the acylation of the lysophospholipid. To complete this cycle, the *Acanthamoeba* plasma membrane contains all the necessary enzymes such as phospholipase A2, (that shows Ca^{2+}-independence), acyl CoA:lysolecithin acyltransferase, and acyl CoA synthetase (Victoria and Korn, 1975). In addition, phospholipase A1 and lysophospholipases are present in the plasma membrane of *Acanthamoeba* (Victoria and Korn, 1975). The phospholipase A activity is dependent on the population density of the cultures (Hax *et al.*, 1974). An increased phospholipase A activity in the older cultures of *Acanthamoeba* due to an increased population density induces aggregation of these cells in a way that communicative contacts are formed (Hax *et al.*, 1974). Furthermore, at pH 5.5 the phospholipase A activity is stimulated by cyclic AMP but at pH 8.0 no stimulation occurs (Hax *et al.*, 1974). The plasma membrane lysophospholipase may serve to protect the cell from the lytic effect of lysophospholipids, either of exogenous or endogenous origin. The plasma membrane of *Acanthamoeba* has the enzymatic capability of modulating the fatty acyl composition of phospholipids by deacylation and acylation (Robertson and Lands, 1964). However, the plasma membrane is incapable of synthesizing lecithin *de novo* by the cytidine dinucleotide pathway. The synthesis of lecithin appears to be confined almost entirely to the microsomal fraction of *Acanthamoeba*. This is in contrast to other protozoa, such as *Tetrahymena*, in which mitochondria are the main site of lecithin synthesis (Smith and Law, 1970). Overall, these studies suggest that particle uptake in *Acanthamoeba* is a complex process that may play a significant role, both in the food uptake as well as in the pathogenesis of *Acanthamoeba* infections (discussed in section D6.4). Future studies are needed to determine how *Acanthamoeba* discriminates between pinocytosis and phagocytosis, why they use one or the other, and whether there are any differences in this respect, between pathogenic and non-pathogenic *Acanthamoeba*.

7. Metabolism

The metabolism is a set of chemical reactions, which allow organisms to grow and reproduce, maintain their structures, and respond to their environments. The metabolism is usually divided into two categories: catabolic reactions that break down large molecules to harvest energy in cellular respiration, while anabolic reactions that use energy to construct components of cells, such as proteins and nucleic acids. The chemical reactions of metabolism are organized into the metabolic pathways, in which one chemical is transformed into another by a sequence of enzymes. The structures that make up eukaryotes are made from three basic classes of molecules: amino acids, carbohydrates and lipids (the latter are often called fats). As these molecules are vital for life, metabolism is focussed on making these molecules, in the construction of cells, or breaking them down and using them as a source of energy, in the digestion and its use as a food source (Fig. 7).

Sensitivity of *Acanthamoeba* cell respiration to cyanide varies markedly during the exponential growth. Cyanide was found to stimulate respiration if added to the early exponential phase cultures but inhibits respiration if added to the late exponential phase cultures. Although *Acanthamoeba* has an absolute O_2 requirement for cell division, it can survive anaerobic conditions for a period of several hours.

8. *Acanthamoeba* is a heterotroph (chemo-organotroph)

A heterotroph, or chemo-organotroph [Greek "heterone" means "another" and "trophe" means "nutrition"), is an organism that requires organic substrates to get its carbon for growth and development. A heterotroph is known as a

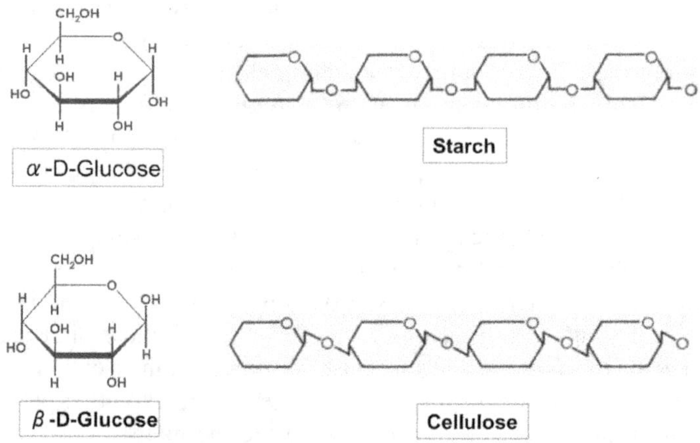

Figure 7. The structure of starch (polymer of α-glucose) and cellulose (polymer of β-glucose) are similar, i.e., formed by long glucose unit chains. Glycogen (polymer of α-glucose) however has a longer chain and is more highly branched as shown in Section A, figure 9B. The orientation of the units in cellulose results in a linear chain, which allows other chains to closely pack by hydrogen bonding. This increases the strength of cellulose (compared with starch) and accounts for its low solubility in water, hence it can be used to make clothes, ropes etc.

consumer in the chain. This is in contrast with autotrophs, which use inorganic carbon dioxide or bicarbonate as sole carbon source. All animals are heterotrophic, as well as fungi and many bacteria, but plants are mainly autotrophs except insectivorous which also are heterotrophs. Heterotrophs are unable to synthesize organic carbon-based compounds independently from the inorganic environment's sources. Therefore heterotrophs must obtain their nutrition from another heterotroph or an autotroph. Within heterotroph, there may be photoheterotroph, i.e., obtains energy from light and must obtain carbon in an organic form, and chemoheterotroph, i.e., obtains energy from the consumption of organic molecules and an organic form of carbon. In simple terms, a heterotroph is an organism that is incapable of making its own food from light or inorganic compounds, and feeds on organisms or the remains of other organisms to get its necessary energy to survive.

Being one of the most ubiquitous protozoa, *Acanthamoeba* has the ability to occupy diverse environments and this function requires properties to rapidly adapt to changes in the environment. Under favorable conditions, *Acanthamoeba* exhibits cellular movements, increased phagocytosis, cell division, and other cellular functions with increasing energy demands. To this end, *Acanthamoeba* possesses a large number of mitochondria, which exhibit properties similar to those from the mammalian cells (Klein and Neff, 1959). The energy is obtained by oxidizing organic compounds such as sugars, lipids and proteins. Enzymes (glycosidases) degrade complex sugars into simple molecules; while phospholipases degrade lipids, and proteases degrade proteins. *Acanthamoeba* synthesize fatty acids from acetate and contain glycolytic, glyoxylate, and the Krebs cycle enzymes. *Acanthamoeba* form UDP-glucose and use it as a glucosyl donor in the synthesis of glycogen, cellulose, glycolipids, and glycoproteins (reviewed in Weisman, 1976). *Acanthamoeba* is shown to grow in the minimal medium containing only arginine, methionine, leuine, isoleucine, and valine as sole nitrogen source, as long as glucose is present as the carbon source (Dolphin, 1976). However, if acetate is the carbon source then glycine must also be added to the minimal medium (Dolphin, 1976; Ingalls and Brent, 1983).

The majority of important chemical reactions such as the Krebs cycle required for the oxidation of sugars as well as for the fat oxidation occur within the mitochondria, thus extracting energy in the form of ATP. Under aerobic conditions, O_2 serves as the terminal acceptor of electrons (see below), while in anaerobic conditions, nitrate, sulfate, or fumarate can serve as terminal electron acceptors. Overall, this results in the synthesis of ATP as the energy source. As an example, glucose oxidation is described below:

$$C_6H_{12}O_6 + 6O_2 \longrightarrow 6CO_2 + 6H_2O + \text{energy}$$

The net result of a complete oxidation of a single glucose molecule is the production of 38 ATP molecules. If the substrate is protein, then ammonia is also produced (Fig. 8).

Like other eukaryotes, glucose is oxidized into CO_2 and water into the following three biochemical pathways including the glycolysis pathway, the Krebs cycle and the oxidative phosphorylation as described below.

8.1. Glycolysis pathway (also called Embden- Meyerhof-Parnas pathway)

As indicated above, glycolysis in *Acanthamoeba* is most likely similar to other eukaryotes. Glycolysis is used to extract carbon and energy from glucose. Glucose is transported into the cell and phosphorylated by phosphoenolpyruvate and transforms into glucose-6-phosphate by group translo-

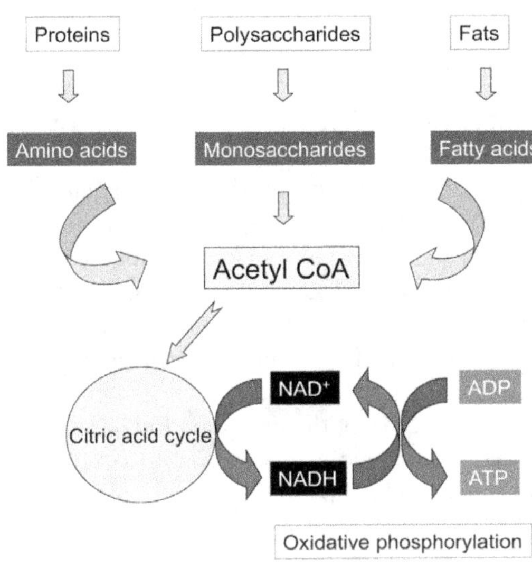

Figure 8. Summary of catabolic reactions.

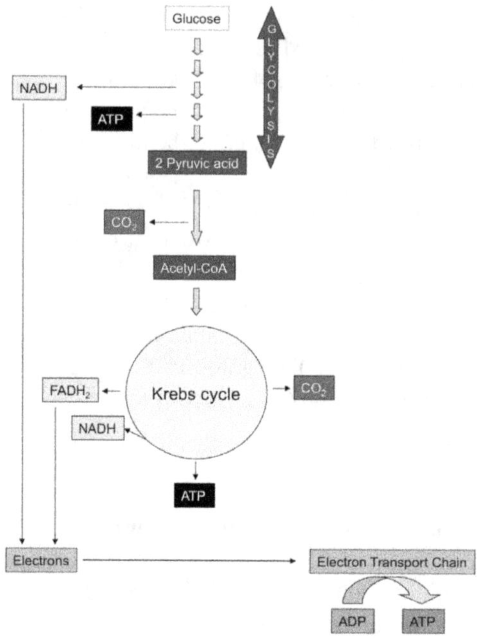

Figure 9. The summary of glucose metabolism.

cation. Glucose can also be transported into the cytoplasm by active transport without being phosphorylated. Glucose molecules transported via active transport are then phosphorylated by hexokinase enzymes, which transfer a phosphate group from an ATP molecule to the glucose resulting in 6C (carbon) glucose-6-phosphate. Glucose-6-phosphate is then phosphorylated again to 6C fructose-1,6-bihosphate (Fig. 9). Next, 6C fructose-1,6-biphosphate into two 3C pyruvate molecules (Fig. 9). In the process, two ATP are used while four ADP molecules are phosphorylated to make four ATP, i.e., net gain of 2ATP and two molecules of nicotinamide adenine dinucleotide (NAD^+) are reduced to NADH.

8.2. Krebs cycle (also called citric acid cycle or tricarboxylic acid cycle)

Generally in eukaryotic cells, before pyruvic acid can enter the Krebs cycle, it must be converted into acetyl-CoA (Fig. 10). In brief, enzymes remove one carbon from pyruvic acid as CO_2, and join it with coenzyme-A resulting in acetyl-coenzyme-A (acetyl-CoA), a process called decarboxylation (Fig. 10). Overall, two molecules of pyruvic acid will result in two molecules of acetyl-CoA and two molecules of CO_2, and two molecules of NADH. Finally acetyl-CoA enters into the Krebs cycle (Fig. 11). It occurs in the cytoplasm in prokaryotes and in the matrix of the mitochondria in eukaryotic cells. The overall outcome is that, for two molecules of acetyl-CoA, two ATP molecules are produced and six molecules NADH and two molecules of $FADH_2$ are produced.

However, the pathway for the conversion of the acetyl-CoA to glucosyl residues is not clearly understood in *Acanthamoeba*. The acetyl-CoA can be

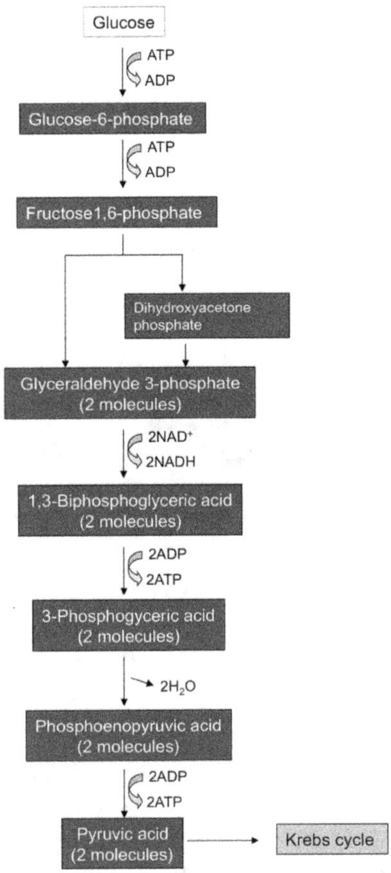

Figure 10. Summary of Glycolysis. Glucose is cleaved into two molecules of pyruvic acid. Four molecules of ATP are formed and 2 ATP are used, so net gain of 2 ATP. In addition, 2 molecules of NAD^+ are reduced to NADH.

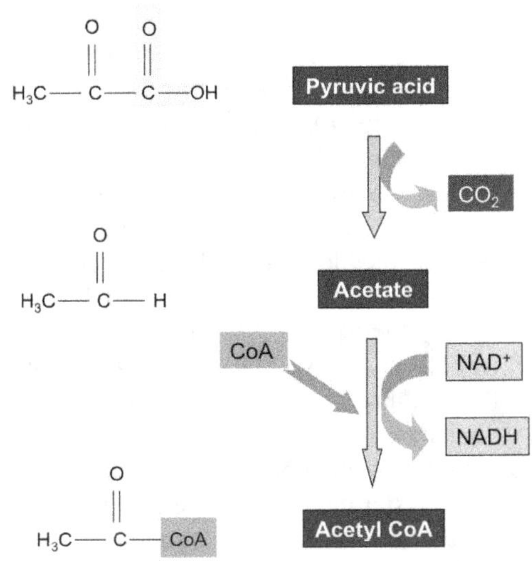

Figure 11. Formation of acetyl-CoA.

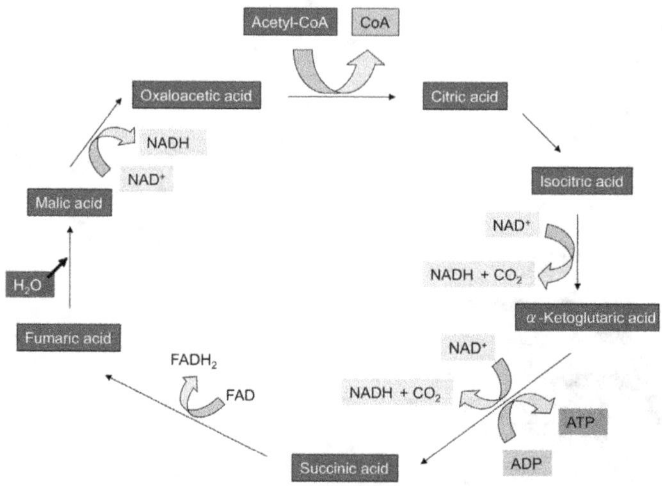

Figure 12. The Krebs cycle. Acetyl-CoA enters the Krebs cycle by joining with oxaloacetic acid to form citric acid and CoA.

metabolized to malate and oxaloacetate by the action of glyoxylate and the Krebs cycle enzymes, but as yet, no enzymes favouring the conversion of either of these dicarboxylic acids to phosphoenolpyruvate or pyruvate have been detected in *Acanthamoeba*.

8.3. Electron transport

The electron transport system consists of a number of membrane-bound molecules (inner membrane of mitochondria in eukaryotes and the cytoplasmic membrane in prokaryotes) that pass electrons from one to another, ultimately to a final electron acceptor, while pumping protons across the membrane (Fig. 12). This process is associated with a protein complex, ATP synthase that generates ATP. The NADH and $FADH_2$ donate electrons, which are used to generate energy, which is captured to synthesize ATP (Fig. 13). Generally in eukaryotes, the process initiates with NADH oxidation to NAD^+ transferring a proton and two electrons to a flavoprotein (flavin monnucleotide, FMN). Flavoproteins are integral-membrane proteins that alter between oxidized and reduced forms. Flavoprotein transfers two electrons to the next protein, iron-sulfur protein (FeS) while releasing the two protons outside the cytoplasmic membrane. The iron-sulfur protein transfers electron to lipid ubiquinones, called so as they are ubiquitous in the cell (also called Co-enzyme Q in mitochondria) which also picks up two protons as well, from the cytoplasmic side of the membrane. Under aerobic conditions, the ubiquinone carrying protons and electrons interact with cytochrome and donate two electrons to cytochrome while releasing two protons outside the cytoplasmic membrane. The cytochrome oxidase finally oxidizes cytochrome and transfer electrons to the terminal electron acceptor, i.e., oxygen (nitrate, sulfate, or organic compounds can be electron acceptors under anaerobic conditions), which is converted to water by cytochrome oxidase. This process

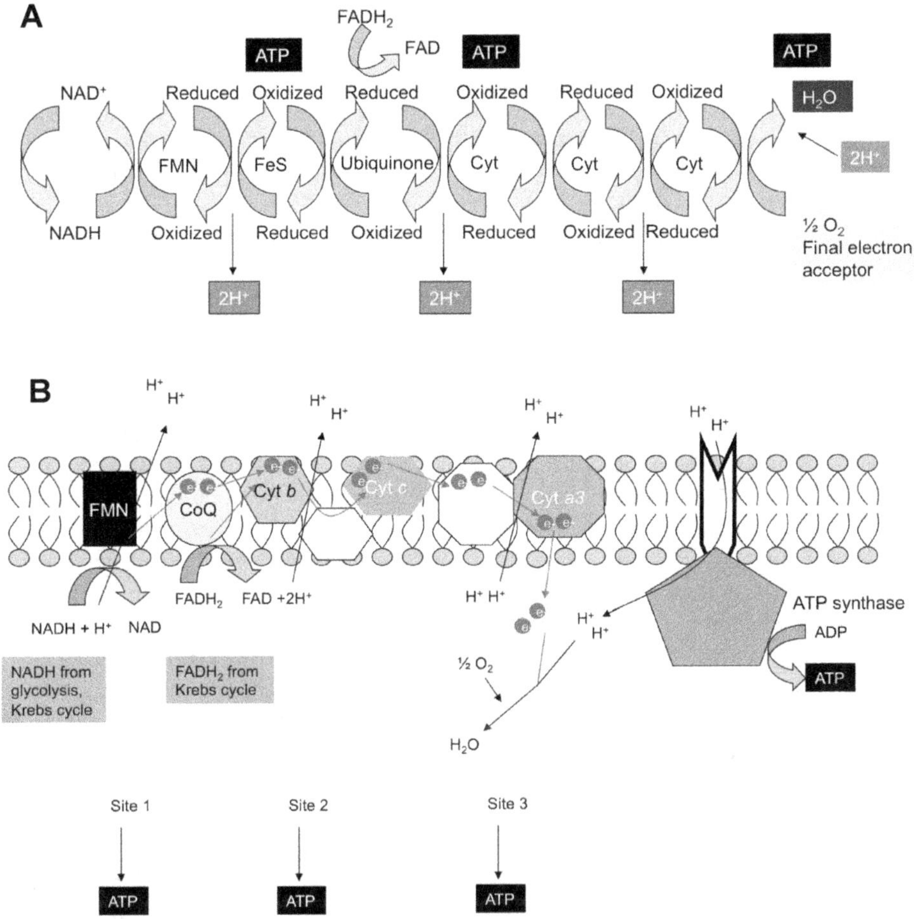

Figure 13. Electron transport chain (**A**). ATP production is indicated at approx. point in the chain (**A**) and a possible arrangement of an electron transport chain (**B**).

generates a proton gradient outside the cytoplasmic membrane, which re-enter the cytoplasm through a complex called ATP synthase inducing the phosphorylation of ADP to ATP. It takes around 2 protons to move through the ATP synthase to drive phosphorylation of an ADP to ATP. Although in eukaryotic organisms, cytochrome $a + a_3$ is normally the major terminal oxidase of the mitochondrial respiratory chain, mitochondria of *Acanthamoeba* in addition to the conventional cytochrome pathway possesses, like plant mitochondria, a cyanide-resistant alternative oxidase pathway exists. The steady-state activity of these two quinol-oxidizing pathways of *Acanthamoeba* mitochondria, is fixed by ubiquinone (Q) pool redox state independently of the reducing substrate [succinate or exogenous NADH], indicating that the active Q pool is homogenous (Jarmuszkiewicz *et al.*, 2002).

The branched mitochondrial electron-transport chain in *Acanthamoeba* terminates in three distinct oxidases (Edwards *et al.*, 1977; Edwards and Lloyd, 1977; 1978a, b), namely two oxidases with high affinities for oxygen [(i) major cytochrome *c* oxidase of the main phosphorylating electron-transport chain that is azide- and cyanide-sensitive and (ii) the salicylhydroxamic acid-sensitive alternative oxidase that is azide- and cyanide-insensitive)] and a third oxidase with a low affinity for oxygen, sensitive to inhibition by cyanide but not by azide or salicylhydroxamic acid and is responsible for the oxidation of NADH added externally to the mitochondria (Lloyd *et al.*, 1979). Later studies revealed that *Acanthamoeba* contains cytochrome a_3, cytochrome a_{614}, *b*- and *c*-type cytochromes (Lloyd *et al.*, 1981). Also unusual is that whereas in mammalian mitochondria cytochrome *c* oxidation is very slow and indistinct at -78°C, cold conditions did not disturb the cytochrome pathway in *Acanthamoeba* and the electron transport is evident in the mitochondria of *Acanthamoeba* at -80°C. Furthermore, in the mitochondria from low-temperature grown *Acanthamoeba*, the content of the alternative oxidase was increased (Lloyd *et al.*, 1981; Jarmuszkiewicz *et al.*, 2001).

Dibromothymoquinone acting as an analogue of ubiquinone inhibits mitochondrial electron transport at the ubiquinone-cytochrome *b* region of the respiratory chain (Trebst *et al.*, 1970) and in *Acanthamoeba* this compound (i.e., dibromothymoquinone) decreases electron flux through both the azide- and salicylhydroxamic acid-sensitive pathways but not through the third oxidase. Notably, the inhibitor sensitivities of these oxidases are similar to those of the three terminal oxidases of *Trypanosoma* (Hill, 1978). The relative electron flux through these oxidases changes as *Acanthamoeba* grows in a batch culture. For example, in the early phase of exponential growth all three oxidases are potentially active; later in growth electron-transport occurs mainly by way of the main phosphorylating cytochrome pathway but still with the possibility of some flux between alternative oxidase. In the stationary phase cultures, the oxidase with the low affinity for oxygen is not detectable at low oxygen tensions and electrons flow through the cytochrome *c* oxidase and/or the salicylhydroxamic acid-sensitive oxidase (Edwards *et al.*, 1977; Edwards and Lloyd, 1977; 1978a, b; Lloyd and Edwards, 1979; Lloyd *et al.*, 1979; Jarmuszkiewicz *et al.*, 2002).

The depletion of iron from the growth medium leads within two to three generations to respiration of trophozoites of *Acanthamoeba* from cyanide-stimulated to the cyanide-inhibited pattern (Hryniewiecka *et al.*, 1980). While addition of iron restores the activity of the alternative respiratory pathway within one hour (Hryniewiecka *et al.*, 1980).

9. Isolation of *Acanthamoeba* from the environmental samples

In natural environments, *Acanthamoeba* feeds on yeasts, other protozoa, bacteria, small organisms and organic particles. Perhaps, any of these can be used as growth substrates to isolate *Acanthamoeba* in the laboratory, but there are some technical problems. For example, the use of yeast and protozoa as growth substrates is problematic due to complexity in their preparations, and their possible overwhelming growth and the difficulty in eradicating yeast to obtain pure axenic *Acanthamoeba* cultures. Organic matters such as glucose, proteose peptone or other substrates provide rich nutrients for unwanted organisms i.e., yeasts, fungus, other protozoa and bacteria. To overcome these technical problems and to maximize the likelihood of *Acanthamoeba* isolation from the environmental (as well as clinical) samples, protocols have been developed using simple plating assays as described below. Both of the following methods can be used to obtain large number of *Acanthamoeba* trophozoites, which may be tested for further biochemical studies.

9.1. Xenic cultivation (non-nutrient agar plates seeded with Gram-negative bacteria)

The term "xenic cultivation" is used when *Acanthamoeba* (or other organisms) are cultured in association with another organism, e.g., bacteria, yeast etc. If there is only one organism in addition to *Acanthamoeba*, it is referred to as "monoxenic cultivation". This method has been used extensively in the isolation of *Acanthamoeba* both from the environmental and the clinical samples, worldwide. For *Acanthamoeba*, the basis of this method is the use of Gram-negative bacteria (*Escherichia coli* or *Enterobacter aerogenes*, the latter was formerly known as *Klebsiella aerogenes*; both are commonly used) that are seeded on the non-nutrient agar plate as a food source for *Acanthamoeba*. The non-nutrient agar contains minimal nutrients and thus inhibits the growth of unwanted organisms. Briefly, non-nutrient agar plates containing 1% (w/v) Oxoid no.1 agar in Page's amoeba saline (PAS) [2.5 mM NaCl, 1 mM KH_2PO_4, 0.5 mM Na_2HPO_4, 40 µM $CaCl_2.6H_2O$ and 20 µM $MgSO_4.7H_2O$] supplemented with 4 % (w/v) malt extract and 4 % (w/v) yeast extract are prepared, and the pH adjusted to 6.9 with KOH. Approximately, 5 ml of the late log phase cultures of Gram-negative bacteria (*Escherichia coli* or *Enterobacter aerogenes*) are poured onto the non-nutrient agar plates and left for 5 min, after which, the excess culture fluid is removed and plates are left to dry before their inoculation with an environmental sample or a clinical specimen. Once inoculated, plates are incubated at 30°C and observed daily for the presence of *Acanthamoeba* trophozoites. Depending on the number of *Acanthamoeba* in the acquired sample, trophozoites can be observed within a few hours (up to 12 hours). However if no *Acanthamoeba* are detected after this time, plates should be monitored for up to 7 days. Once bacteria are consumed, amoebae differentiate into characteristic cysts (Fig. 13). The confirmatory evidence of *Acanthamoeba* identity comes from the polymerase chain reaction or the immunofluorescence assays (described in section C2.9). The precise understanding of bacterial preference by *Acanthamoeba*, i.e., why *Escherichia coli* or *Enterobacter aerogenes* are used most commonly as food substrate, and whether bacterial preferences vary between *Acanthamoeba* isolates belonging to different genotypes/species is not known (Weekers *et al.*, 1993; Khan *et al.*, 2001; Khan and Paget, 2002). For example, *Chromatium vinosum* and *Serratia marcescens* did not or hardly supported *Acanthamoeba* growth, but why and how are questions for future studies.

9.2. "Axenic" cultivation

Acanthamoeba can be grown "axenically" in the absence of external live food organisms. This is typically referred to as "axenic culture" to indicate that no other living organisms are present. Notably, *Acanthamoeba* cultures may never be truly "axenic" as they may contain live bacteria, viruses or other microbes surviving internally as endosymbionts. However, strains such as the Neff strain have been grown in "axenic" media since 1959 with no such endosymbionts being detected. For simplicity the term "axenic" is used for *Acanthamoeba* grown in the liquid growth medium. Under laboratory conditions, axenic growth is achieved using liquid growth medium, known as PYG medium [proteose peptone 0.75% (w/v), yeast extract 0.75% (w/v) and glucose 1.5% (w/v)] at 30°C. To achieve axenic cultivation from xenic cultures, non-nutrient agar plates overlaid with bacteria are placed under ultraviolet light for 15 – 30 min to kill the bacterial lawn. A small piece of non-nutrient agar (stamp-sized) containing *Acanthamoeba* cysts are placed onto plates containing ultraviolet light-killed bacteria. When amoebae begin to grow, a stamp-sized piece of the agar containing trophozoites or cysts is transferred into 10 ml of sterile growth medium containing antibiotics, i.e., penicillin and streptomycin. The amoebae switch to the axenic growth medium as a food source, and their multiplication can be observed within several days. Once multiplying in the growth medium, *Acanthamoeba* is typically grown aerobically in tissue-culture flasks with filter caps at 30°C in static conditions. The trophozoites adhere to the flask walls and are collected by chilling the flask on ice for 15 – 30 min, followed by centrifugation of the medium containing the cells.

9.3. Chemically-defined medium

A chemically defined medium containing 11 amino acids, 3 vitamins, 6 inorganic salts and glucose, yielding maximum cell densities of 1.5 – 2.5 x 10^7 cells per ml, has been developed for *Acanthamoeba* with a mean generation time of 10 hours. A medium containing six amino acids *viz.* arginine, methionine, leucine, isoleucine, valine and glycine along with other components could also support good albeit slower growth (mean generation time of 27 hours) of *Acanthamoeba*. Interestingly, acetate does not serve as a suitable carbon source for *Acanthamoeba* (Shukla *et al.*, 1990).

10. Methods of encystation

Both xenic and axenic methods have been developed to obtain *Acanthamoeba* cysts. For xenic cultures, *Acanthamoeba* are inoculated onto non-nutrient agar plates seeded with bacteria as indicated above. Plates are incubated at 30°C until the bacteria are cleared and trophozoites have transformed into cysts. Cysts can be scraped off the agar surface using phosphate buffered saline and used for assays. This resembles the most likely natural mode of encystation and can be effective, achieving up to 100% cysts. However, one major limitation may be the presence of bacterial contaminants that could hamper molecular and biochemical studies. For axenic encystation, amoebae are grown in the growth medium for 17 – 20 hours as described above. After this incubation, 8% glucose in RPMI 1640 is added to stimulate encystation. Alternatively $MgCl_2$ can be added to induce encystation. Plates are incubated at 30°C for up to 48 hours. To confirm transformation of trophozoites into cysts, sodium dodecyl sulfate (0.5% final concentration) is added. Trophozoites are SDS-sensitive and are lysed immediately, while cysts remain intact (Cordingley *et al.*, 1996; Dudley *et al.*, 2005). This method allows the simple counting of cysts using a hemocytometer and is useful in studying the process of encystation.

11. Storage

For short-term storage, *Acanthamoeba* cultures maintained on non-nutrient agar plates. Plates inoculated with *Acanthamoeba* can be kept at 4°C under moist conditions for several months, or as long as plates are protected from drying out. Cysts can be re-inoculated into the growth medium in the presence of antibiotics (penicillin and streptomycin) to obtain axenic cultures as described above. Alternatively, *Acanthamoeba* trophozoites can be stored as axenic cultures for long-term storage. Briefly, exponentially growing amoebae are resuspended at a density of 3 – 5 x 10^6 parasites per ml in the freezing medium (PYG containing 10% dimethylsulfoxide) (John and John, 1994). Cultures are transferred to –20°C for 60 min, followed by their storage at –70°C or in the liquid nitrogen indefinitely. *Acanthamoeba* cultures can be revived by thawing at 37°C, followed by their immediate transfer to the growth medium in a T-75 flask at 30°C. The inclusion of 20% heat-inactivated foetal bovine serum in the freezing medium has been shown to improve the revival viability of *Acanthamoeba* in the long-term storage. Alternatively clay plugs or filter paper can be sued to store cysts for years (John *et al.*, 1994).

12. Determination of the viability of trophozoites and cysts

Following treatment with the antimicrobial compounds or other inhibitors, viability of trophozoites and cysts can be determined using various methods as described below.

A simple but confirmatory assay is to inoculate treated trophozoites and cysts onto non-nutrient agar plates seeded with bacteria. As described above, bacteria are preferential food substrate for *Acanthamoeba* and any viable cells would re-emerge as trophozoites (Fig. 13). A major limitation with this assay is that it is time-consuming, and it may take up to several days (around 7 days), before it can be concluded that cells are non-viable. Alternatively, treated *Acanthamoeba* can be inoculated into the growth medium and their growth can be used as a criterion for viability.

Alternatively, Trypan blue exclusion test provides a useful indicator to determine the viability of the trophozoites and cysts. The Trypan blue is a stain (derived from toluidine) that selectively colours dead tissues or cells blue. Live cells with intact cell membranes are not coloured. Since cells are very selective in the compounds that pass through the membrane, in a viable cell, Trypan blue is not absorbed; however, it traverses the membrane in a

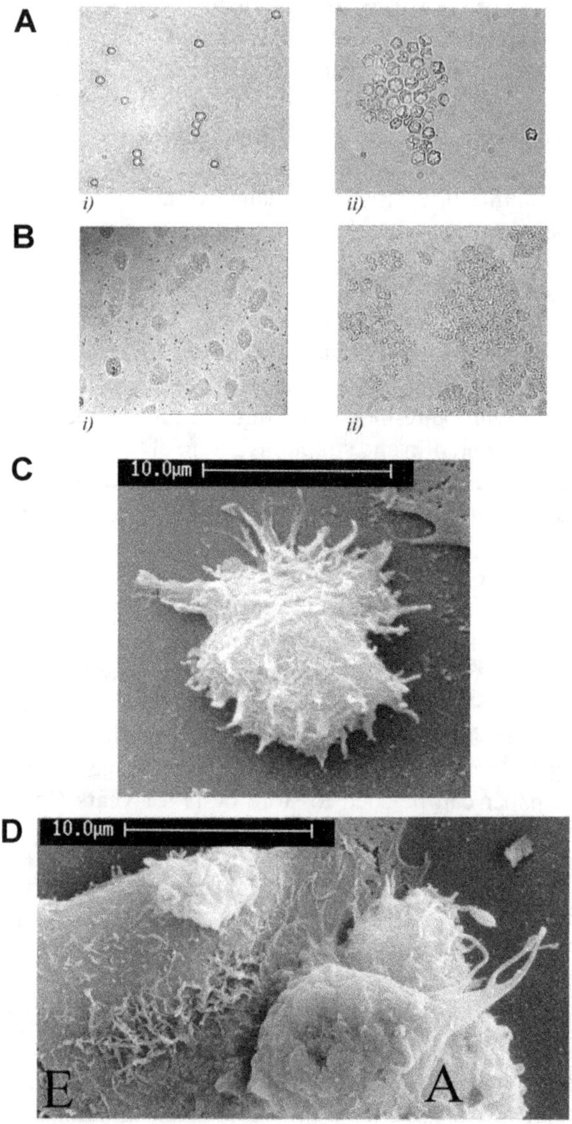

Figure 14. A) *Acanthamoeba* cysts under phase-contrast microscope. ***i)*** Non-nutrient agar plates exhibiting *Acanthamoeba* cysts. ***ii)*** *Acanthamoeba* cysts were collected from non-nutrient agar plates using PBS and observed under the phase-contrast microscope. Note cysts formed clusters in PBS, X400. **B)** *Acanthamoeba* trophozoites on non-nutrient agar plates observed under phase-contrast microscope. Note the characteristic contract vacuole in *Acanthamoeba* trophozoites, X400. **C)** *Acanthamoeba* trophozoite binding to glass cover slips observed under scanning electron microscope. Note large number of acanthopodia on the surface of *Acanthamoeba* trophozoites belonging to the T4 genotype. **D)** *Acanthamoeba* binding to corneal epithelial cells. *Acanthamoeba castellanii* (T4 isolate) were incubated with corneal epithelial cells, followed by several washes and observed under scanning electron microscope. Note that parasites were able to exhibit binding to the host cells and binding was mediated by acanthopodia. A is amoeba, E is corneal epithelial cell. Bar = 10 µm.

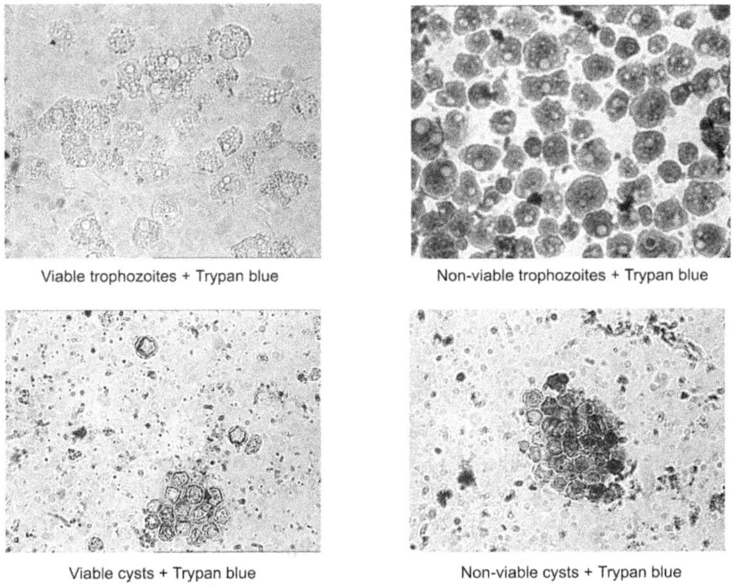

Figure 15. Trypan blue staining of viable and dead trophozoites and cysts of *Acanthamoeba*. X400.

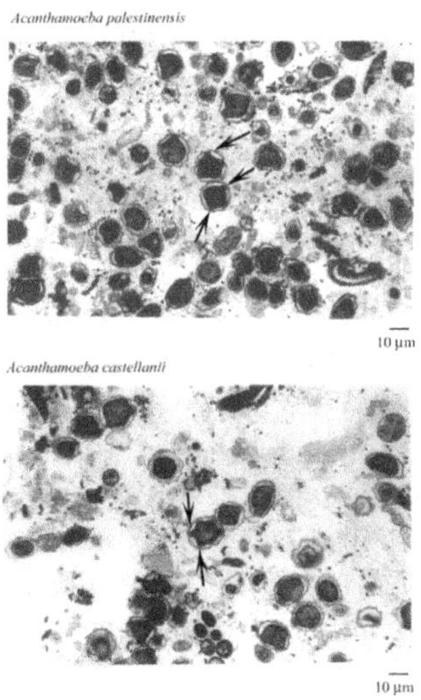

Figure 16. *Acanthamoeba* cysts exhibiting characteristic arm-like structures.

dead cell. Hence, dead cells are shown as distinctive blue colour under a microscope. Since live cells are excluded from staining, this staining method is also described as a dye exclusion method. Overall, when incubated with cells, this dye (0.5% concentration) is expelled or does not enter live cells (which appear non-stained) but dead cells are stained blue (Fig. 14). The Trypan blue exclusion test may thus be a good method of the estimation of the viability of trophozoites and cysts of *Acanthamoeba* (Mazur and Zozwiak 1989).

In addition to the aforementioned test, eosin colour test has been shown to provide an estimate of the viability of cysts. Cysts stained with eosin are dead, as supported by their non-viability on non-nutrient agar plates seeded with bacteria, i.e., eosin stained cysts did not excyst. As for Trypan blue exclusion test, the eosin test may also be a good method of the estimation of the viability of *Acanthamoeba* trophozoites and cysts (Mazur and Zozwiak 1989).

13. Speciation and genotyping

Following the discovery of *Acanthamoeba*, several isolates belonging to the genus *Acanthamoeba* with distinct morphology were isolated and given different names based on the isolator, source, or other criteria. In an attempt to organize the increasing number of isolates belonging to this genus, it was classified based on morphological characteristics of the cysts, which were the most appropriate criteria at the time in the 1970s (Pussard and Pons, 1977). Based on these characteristics, the genus *Acanthamoeba* was classified into three groups based only on two obvious characters i.e., cyst size and the number of arms within a single cyst (Figs. 15 and 16). Based on this scheme, the genus *Acanthamoeba* (18 species at the time) was divided into three groups. Subsequently, this classification scheme gained acceptance (De Jonckheere, 1987; Page, 1988).

Group 1: Four species were placed in this group including *A. astronyxis*, *A. comandoni, A. echinulata and A. tubiashi*. These species exhibit large trophozoites, while in the cyst forms ectocyst and endocyst are widely separated and exhibit the following properties.

1. Less than 6 arms with the average diameter of cysts at ≥18 μm, *A. astronyxis*.
2. 6-10 arms and the average diameter of cysts at ≥25.6 μm, *A. comandoni*.
3. 12-14 arms and the average diameter of cysts at ≥25 μm, *A. echinulata*.
4. The average diameter of cysts at ≥22.6 μm, *A. tubiashi*.

Group 2: This group included 11 species, which are the most widespread and commonly isolated *Acanthamoeba*. Ectocyst and endocyst may be close together or widely separated. Ectocyst may be thick, thin and endocyst may be polygonal, triangular or round with the mean diameter of less than 18 μm. The species included in this group were *A. mauritaniensis, A. castellanii, A. polyphaga, A. quina, A. divionensis, A. triangularis, A. lugdunensis, A. griffini, A. rhysodes, A. paradivionensis* and *A. hatchetti*.

Group 3: Five species were included in this group 3, including *A. palestinensis, A. culbertsoni, A. royreba, A. lenticulata, A. pustulosa*. The ectocyst in this group is thin and endocyst may have 3-5 gentle corners with the mean cysts diameter at <18 μm.

Later, *A. tubiashi* in group 1 and *A. hatchetti* in group 2 were added (Visvesvara, 1991).

13.1. Use of 18S rRNA gene sequences (18S rDNA)

From above it is obvious that identification of the various species of *Acanthamoeba* by morphological features alone is problematic. In addition, several studies have demonstrated inconsistencies and/or variations in the cyst morphology belonging to the same isolate/strain. For example, it was discovered that the ionic strength of the growth medium could alter the shape of cyst walls (Sawyer, 1971), thus substantially reducing the reliability of the cyst morphology as a taxonomic characteristic. Furthermore, this scheme had limited value in associating pathogenesis with a named species. For example, several studies demonstrated that strains/isolates within *A. castellanii* can be virulent, weakly-virulent or avirulent. This discrepancy in assigning an unambiguous role to a given species presented a clear but urgent need to reclassify this genus. The discovery of advanced molecular techniques led to the pioneering work of the late Dr. T. Byers (Ohio University, USA) in the classification of the genus *Acanthamoeba* based on rRNA gene sequences. Because life evolved most likely through self-replicating RNA as the genetic material or as common ancestor and evolved into diverse forms, it is reasonable to study the evolutionary relationships through such molecules, i.e., rRNA. In addition, this is a highly conserve, precise, reliable, and informative scheme. Each base represents a single character providing an accurate and diverse systematic. Sequences of either nuclear 18S rRNA genes

Table 2. Known *Acanthamoeba* genotypes and their associations with human diseases, i.e., keratitis and granulomatous encephalitis.

Acanthamoeba genotypes	Human disease association
T1	Encephalitis
^T2a	Keratitis, Encephalitis
^T2b – ccap1501/3c-alike sequences	NA
T3	Keratitis
T4*	Encephalitis, Keratitis
T5	Keratitis
T6	Keratitis
T7	NA
T8	NA
T9	NA
T10	Encephalitis
T11	Keratitis
T12	Encephalitis
T13	NA
T14	NA
T15	NA

*this genotype has been most associated with both diseases
^basis of T2 division into T2a and T2b has been proposed by Maghsood *et al.,* (2005)
NA – no disease association has been found yet

(18S rDNA) or mitochondrial 16S rRNA genes (16S rDNA) are suitable for classifying isolates. Based on rDNA sequences, the genus *Acanthamoeba* is divided into 15 different genotypes (T1 – T15) (Table 2), (Schuster and Visvesvara, 2004). Each genotype exhibits 5% or more sequence divergence between different genotypes. Note, it is recently (Maghsood *et al.*, 2005) proposed to subdivide T2 further into two groups, i.e., T2a and T2b. This is due to the sequence dissimilarity of 4.9% between these two groups, which is very close to the current cut-off limit of 5% between different genotypes. This should help differentiate pathogenic and non-pathogenic isolates within this genotype. With the clear advantage of rDNA sequences over morphology-based classification, I have made an attempt here to refer to the genotype rather than species name, wherever possible in this book.

Based on this classification scheme, the majority of human infections due to *Acanthamoeba* have been associated with the isolates of the T4 genotype. For example, more than 90% of the keratitis cases have been linked with this genotype. Similarly, T4 has been the major genotype associated with the non-keratitis infections such as *Acanthamoeba* granulomatous encephalitis and cutaneous infections. At present, it is unclear why T4 isolates are most abundant in human infections and studies are in progress to address these issues. Notably,, recent findings suggest that the abundance of T4 isolates in human infections is most likely due to their greater virulence and/or properties that enhance their transmissibility as well as their decreased susceptibility to chemotherapeutic agents (Maghsood *et al.,* 2005). Future studies will identify virulence traits and genetic markers limited only to certain genotypes (e.g., T4 genotype), which may help clarify these issues. A current list of genotypes and their association with the human infections is presented in Table 2. With the increasing research interests in the field of *Acanthamoeba* and the worldwide availability of advanced molecular techniques, undoubtedly additional genotypes will be identified. These studies will help clarify the role of *Acanthamoeba* in the ecosystem, and its symbiosis with microbial pathogens, as well as in causing primary and secondary human infections.

From: *Acanthamoeba*: Biology and Pathogenesis. By: Naveed Khan.

Section C

Acanthamoeba Infections

1. Human infections

Acanthamoeba causes two well-recognized diseases in humans: a rare granulomatous encephalitis involving the central nervous system that is limited typically to immunocompromised people and almost always results in death, and a painful keratitis that can result in blindness. These infections are becoming increasingly important in human health. This is due an exponential increase in the number of immunocompromised individuals and contact lens wearers.

2. *Acanthamoeba* keratitis

First discovered in the early 1970s, *Acanthamoeba* keratitis has become a significant ocular infection (Fig. 1). Keratitis is a condition in which the eye's cornea, i.e., the front part of the eye, becomes inflamed (Fig. 2). *Acanthamoeba* keratitis is often marked by intense pain and usually involves impaired eye sight. A key predisposing factor in *Acanthamoeba* keratitis is the use of contact lenses exposed to contaminated water, but the precise mechanisms associated with this process are not fully understood. Overall this is a multifactorial process that involves (i) contact lens wear for extended periods of time, (ii) lack of personal hygiene, (iii) inappropriate cleaning of contact lenses, (iv) biofilm formation on contact lenses, and (v) the exposure to contaminated water.

Figure 1. *Acanthamoeba* keratitis has become a significant problem in recent years, especially in contact lens wearers exposed to contaminated water.

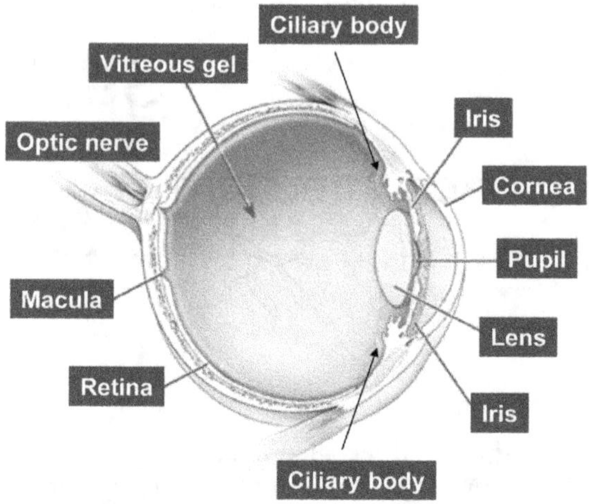

Figure 2. Anatomy of the eye and cellular structure of the cornea.

The sequence of events in *Acanthamoeba* keratitis involves breakdown of the epithelial barrier, stromal invasion by *Acanthamoeba*, keratocyte depletion, induction of an intense inflammatory response and finally stromal necrosis (Garner, 1993; Vemuganti *et al.,* 2004). One of the modes of keratocyte loss in *Acanthamoeba* keratitis is by apoptosis, in addition to the necrotic process and phagocytic activity of the parasite. The death of inflammatory cells also appears to be mediated by apoptosis (Vemuganti *et al.*, 2000). Below, I have discussed the possible steps involved in contracting *Acanthamoeba* keratitis.

2.1. *Acanthamoeba* binding to worn and unworn contact lenses

It is interesting that *Acanthamoeba* trophozoites adhere to the unworn soft contact lenses and show the presence of acanthopodia, lobopodia, and filopodia. Moreover, *Acanthamoeba* trophozoites show binding to almost all types of lenses including low-water, and high-water content soft contact lenses and rigid gas-permeable contact lenses. However, a higher binding is reported to soft contact lenses with higher water content, compared with the rigid gas-permeable contact lenses. Notably, *Acanthamoeba* cysts also showed attachment to the soft lenses but not to the rigid gas-permeable lenses. Attachment of *Acanthamoeba* cysts is significantly lower than that of trophozoites to each lens type tested. Moreover, the adherence of *Acanthamoeba* cysts and trophozoites to disposable lenses is less than that to daily wear and extended wear lenses. However, any adherence of *Acanthamoeba* to the soft contact lenses is a potential source of corneal infection. Overall, these findings suggest that contact lens may provide means to introduce *Acanthamoeba* from the environment onto the cornea.

For worn contact lenses, it is shown that extended wear of contact lenses results in their increased affinity for *Acanthamoeba*. In support, it is shown that *Acanthamoeba* exhibits significantly higher binding to worn contact

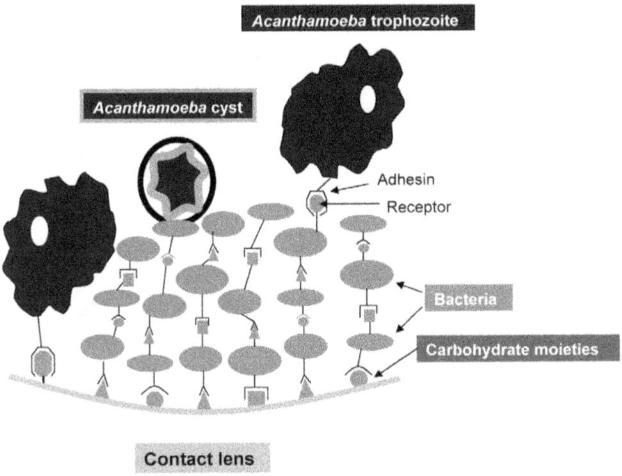

Figure 3. Diagrammatic representation of biofilm formation on contact lens surface. Initial bacterial co-aggregation involves protein-saccharide interactions leading to microbial colonization including *Acanthamoeba*. Note carbohydrate residues (α-mannose) on contact lens may directly act as receptors for *Acanthamoeba* trophozoites.

lenses as compared with the unworn contact lenses. Tests to determine the molecular differences between worn and unworn lenses revealed that worn contact lenses showed the presence of saccharides including mannose, glucose, galactose, fucose, *N*-acetyl-glucosamine, *N*-acetyl-galactosamine, *N*-acetyl-neuraminic acid (sialic acid), and proteins, glycoproteins, lipids, mucins, polysaccharides, calcium, iron, silica, magnesium, sodium, lactoferrin, lysozyme and immunoglobin molecules on the surface of contact lenses, after only 30 min of contact lenses being worn (Gudmundsson *et al.*, 1985; Klotz *et al.*, 1987; Tripathi and Tripathi, 1984). A possible explanation of increased affinity of worn contact lenses for *Acanthamoeba* is that these molecules may act as receptors for *Acanthamoeba* trophozoites and/or enhance parasite ability to bind to contact lenses. For example, *Acanthamoeba* expresses a mannose-binding protein on its surface membranes, which specifically binds to mannose residues. This may explain the ability of *Acanthamoeba* to exhibit higher binding to worn contact lenses compared with the unworn contact lenses (Nagington *et al.*, 1974; Beattie *et al.*, 2003; John *et al.*, 1989; 1991; Kilvington and Larkin, 1990).

2.2. Biofilm formation

The formation of a biofilm on contact lenses may prove a critical step in contracting *Acanthamoeba* keratitis. The basis of biofilm formation is bacterial contamination of the contact lens and/or storage case, which is highly common in contact lens wearers (Fig. 3). For example, bacterial contamination of storage cases for contact lenses of 102 asymptomatic lens wearers revealed that 83 were contaminated with bacteria. Interestingly, contact lens cases with the highest bacterial counts showed the presence of *Acanthamoeba* (i.e., 7 out of 83) (Larkin *et al.*, 1990). The biofilm may play a role in contracting *Acanthamoeba* by increasing attachment affinity for amoebae as well as promoting *Acanthamoeba* growth. In support, several lines of evidence suggest that the biofilm formation on contact lenses provides increased attachment affinity for *Acanthamoeba*. This is shown by increased *Acanthamoeba* binding to biofilm-coated contact lenses as opposed to contact lenses without biofilms (Beattie *et al.*, 2003; Simmons *et al.*, 1998; Tomlinson *et al.*, 2000). For example, biofilm-coating with *Pseudomonas aeruginosa* gave a significantly increased attachment of *Acanthamoeba* trophozoites to the contact lens. Interestingly, sodium salicylate reduced *Acanthamoeba* trophozoite attachment to the hydrogel lens, but the precise mechanisms of this inhibition are unclear. These findings suggested salicylate's potential benefit as a component of contact lens care solutions, designed to reduce microbial attachment and the subsequent risk of infection (Tomlinson *et al.*, 2000). In addition to a role in attachment, biofilms provide nutrients for *Acanthamoeba* and may allow amoebae growth. Notably, *Acanthamoeba* possess alginate lyase (alginate is an important component of the biofilm matrix) which may provide access to bacteria within biofilms by breaking down the biofilm matrix. In addition, *Acanthamoeba* is shown to possess polyhydroxybutyrate depolymerase (polyhydroxybutyrate is a major bacteria storage compound), which may facilitate amoebae to utilize bacteria as food source. In addition, biofilms may enhance *Acanthamoeba* persistence in the presence of antimicrobials during contact lens cleaning and/or storage in the presence of antimicrobials. Overall, these findings suggest that biofilm-coated contact lens increase the likelihood of contracting *Acanthamoeba*, if exposed to amoebae-contaminated environment.

2.3. Corneal injury

Once an *Acanthamoeba*-contaminated lens is placed over the cornea, amoebae transmit to the cornea. *Acanthamoeba* transfer onto the cornea may also be dependent on the physiological status of the cornea. For example, *in vivo* studies revealed that animals with intact corneas (i.e., with intact epithelial cells) do not develop *Acanthamoeba* keratitis (Niederkorn *et al.*, 1999). It was shown further that the corneal trauma plays an important role in contracting *Acanthamoeba* keratitis, *in vivo*.

The presence of corneal trauma seems to be particularly important in *Acanthamoeba* keratitis cases in non-contact lens wearers. In the non-contact lens wearers, it is observed that the corneal trauma followed by the exposure to contaminated water, soil, vector (inert objects or biological) is sufficient to contract amoebae, resulting in *Acanthamoeba* keratitis (Chang and Soong, 1991; Sharma *et al.*, 1990).

The requirement of corneal trauma may be explained by the fact that the expression of *Acanthamoeba*-reactive glycoprotein(s) on the damaged corneas is 1.8 times higher, compared with the healthy ones, suggesting that corneal injury contributes to *Acanthamoeba* keratitis (Jaison *et al.*, 1998). Future studies will determine whether corneal injury simply exposes mannose-containing glycoprotein(s), providing additional binding sites for *Acanthamoeba*, or whether the expression of mannose-glycoprotein(s) is generally higher on the healing corneal epithelial cells. It is important to note that *Acanthamoeba* must be present in the trophozoite stage to bind to the human corneal epithelial cells. Recent studies have shown that *Acanthamoeba* cysts do not bind to the human corneal epithelial cells, indicating that cysts must transform into trophozoites to initiate the infection (Dudley *et al.*, 2005; Garate *et al.*, 2006).

2.4. Epidemiology

The first case of *Acanthamoeba* keratitis was diagnosed in 1973 in a man with a history of ocular trauma and exposure to contaminated water. Originally thought to be a rare infection, *Acanthamoeba* keratitis has now become increasingly important in human health. This is due to increased awareness and the availability of diagnostic methods. Over the last few decades, it has become clear that the users of contact lenses are at increased risk of corneal infections. For example, contact-lens wearers are 80-fold more likely to contract corneal infection than those who do not wear contact lenses (Alvord *et al.*, 1998; Dart *et al.*, 1991). The incidence rate of microbial keratitis in users of extended-wear contact lenses is determined at 20.9 per 10,000 wearers per annum in the USA (Poggio *et al.*, 1989). Similar findings have been reported from Sweden (Nilsson and Montan, 1994), Scotland (Seal *et al.*, 1999) and the Netherlands (Cheng *et al.*, 1999). At present there are approximately 120 million people wearing contact lenses, throughout the world (Barr, 1998). Furthermore, with their wider potential applications beyond vision correction, such as ultraviolet light protection and cosmetic purposes, this number will undoubtedly rise. With an increasing number of people wearing contact lenses, it is important to assess any associated risks, and to make both existing and new users aware of such risks. Among other microbial agents, bacteria including *Pseudomonas*, *Staphylococcus*, protozoa including *Acanthamoeba*, and fungi including *Fusarium* are the major causes of corneal infections in the users of contact lenses (Giese and Weissman, 2002). The incidence rate of *Acanthamoeba* keratitis varies between different geographical locations as well as between different surveys. For example, in some surveys, the incidence rate of *Acanthamoeba* keratitis was reported as 0.33 per 10,000 contact lens wearers in Hong Kong, 0.05 per 10,000 in Holland, 0.01 per 10,000 in the USA (Stehr-Green *et al.*, 1989), 0.19 per 10,000 in England (Radford *et al.*, 2002), and 1.49 per 10,000 in Scotland (Lam *et al.*, 2002; Seal *et al.*, 1999). It is noteworthy that variations in the incidence rate of *Acanthamoeba* keratitis do not reflect the geographical distribution of *Acanthamoeba*, and are most likely due to variations in the extended wear of soft contact lenses, varied awareness of the potential risks associated with the contact lens wear, enhanced detection, and/or local conditions that promote growth of pathogenic *Acanthamoeba* only e.g., water hardness or salinity, or conditions that suppress growth of non-pathogenic *Acanthamoeba*.

2.5. Pathophysiology

The onset of symptoms associated with *Acanthamoeba* keratitis can take from a few days to several weeks, depending on the inoculum size of

Table 1. Symptoms associated with *Acanthamoeba* granulomatous encephalitis and *Acanthamoeba* keratitis.

***Acanthamoeba* granulomatous encephalitis**	***Acanthamoeba* keratitis**
Headache	Conjunctivitis
Fever	Corneal ulceration
Sleep disturbances	Conjunctival hyperemia
Hemiparesis	Lid edema
Mental status changes	Epithelial defects
Visual disturbances	Considerable production of tears
Ataxia	Photophobia
Nausea and vomiting	Severe pain
Hallucinations	Stromal abscess
Personality changes	A characteristic corneal ring stromal infiltrate
Photophobia	Increased intraocular pressures
Seizures	Hypopyon
Coma	Cataract formation
Skin infection may predate the onset of the central nervous system manifestations and may include ulcers, nodules, or subcutaneous abscesses.	Anterior uveitis of fluctuating severity. Anterior uveitis occurs when the iris and/or the ciliary body become inflamed.

Acanthamoeba and/or the extent of the corneal trauma. During the course of infection, symptoms may vary depending on the clinical management of the disease (Table 1). Most commonly *Acanthamoeba* keratitis is associated with considerable production of tears; eyelid ptosis (i.e., abnormally low position or drooping of the upper eyelid); conjunctival hyperemia (i.e., engorgement conjunctival blood vessels or increased blood flow to the conjunctiva); epithelial defects (damage to the outermost layer of the cornea); photophobia (excessive sensitivity to the light) which leads to the inflammation with redness, stromal infiltration, edema, stromal opacity, together with excruciating pain due to radial neuritis, epithelial loss and stromal abscess formation with vision-threatening consequences (Fig. 4). Other symptoms may involve scattered sub-epithelial infiltrates, abscess in the stroma, anterior uveitis (also referred to as iritis as the iris is the part of the uvea that is usually inflamed, the ciliary body may also be involved and this would be termed iridocyclitis). Anterior uveitis occurs when the iris and/ or the ciliary body becomes inflamed. In addition, stromal perforation, and the presence of scleral inflammation is observed. As *Acanthamoeba* keratitis can progress to scleritis (inflammation of the white outer coating of the eye, known as the sclera), and in severe cases, uncontrolled infection requires the removal of the affected eye. Secondary infection due to bacteria may additionally complicate the clinical management of the disease. Glaucoma (i.e., damage of the optic nerve, usually due to raised intraocular pressure) is commonly reported, and occasionally posterior segment signs such as nerve edema, optic atrophy or neuropathy (i.e., the loss of some or most of the fibers of the optic nerve, "atrophy" usually means shrunken but capable of

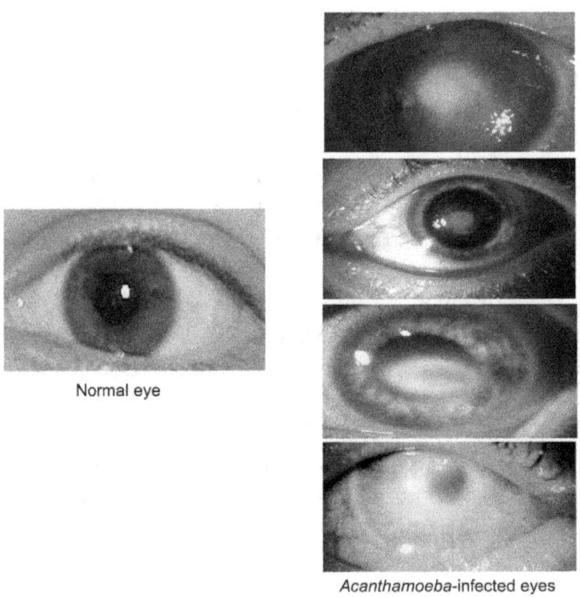

Figure 4. Normal eye and an *Acanthamoeba*-infected eyes exhibiting ulcerated epithelium and corneal opacity in acute *Acanthamoeba* keratitis.

re-growth, so term "optic neuropathy" is also used); and retinal detachment are observed. In untreated eyes, blindness may eventually result as the necrotic region spreads inwards (Niederkorn *et al.,* 1999). Some patients have been presented with a pseudodendritic epithelial lesion (dendritic ulcer morphology) that gradually progressed to the stromal involvement, while in other cases patients presented with the central stromal infiltrate and necrosis, and others exhibited features of a disciform lesion (flat and round lesion) with the development of an immune ring (Johns *et al*., 1987).

2.6. Is *Acanthamoeba* keratitis limited to cornea?

Unlike bacterial keratitis, *Acanthamoeba* keratitis rarely progresses beyond the cornea to produce intraocular infections and endophthalmitis (an inflammation of the internal coats of the eye). Following corneal epithelial damage, penetration of the Bowman's membrane and desquamating of the stroma occurs, but it rarely invades the corneal endothelium and the anterior chamber. This is in contrast to the *in vitro* studies, which suggested that *Acanthamoeba* trophozoites kill corneal endothelial cells. Only a handful of cases of *Acanthamoeba* endophthalmitis have been reported in which *Acanthamoeba* has been shown to be the causative agent of intraocular infection. Furthermore, it is questionable whether these cases of endophthalmitis were due to the natural progression of the trophozoite beyond the corneal endothelium. A likely explanation is that *Acanthamoeba*, unlike other ocular pathogens, does not progress and produce intraocular infections. To determine whether *Acanthamoeba* exhibits properties to penetrate Descemet's membrane (i.e., basement membrane that lies between the corneal stroma, and the endothelial layer of the cornea) and the corneal endothelium to enter the anterior chamber and survive the aqueous humor. Using *in vitro* assays, it was shown that the amoebae penetrated Descemet's membrane within 24 hours. Penetration was dependent on serine protease activities of *Acanthamoeba* as demonstrated by inhibition using a serine protease inhibitors or a chicken monoclonal antibody against the *Acanthamoeba* serine protease. Although the aqueous humor induced encystation in *Acanthamoeba*, however cysts remained viable. To determine the capacity of *Acanthamoeba* to survive in the anterior chamber, trophozoites were injected into the anterior chamber of hamster eyes, *in vivo*, and the number of amoebae determined, post-injection. Injection of amoebae into the anterior chamber induced a robust neutrophil infiltrate, which was associated with complete clearance by day 15, post-injection. Overall, these findings suggested that *Acanthamoeba* is capable of penetrating Descemet's membrane and entering the anterior chamber. However, a robust neutrophil response is associated with the disappearance of intraocular trophozoites and suggests that cells of the innate immune apparatus are important in preventing *Acanthamoeba* keratitis from progressing to become an intraocular infection (Clarke *et al*., 2005).

2.7. Why is it so painful?

Histopathological and clinical data indicate that *Acanthamoeba* trophozoites cluster around corneal nerves, producing radial keratoneuritis (inflammation

Table 2. Most widely used criteria in the diagnosis of *Acanthamoeba* granulomatous encephalitis and *Acanthamoeba* keratitis.

Acanthamoeba **granulomatous encephalitis**	***Acanthamoeba*** **keratitis**
Presence of skin lesions, ulcers, nodules, or subcutaneous abscesses together with the neurological manifestations should be considered as a high index of suspicion of amoebic encephalitis.	Use of contact lens should be considered as a high index of suspicion
Magnetic resonance imaging or computerized tomography of the brain shows multiple ring-enhancing lesions in the cerebral cortex.	Dendriform pattern are noted on the epithelium of the cornea
Cerebrospinal fluid examination reveals an increased number of white blood cells, elevated protein levels, and decreased glucose levels.	Demonstration of trophozoites or cysts in corneal scrapings in wet-mount preparations using microscopy or immunofluorescence assays.
Demonstration of trophozoites or cysts in affected areas in wet-mount preparations using microscopy or immunofluorescence assays.	Demonstration of *Acanthamoeba* by culturing the clinical specimens on bacterial-seeded non-nutrient agar plates.
Use of indirect immunofluorescence assays to demonstrate *Acanthamoeba* as the causative agent.	Demonstration of *Acanthamoeba* by culturing contact lenses, storage case and the storage solution on bacterial-seeded non-nutrient agar plates.
Demonstration of *Acanthamoeba* by culturing the clinical specimens on bacterial-seeded non-nutrient agar plates.	Polymerase chain reaction of biopsy specimens.
Polymerase chain reaction of biopsy specimens.	Use of tandem scanning confocal corneal microscopy has been described as a non-invasive method
Matrix-assisted laser desorption-ionization time-of-flight mass spectrometry (MALDI-TOF MS)	Matrix-assisted laser desorption-ionization time-of-flight mass spectrometry (MALDI-TOF MS)

of the corneal nerves) in *Acanthamoeba* keratitis patients. The severe pain associated with *Acanthamoeba* keratitis is not commensurate with the degree of tissue destruction and is not seen with other types of infectious keratitis including Herpes simplex virus-1 keratitis and *Pseudomonas aeruginosa* keratitis. To this end, *in vitro* studies have shown the predilection of *Acanthamoeba* trophozoites for responding chemotactically to neural-crest-derived cells. Perhaps, a combination of chemotactic responses and *Acanthamoeba* trophozoites-mediated nerve cell death by direct cytolysis and apoptosis may account for the cause of extreme pain associated with this infection.

2.8. Diagnosis of *Acanthamoeba* keratitis

2.8.1. Clinical diagnosis

The correct diagnosis of *Acanthamoeba* keratitis is problematic as it often is misdiagnosed for bacterial, viral or fungal keratitis. Of interest, the most common bacterial pathogens in keratitis are *Staphylococcus aureus* and for contact lens wearers it is *Pseudomonas aeruginosa*. *Fusarium* spp. is a common cause of keratitis among fungal pathogens; while the causes of viral keratitis include, Herpes simplex virus-1 and Herpes zoster virus. Thus a correct diagnosis for *Acanthamoeba* keratitis requires awareness of its clinical symptoms, and familiarity with the morphological characteristics of the parasite may prove very useful for rapid diagnosis.

The clinical diagnosis of *Acanthamoeba* keratitis includes both clinical syndromes and/or demonstration of the presence of amoebae (Table 2) (Martinez and Visvesvara, 1991). In the majority of cases, this infection is misdiagnosed as the Herpes simplex virus or adenovirus infection. The clinical symptoms are indicated above, but the use of contact lenses by the patient, together with the excruciating pain, is strongly indicative of *Acanthamoeba* keratitis.

The use of slit lamp biomicroscopy *in vivo* may be helpful in demonstrating epithelial defects, and stromal involvement. However the use of the tandem scanning confocal microscope *in vivo* provides superior lateral and axial resolution and serial optical sectioning capability, which is useful for the identification of corneal pathogens in the early stages of infection, however it does not provide identification specific for *Acanthamoeba*. Because this technique is a non-invasive diagnostic imaging method, it has advantage over conventional optical microscopes in that it can image layers within the substance of a specimen of substantial thickness. Therefore, it is effective in imaging the cornea. In preparations for microscopic observations a drop of topical anesthetic is placed on to the patient's cornea and 2% hydroxymethylcellulose is placed on the end of the objective lens. The imaging time used is less than 5 min. However the resolution of *in vivo* confocal microscopy of the cornea is not equal to routine light microscopy of fixed histological specimens. Thus, visual identification of *Acanthamoeba* is not reliable. Also small eye movements interfere with the confocal microscopy imaging. The eye must remain still and for this reason, it may not be a diagnostic option for children or patients with nystagmus (restless).

Notably, *in vivo* studies using tandem scanning confocal microscope demonstrate bacterial pathogens as highly refractile bodies in the epithelium and superficial stroma, while branching fungal hyphae can be highly visible, and *Acanthamoeba* cysts are observed in the subepithelial stroma. To this end, the use of confocal microscopy has an advantage over the conventional optical microscopy in that it can image layers within the substance of a specimen of substantial thickness, so it is effective in imaging the cornea thus it can aid in the diagnosis of *Acanthamoeba* keratitis. A limitation of such microscopic identification is that it is based on morphological characteristics of a pathogen, which requires skill and the use of robust keys for identification. The examiners must have familiarity with the morphological characteristics of *Acanthamoeba* otherwise diagnosis may require histological examination of material obtained by corneal biopsy or keratoplasty. Overall, these findings suggested that tandem scanning confocal microscope may provide an excellent *in vivo*, rapid, non-invasive method for identifying the agent of corneal infection, thereby facilitating prompt and appropriate treatment (Epstein *et al.,* 1986; Winchester *et al.,* 1995; Chew *et al.*, 1992; Palmer *et al.*, 2006).

2.9. Laboratory identification and differentiation of *Acanthamoeba* spp.

Clinical diagnosis may provide an indication of the causative agent, however the confirmatory evidence comes from the laboratory identification, usually by the isolation of *Acanthamoeba* from either the contact lens case or the corneal biopsy. With the devastating nature of *Acanthamoeba* keratitis, it is not only important to develop assays to identify *Acanthamoeba* at the genus-level, but also to rapidly differentiate pathogenic *Acanthamoeba* from non-pathogenic. Although, the ability of *Acanthamoeba* to produce disease in an animal model is a well-accepted measure to establish an isolate's virulence, these models are not routinely available in many laboratories. Furthermore, a requirement of a large number of axenically grown parasites suggests that these methods may not be appropriate for the rapid identification of pathogenic *Acanthamoeba* (DeJonckheere 1980). To this end, the ability of an *Acanthamoeba* isolate to produce cytopathic effects on mammalian cell cultures; growth at high temperatures (37 – 40°C); and to grow readily in axenic medium may provide an indicator for their pathogenic potential. These methods can routinely be carried out in the laboratory and may provide a useful tool to indicate the pathogenicity of an isolate, which should be confirmed using *in vitro* and *in vivo* assays (DeJonckheere 1980). Currently, the following laboratory methods are used to identify *Acanthamoeba* at the genus-level and some may be useful to differentiate pathogenic and non-pathogenic *Acanthamoeba*.

2.9.1. Culture of Acanthamoeba

Despite the development of microscopic and molecular-based approaches, the cultivation of *Acanthamoeba* from the corneal biopsy specimens or from contact lenses or lens cases remains the most widely used assay, because it is simple, inexpensive, and there is no loss of cells during centrifugation or washing steps. In addition, this method can provide a large numbers of *Acanthamoeba*, which could be used for typing, sequencing, epidemiological studies or pathogenicity assays. Briefly, specimens (contact lenses or corneal biopsy specimens) are inoculated onto non-nutrient agar plates seeded with Gram-negative bacteria, e.g., *Escherichia coli* or *Enterobacter aerogenes*. Plates are incubated at 30°C and observed daily for the presence of amoebae (described in section C2.9). *Acanthamoeba* can be identified at the genus-level, based on the morphological characteristics of the trophozoite and the cyst using a phase-contrast microscope. However, it may take several days before viable cells can be observed. In addition, this method is limited in differentiating between pathogenic and non-pathogenic *Acanthamoeba*.

2.9.2. Microscopic identification based on morphological characteristics

Microscopy is a powerful tool for identification of most protozoa. Since microscopic identification is based on the morphological characteristics, skill and use of robust keys for identification are required. Examiners must have familiarity with the morphological characteristics of *Acanthamoeba* spp. otherwise diagnosis may require histological examination of material obtained by corneal biopsy (Epstein *et al*., 1986).

Samples such as corneal biopsy can be directly placed on a glass slide and observed under a microscope for the presence of *Acanthamoeba*. In addition, the identification of amoebae on the surface of the patient's contact lens by light microscopy can provide rapid diagnosis of *Acanthamoeba* keratitis. This is a simple, fast and an inexpensive technique. Furthermore, the contact lens case should be placed in a 50-ml centrifuge tube containing Page's amoeba saline (described in section C2.9) and tube shaken vigorously. Next, the contact lens case should be removed, followed by centrifugation at 1,000 x g for 5 min. The supernatant should be aspirated leaving behind approximately 100 μl, vortexed and placed on a slide to observe under the microscope. The use of Calcofluor white, a chemofluorescent dye with an affinity for the polysaccharide polymers of amoebic cysts is shown to be helpful for the diagnosis of *Acanthamoeba* but may lack accuracy as this dye is not specific for *Acanthamoeba* and may bind to other microbes (Wilhelmus *et al*., 1986). Although, the scanning electron microscopy clearly shows the morphological characters of *Acanthamoeba* such as acanthopodia, however, this technique requires a large number of cells, it is expensive, labour-intensive and thus limited in its clinical use. All the techniques described above rely on microscopy and as such open to error, i.e., require skill and familiarity with the morphological characteristics of *Acanthamoeba*. Misdiagnosis of *Acanthamoeba* keratitis can lead to significant delay in proper diagnosis and treatment. Furthermore, these methods are not useful in the differentiation of pathogenic and non-pathogenic *Acanthamoeba*.

2.9.3. Isoenzyme analysis

Isoenzyme typing has previously been used for the identification to the species level of protozoa. Isoenzyme analysis has been useful in *Acanthamoeba*, *Naegleria*, *Giardia*, *Vahlkampfia* and *Entamoeba* (Moss *et al.,* 1992). Isoenzymes can be run either on cellulose acetate or starch gels. Cellulose acetate gels have the advantage over starch gels that they can be run faster thus avoiding the need to keep them cold. With some isoenzymes, cellulose acetate gels do not give such good banding patterns and thus reduce differentiation. Starch gels in comparison take longer to run, but are cheaper and give good differentiation. Although this technique is excellent for the classification of *Acanthamoeba* into specific genotypes/species, however this technique can

Figure 5. Indirect immunofluorescence assays. Serum or mucosal secretions from a suspected patient is incubated with pathogen (e.g., *Acanthamoeba*)-coated slides, at various titres. A strong reaction suggests *Acanthamoeba* as the causative agent. Note that the levels of anti-*Acanthamoeba* antibodies in normal populations may be in the range of 1:20 to 1:60, however, patients with severely impaired immune system may not develop a high titre, thus other clinical findings should be taken into account for the correct diagnosis.

only be applied to homogenous populations of *Acanthamoeba* and therefore would not be practical when analyzing environmental or clinical samples that contain more than one species. In such cases, isoenzyme patterns must not be used as the sole criterion to establish a new species nor should a morphological analysis be used alone.

For example, several groups have used analysis of isoenzyme electrophoretic patterns to test the morphological classification scheme. These studies discovered extensive diversity among isolates of *Acanthamoeba* due to axenic and monoaxenic cultivation of *Acanthamoeba*. Thus, although isoenzyme patterns have highlighted ambiguities in the morphology-based classification scheme but this approach itself does not address the phylogeny of *Acanthamoeba* due to its lack of reproducibility. Another drawback to this technique is the requirement of large number of axenically grown parasites which is both time consuming and laborious.

2.9.4. Antibody-based assays

2.9.4.1. Immunofluorescence assays

Traditionally, identification of infectious agents included culture techniques, but these assays are labor-intensive, insensitive and may not be useful in the differentiation of pathogens from non-pathogens. Immunoassays became popular with the development of immunofluorescence assays. In these, either specific microbial antigen are detected in the clinical specimens using specific antibodies (i.e., immunofluorescence assays); or antibodies in serum and mucosal secretions are detected against specific microbial antigens (i.e., indirect immunofluorescence assays) (Fig. 5). The lower limits of sensitivity for detecting antigens by immunofluorescence assays in most studies are approximately 100 pg – 1 ng. This level of sensitivity may be sufficient to detect bacterial, viral or parasitic antigens directly in clinical specimens (Crafts *et al.*, 1982; Nash *et al.*, 1987; Villa *et al.*, 1989).

The basic requirement in these assays is antibodies, which can be polyclonal, monoclonal or merely fragments of antibodies as described below. Although polyclonal antibodies have been successfully used as diagnostic reagents in immunoassays, these tests have been difficult to standardize mainly due to broad specificity of the anti-serum which can lead to undesirable cross reactions with other antigens (Underwood 1988; Van

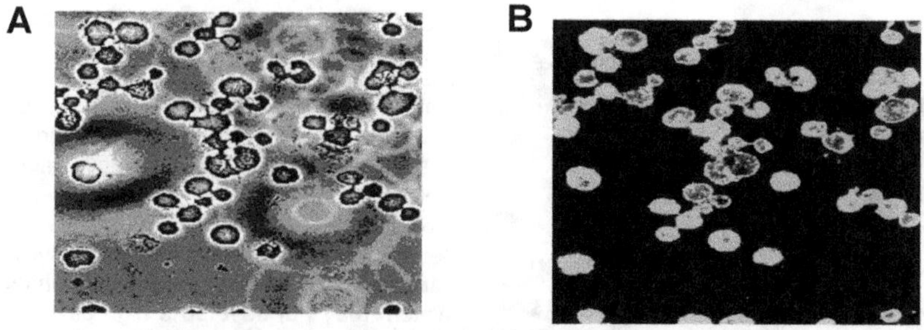

Figure 6. Antibody fragments isolated from a phage display antibody library and tested using indirect immunofluorescence assays. **(A)** Under a light microscope. **(B)** Indirect immunofluorescence reactivities of an antibody fragment with *Acanthamoeba*.

Regenmortel 1992). On the other hand monoclonal antibodies (Kohler *et al.*, 1975) have defined specificity and sensitivity when used in immunoassays. In immunoassays using monoclonal antibodies, microbial antigens can be directly tested within the clinical specimen. In general, monoclonal antibodies to microbial antigens have been produced by *in vivo* techniques following the fusion of myeloma cells with antibody secreting B-cells. The resultant continuous cell line (hybridoma) produces large quantities of homogeneous single epitope antibody. Antibodies from a given hybridoma are immunologically identical and react with one specific epitope on the antigen against which they are raised. Problems with this method are that antigens have to be immnunogenic, animals are required, and it could take several months before a useful antibody is isolated, and this is an expensive method. This has led to the development of phage antibody display libraries. Phage display technology has been developed recently as an alternative to isolating specific antibodies (Winter *et al.*, 1994). Antibody fragments are generated by the random pairing of large diverse repertoires of variable heavy and light chain genes, derived by PCR from naive human lymphocytes, and cloned for expression of individual specificities on the surface of the filamentous bacteriophage. A library contains antibody specificities varying from 10^7 to 10^{12} depending on how the library is constructed. This approach for generating antibodies has the major advantage that epitopes do not have to be immunogenic, i.e. antibody fragments can be isolated that recognize native cell surface structures. The first phage antibody display library was constructed in 1985. Using this technology, antibody fragments, showing a high level of specificity for *Acanthamoeba* spp. have been isolated (Fig. 6), from a naive bacteriophage display library suggesting that bacteriophage antibody display technology is useful in the isolation of novel antibodies for use as research and diagnostic tools in the clinical microbiology. Furthermore, this approach may be of value in isolating antibody fragments specific for a given genotype of *Acanthamoeba* (Fig. 7).

These antibodies can then be used in the development of immunoassays for the identification and differentiation of *Acanthamoeba* spp. Immunoassays can be used to identify a specific genotype (or a species) of *Acanthamoeba* in a complex mixture with other species and provide a sensitive and specific

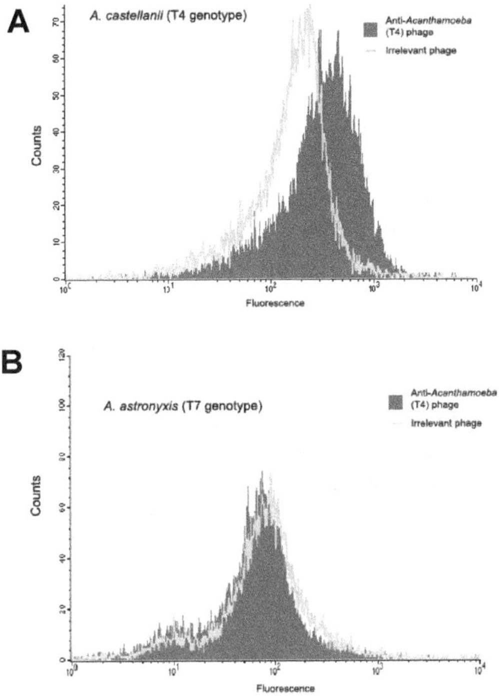

Figure 7. Flow cytometry analysis of specificities of *Acanthamoeba castellanii* (T4 genotype)-specific phage clone with (A) *A. castellanii* (T4 genotype); (B) *A. astronyxis* (T7 genotype). *Acanthamoeba* (10^6) were incubated with phage clone (10^{12}), washed and resuspended in sheep anti-M13 antibody. Following this, cells were washed, resuspended in fluorescein isothiocyanate conjugated anti-sheep IgG (10 µg / ml) and analyzed by flow cytometry. Note that mean channel fluorescence is higher in pathogenic *A. castellanii* (T4 genotype) than weak-pathogenic *A. astronyxis*.

means for the detection. The use of indirect immunofluorescence staining of corneal scrapings provides a simple means of accomplishing this goal with a high degree of accuracy (Epstein *et al.*, 1986). Both immunofluorescence and indirect immunofluorescence assays make use of antibodies generated against the whole *Acanthamoeba* or their plasma membranes. In addition to the traditional generation of antibodies, the use of phage antibody display technology has been used in the isolation of antibody fragments that can be used for the identification and differentiation of pathogenic *Acanthamoeba* using the immunofluorescence assays. Furthermore, specific antibodies can be used for the enzyme-linked immunosorbent assay.

2.9.4.2. Flow cytometry

Flow cytometry has been successfully used to identify *Acanthamoeba* spp. This technique has been proven efficient for the environmental samples, and may also be valuable for the clinical diagnosis of *Acanthamoeba* spp. In a flow microcytometric system, cells are incubated with a fluorescein-conjugated antibody and passed in front of a focused argon-ion laser beam. Cells are then distinguished by the degree of the scattered light (size) or fluorescence. Scatter and fluorescent signals are generated in the instrument as cell pass in a single file through a transparent flow channel and cross a 488 nm argon-ion laser excitation beam. The fluorescein-conjugated immunoglobulins that are bound to the cells re-emit fluorescent light in the green wavelength. This re-emitted fluorescence is then filtered to remove extraneous fluorescence, detected and amplified by photomultiplier tubes and then displayed in

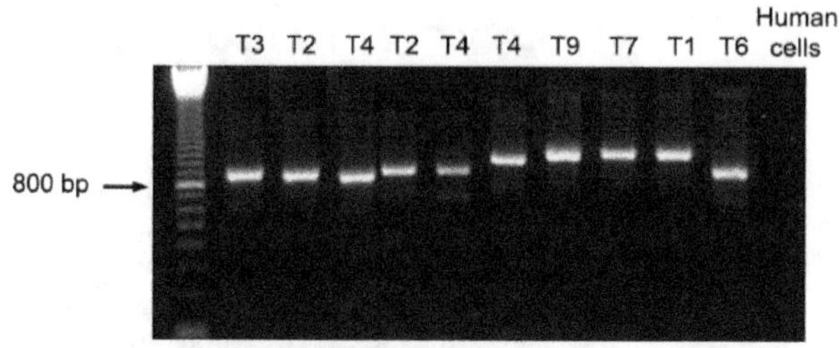

Figure 8. Agarose gel showing PCR amplification of 18S rDNA of *Acanthamoeba* using genus-specific primers. Note that primers amplified various genotypes of *Acanthamoeba* but not human DNA, suggesting their use in the diagnosis of *Acanthamoeba* in the environmental and clinical settings.

the form of cytogram or histogram patterns. The cluster of dots seen in the cytogram patterns is representative of cell size (scatter) and degree of fluorescence. These data are stored in a multichannel distribution analyzer and the histogram distribution of the cells indicates fluorescent frequency. This technique reduces the time presently required to screen environmental sites for the presence of amoebae. Cytofluorograph analysis permits detection of amoebae and may aid in the differentiation of *Acanthamoeba* spp. using antibodies specific for pathogenic *Acanthamoeba* spp. Other advantages of flow cytometry over fluorescent microscopy include a high degree of statistical precision due to the large numbers measured, high immunofluorescent titres, and elimination of subjectivity and fluorescence fading (Muldrow *et al.*, 1982).

2.9.5. Polymerase chain reaction

In the past few decades, genetic and molecular methods for characterizing pathogen strains have taken a major place in modern approaches, in diagnostic and epidimiological studies of parasitic and other infectious diseases (Tibayrenc 1996). These techniques are more sensitive than the conventional morphological and isoenzyme methods. The use of PCR as a diagnostic tool has increased during the last two decades, because it is a sensitive method and amplification can be achieved from a single cell (Abbaszadegan *et al.*, 1991). This method is increasingly being used to identify parasitic organisms however the scarcity of detailed molecular analysis of many parasites limits its use. Usually PCR can be performed within 2 – 3 hours, and samples can be directly assayed without the prior need of cultivation making it a rapid assay. To devise a PCR-based assay, a target sequence is required. If a genus-specific assay is needed, the target sequence should be highly conserved (such as ribosomal DNA sequence) to allow primers to be designed from non-polymorphic regions (Fig. 8). Following the identification of a conserved sequence, the next step is to design primers from that sequence and to optimize the conditions for PCR (concentration of magnesium chloride, nucleotide triphosphates and primer concentration). Finally, it is important to select the proper temperature for annealing and the duration of time for each step. The PCR-based assays have been successfully employed for a number of protozoa, including *Plasmodium* (Waters *et al.*, 1989; Mathiopoulos *et al.*, 1993), *Leishmania* (Uliana *et al.*, 1991), *Toxoplasma* (Cazenave *et al.*, 1991; Guay *et al.*, 1993; MacPherson *et al.*, 1993; Verhofstede *et al.*, 1993), *Entamoeba histolytica* (Garfink *et al.*, 1989; Bracha *et al.*, 1990), *Naegleria* (Sparagano *et al.*, 1993), *Cryptosporidium parvum* (Laxer *et al.*, 1991; 1992) and *Giardia* (Mahbubani *et al.*, 1991; 1992). Genetic characterization of pathogens allow their identification at the strain, subspecies or species levels and explore the impact of genetic diversity on relevant biomedical parameters such as virulence or resistance to drugs (Tibayrenc 1995; Tibayrenc 1996).

Polymerase chain reaction-based methods using the 18S rRNA gene sequences have been developed for the rapid detection of *Acanthamoeba*. This method is highly specific and can detect fewer than five cells (Khan *et al.*, 2001; Lehmann *et al.*, 1998; Schroeder *et al.*, 2001). A real-time PCR assay for the detection of *Acanthamoeba* in keratitis patients has also been developed. This is a TaqMan assay specific for *Acanthamoeba* species (Rivi´ere *et al.*, 2006). However, this assay can identify only genotypes T1 and T4 and lacks the ability to identify the two other genotypes T7 and T10 known to cause granulomatous encephalitis. A multiplex real-time PCR assay has recently been developed that can not only detect several genotypes of *Acanthamoeba* but also *Balamuthia* and *Naegleria fowleri* in human specimens (Qvarnstrom *et al.*, 2006). The multiplex assay is of value for the simultaneous detection of all three pathogenic free-living amoebae in the clinical specimens as well as in the environmental samples.

2.9.6. Matrix-assisted laser desorption-ionization time-of-flight mass spectrometry (MALDI-TOF MS)

In addition, matrix-assisted laser desorption–ionization time-of-flight mass spectrometry may be of potential value in the rapid identification of *Acanthamoeba* in the clinical specimens. In this, biomolecules to be analyzed are given a charge and then separated on the basis of their mass:charge ratio. Biomolecules can be charged by mixing with an organic chemical that, when illuminated with laser light, ionises associated biomolecule and gives them a charge. Charged molecules are then allowed to move down a tube to a detector. Time taken to move is proportional to mass of the biomolecule. Matrix-assisted laser desorption–ionization time-of-flight mass spectrometry is sensitive, accurate to hundredths of a Dalton and provides a mass fingerprint for proteins of interest. If the genome of the organism under investigation is known, theoretical mass spectra of all proteins encoded by genome is determined. Mass fingerprint of the sample protein can be compared with that of theoretical mass spectra of the whole organism to identify the protein.

This method has been used to identify and characterize protists at the strain-level, based on protein profiles within 15 min. Using matrix-assisted laser desorption–ionization time-of-flight mass spectrometry, at least 20 biomarker fingerprints of *Acanthamoeba*, *Balamuthia* and *Naegleria fowleri* with molecular masses ranging from 2 to 14 kDa have been detected. Because the entire procedure can be performed in 15 min, it too may be extremely useful in the rapid identification of these amoebae in biological samples (Moura *et al.*, 2003b).

2.10. T4 genotype is the predominant cause of keratitis and granulomatous encephalitis

Based on 18S rRNA gene sequencing, most of the clinical isolates of *Acanthamoeba* (from keratitis, encephalitis and cutaneous infections) as well as environmental isolates have been typed as the T4 genotype. In one survey, out of 249 isolates examined, 72% (179 / 249) were typed as genotype T4. Of these 179 isolates identified as T4, 53% came from the environment, 94.3% from keratitis patients, and 79.3% from non-*Acanthamoeba* keratitis sources. The non-*Acanthamoeba* keratitis isolates were isolated from the cerebrospinal fluid, the brain, nasal passages, lungs and skin. A few *Acanthamoeba* isolates from the brain have been typed as belonging to rare genotypes (T1, T10, and T12), and at least two of these rare genotypes (T10 and T12) have yet to be isolated from the environment (Booton *et al.*, 2005). Moreover, of the few non-T4 keratitis isolates, three isolates from Iran have been typed as T2 and 2 as T3, one each from the UK as T3 and T12, and one isolate from Austria as T6 (Magshood *et al.*, 2005; Blaschitz *et al.*, 2006). Future studies would differentiate more precisely the group II and/or sequence type T4 by examining variability within the internal transcribed spacer 1 (ITS1) sequence, and analysis of GC contents and microsatellites (K¨oshler *et al.*, 2006).

2.11. Host susceptibility

Several lines of evidence suggest that *Acanthamoeba* keratitis patients exhibit reduced levels of anti-*Acanthamoeba* defence determinants such as sIgA in their tear film (discussed in section E4), or exhibit corneal properties both at the surface and at the molecular level, which could render their corneas more susceptible to *Acanthamoeba* keratitis. In addition, corneal injury is a prerequisite for *Acanthamoeba* keratitis. *In vivo* studies showed that animals that have intact epithelial layers do not develop *Acanthamoeba* keratitis (Niederkorn *et al.,* 1999). The importance of corneal injury is demonstrated further by reports that injury to the surface of the cornea, even with a splash of *Acanthamoeba*-contaminated matter and/or water, can lead to *Acanthamoeba* keratitis in individuals who do not wear contact lenses (Chang and Soong, 1991; Sharma *et al.,* 1990). It has been shown that the expression of *Acanthamoeba*-reactive glycoproteins(s) on the surface of damaged corneal epithelial cell is significantly higher compared with the surface of the normal corneal epithelial cells, suggesting that corneal injury contributes markedly to *Acanthamoeba* keratitis (Jaison *et al.,* 1998).

2.12. Species specificity

Studies have shown the host specificity in *Acanthamoeba* keratitis. For example, successful *Acanthamoeba* keratitis models that mimic the human form of disease were only produced in pigs and Chinese hamsters but not

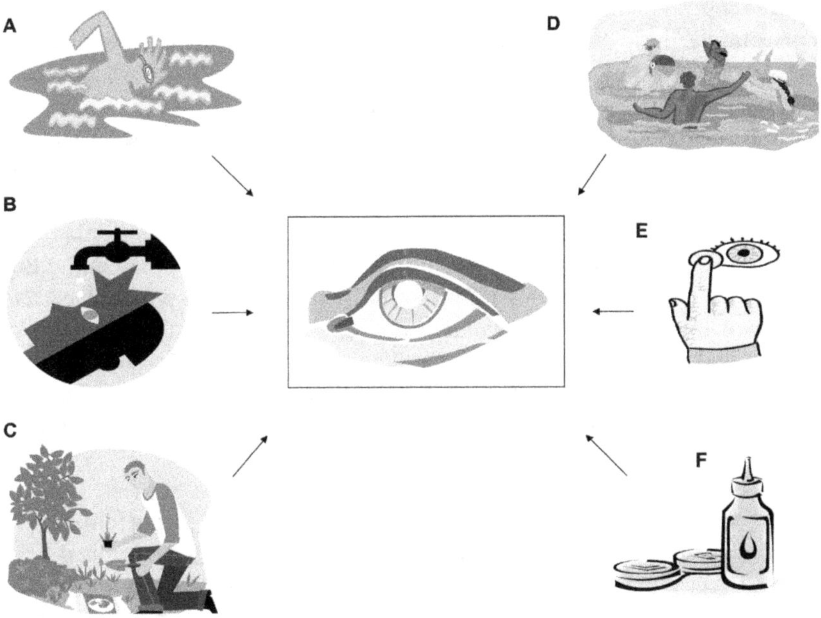

Figure 9. The risk factors contributing to *Acanthamoeba* keratitis. **A)** swimming, especially while wearing contact lenses; **B)** washing eyes during or immediately after contact lens wear; **C)** working with soil and rubbing eyes; **D)** water-related activities (splashing water), especially during or immediately after contact lens wear; **E)** handling contact lenses without proper hand washing; **F)** use of home-made saline (or even chlorine-based disinfectants) for contact lens cleaning.

in rats, mice or rabbits, suggesting that the expression of specific molecular determinants may be limited to certain mammalian species (Niederkorn *et al.*, 1999). Species specificity was demonstrated further by testing normal, intact corneas of 11 mammalian and one avian species. It was interesting to observe the species specificity for *Acanthamoeba* keratitis. *Acanthamoeba* not only failed to produce cytopathic effects but did not even bind to the corneal epithelium of mice, rats, cotton rats, horses, guinea pigs, cows, chickens, dogs, and rabbits. However, parasites adhered, invaded, and produced severe damage to human, pig, and Chinese hamster corneas. *In vitro* experiments determined binding of parasites to the corneas of selected susceptible and non-susceptible species and results revealed scant binding of parasites to mouse, rat, and rabbit and extensive binding to Chinese hamster, pig, and human corneas. These results suggested that *Acanthamoeba* exercises host specificity, at the host cell surface, which is most likely due to the presence of specific molecular determinants expressed on the surface of cornea (Niederkorn *et al.*, 1992).

2.13. Risk factors

Although *Acanthamoeba* keratitis has been observed in patients who did not wear contact lenses, the major risk factors for *Acanthamoeba* keratitis is poor hygiene in the use of contact lenses (Fig. 9). In support, more than 80% of

Acanthamoeba keratitis cases have been reported in wearers of contact lenses. For example, during a 9 month survey in 1986, 24 patients with *Acanthamoeba* keratitis were reported to the Centers for Disease Control (www.cdc.gov). Out of these, 20 patients (83%) wore contact lenses. Of these 2 wore hard lenses (one hard, the other rigid-gas-permeable); 4 wore extended-wear soft lenses; and 14 wore daily-wear soft lenses. Ten of these 20 patients cleaned their lenses with home-made saline solution prepared by mixing salt tablets with bottled, distilled, non-sterile water; 4 used commercially available lens-cleaning solutions followed by a tap water rinse; 1 used commercial bottled saline and 1 cleaned lenses with tap water pumped from a private well. No lens care information was available for 4 patients. With the difficulty of correct diagnosis, it is not surprising that 22 (90%) of the 24 patients were initially diagnosed as having corneal herpes simplex virus (HSV) infections, while in other 2, corneal lesions were attributed to autoimmune diseases (CDC, MMWR 1986).

In a later study, 208 cases of *Acanthamoeba* keratitis in the USA were identified. Out of these 101 (49%) were male and 107 (51%) were female. The median age at the onset of *Acanthamoeba* keratitis was 29, but ranged from 13 to 82. Age distribution did not vary by sex. Age and sex distribution did not vary over time. Most cases (41%) were reported from California (29), Texas (29), Florida (15) and Pennsylvania (12). More than 90% patients had at least one risk factor, i.e., a history of corneal trauma, or exposure to contaminated water or contact lens wear. Consistent with the earlier survey, it was observed that more than 80% wore contact lenses, most of which were daily wear or extended wear soft lenses. Patients aged 50 years and older were more likely to have had a history of trauma than younger patients and males were more likely to have a history of trauma than females (Stehr-Green *et al.*, 1989). Poor personal hygiene, poor handling and care of their lenses or lens storage cases, and non-compliance with disinfection procedures were described as major risk factors (Brennan, 2002). Contact lenses that have been scratched or fragmented through mishandling should not be used. Although disposable soft contact lenses have been marketed as a safer alternative to conventional soft lenses, studies have shown that both extended-wear and daily-wear disposable lenses are associated with higher risks of keratitis than other lens types including conventional extended-wear lenses (Matthews *et al.*, 1992).

Additional risk factors are swimming or washing eyes while wearing contact lenses, and the use of chlorine-based disinfectants for contact lens cleaning should be avoided because *Acanthamoeba* is highly resistant to chlorine (Radford *et al.,* 1995; 1998; Seal *et al.,* 1999). In addition, *Acanthamoeba* exhibits significantly higher binding to silicone hydrogel contact lenses than to the conventional hydrogel contact lenses (Beattie *et al.,* 2003), suggesting that polymer characteristics of the lens or surface treatment procedures may increase the risk of *Acanthamoeba* keratitis. Thus, extended wear of lenses without proper maintenance and recommended replacement, together with lens type can be important risk factors for *Acanthamoeba* keratitis.

Overnight orthokeratology (i.e., the use of rigid gas-permeable contact lenses, normally worn only at night, to improve vision through the reshaping of the cornea) is an important risk factor of *Acanthamoeba* keratitis, especially in school children. Overnight orthokeratology use has been consistently associated with infectious keratitis despite the use of more oxygen-permeable materials and improved lens design, and the risk should not be over-looked. Patient education with informed consent, appropriate lens care, and meticulous follow-up is important. Because this complication is potentially sight-threatening, orthokeratology requires further analysis and evaluation to establish its safety (Hsiao *et al.*, 2005; Tseng *et al.*, 2005; Yepes *et al.*, 2005).

Overall, these characteristics suggest that although the intact cornea is highly resistant to *Acanthamoeba* infection, corneal trauma (microscopic defects) followed by exposure to contaminated water (during swimming, eye washing, water splash), dust, vegetable matter or any foreign particle are important risk factors associated with *Acanthamoeba* keratitis.

Because *Acanthamoeba* is ubiquitously present in water, air, water-air interface and soil (Khan, 2003; Marciano-Cabral and Cabral, 2003; Preston *et al.*, 2001), susceptible hosts should be warned of the risks associated with the wearing of contact lenses while swimming or bathing / washing, and cleaning lenses with home-made saline etc. Proper cleaning of the contact lenses is crucial to prevent this devastating infection. Chlorine-based cleaning solution should not be used because *Acanthamoeba* is highly resistant to chlorine, but a two-step hydrogen peroxide system at the concentration of 3% is highly effective against cysts and trophozoites (Beattie *et al.*, 2002). In addition, strategies and/or control measures to reduce the formation of biofilms and the build-up of carbohydrate moieties on contact lenses should help to prevent this serious infection.

2.14. Treatment

2.14.1. Cryosurgery

To determine the efficacy of the corneal cryosurgery (the application of extreme cold to destroy abnormal or diseased tissue) in the treatment of *Acanthamoeba* keratitis, *in vitro* freeze-thaw-refreeze cryotherapeutic methods were tested on the trophozoites and cysts of *Acanthamoeba*. The trophozoites were killed in all trials. However, the cysts survived all trials. These findings suggested that cryosurgery may not be an effective means to eliminate *Acanthamoeba* cysts from the cornea (Meisler *et al.*, 1986). Later, studies showed that although cryosurgery alone may not eradicate *Acanthamoeba* but may be of value, if it is used as an adjunctive measure to other medical (i.e., chemotherapeutic) and surgical therapy (Binder 1989).

2.14.2. Chemotherapy

Acanthamoeba keratitis is a difficult infection to treat. Early diagnosis followed by aggressive treatment is essential for the successful prognosis of the infection. Although various antimicrobial agents are shown to possess *in vitro* amoebicidal and cysticidal activities, but no agent has been shown to be uniformly effective against all isolates of *Acanthamoeba*. Multiple factors, including the varied clinical presentation and the varied virulence of *Acanthamoeba* isolates/genotypes accounts for the lack of correlation between *in vitro* activity and *in vivo* efficacy (Osato *et al*., 1991).

The recommended treatment regimen includes a biguanide (0.02 % polyhexamethylene biguanide, or 0.02% chlorhexidine digluconate) together with a diamidine (0.1% propamidine isethionate, also known as Brolene or 0.1% hexamidine, also known as Desomedine). The combined treatment with polyhexamethylene biguanide and propamidine isethionate is shown to be highly effective. Using this combination, it is shown that, typically, visual outcome is favorable and the requirement for keratoplasty (corneal transplant) is reduced markedly (Duguid *et al*., 1997).

If bacteria are also associated and/or suspected with the infection, the addition of antibiotics, i.e., neomycin or chloramphenicol is recommended. The initial treatment involves hourly topical application of drugs day and night for 2 – 3 days, followed by hourly topical application during the day for a further 3 – 4 days. Subsequently, application is reduced to 2-hourly application during the day for up to a month. This is followed by application 6 times a day for the next several months for up to a year, clearly presenting appreciable social and economic burdens due to this infection.

Persistent inflammation and severe pain may be managed by topical application of steroids such as 0.1% dexamethasone together with pain killers. However, it should be noted that dexamethasone causes a significant increase in the proliferation of amoebae numbers and increases severity in *Acanthamoeba* keratitis, as demonstrated in an *in vivo* model (McClellan *et al*., 2001). Thus prolonged application of steroids should be carried out with care. At the same time, steroid administration during the course of *Acanthamoeba* keratitis may need to be withdrawn extremely slowly to avoid the recurrence of immunological corneal reaction (Horsburgh *et al*., 1991).

The presence of antibiotics (neomycin or chloramphenicol) limits possible bacterial infection or, at the very least, eliminates food source for *Acanthamoeba*. Co-infection with fungus may complicate the infection, in which case, the inclusion of ketoconazole, miconazole, econasole nitrate, and traconazole is useful. In addition, the development of secondary glaucoma is not uncommon in *Acanthamoeba* keratitis and is a poor prognostic sign in patients with *Acanthamoeba* keratitis, because most progress to light perception or no light perception vision. Histopathologic findings suggest an inflammatory angle-closure mechanism, apparently without direct infiltration of the organism. The secondary glaucoma associated with *Acanthamoeba* keratitis is often severe and frequently requires surgical intervention for intraocular pressure control and vision preservation. Dacryoadenitis (inflammation of the lacrimal glands, i.e., the tear-producing glands), also described as a blocked tear duct may also be an additional complications associated with *Acanthamoeba* keratitis (Kelley *et al*., 2006; Tomita *et al*., 2006). The prolonged treatment should be carried out with care and regular monitoring. For example, it was observed that two patients with *Acanthamoeba* keratitis developed corneal abnormalities following prolonged treatment with topical propamidine isethionate. In both instances, withdrawal of the drug therapy resulted in a gradual clearing of the keratopathy (Johns *et al.*, 1988). Furthermore, in the case of chemotherapeutic resistance or non-responsive to chlorhexidine digluconate, other drugs such as polyhexamethylene biguanide should be considered (Gooi *et al.*, 2008; Vaquero Ruano *et al.*, 2002).

Overall this aggressive, complicated and prolonged management is required because of the ability of *Acanthamoeba* to adapt rapidly to harsh conditions and to switch to a resistant cyst form, and because of the lack of available methods for the targeted killing of both trophozoites as well as cysts. As a last resort a keratoplasty (corneal transplant) may be indicated, especially in drug resistant cases. In the case of penetrating keratoplasty to obtain rehabilitation due to corneal scaring, topical treatment with the above is essential as a first measure. Rejection can occur but is rare. Recurrence of infection does occur. For example, in several cases, patients underwent a corneal transplant but the graft did not alleviate the infection. Thus it is recommended that topical treatment continues for up to a year post operatively as cysts may survive in the acceptor cornea (Perez-Santonja *et al.*, 2003).

3. *Acanthamoeba* granulomatous encephalitis

Acanthamoeba granulomatous encephalitis is a rare infection but it almost always proves fatal. However, this infection is of major concern in view of (i) increasing numbers of immunocompromised persons, (ii) individuals undergoing immunosuppressive therapy and excessive use of steroids, and (iii) the global warming which may add to the ubiquity of these pathogens, and thus a possibility of increased exposure to the susceptible hosts. With the growing HIV pandemic, it is reasonable to predict a further increase in the numbers of *Acanthamoeba* granulomatous encephalitis cases. Perhaps, the most distressing aspect is the limited availability of effective treatment against *Acanthamoeba* granulomatous encephalitis, with a case fatality rate of more than 90%. The gross pathology of the autopsied brain often shows severe edema and haemorrhagic necrosis with severe meningeal irritation and encephalitis. The microscopic findings of the post-mortem biopsies reveal amoebae cysts, predominantly in the perivascular spaces in the parenchyma, indicating the involvement of the cerebral capillaries as the sites of entry into the central nervous system.

The precise mechanisms associated with the pathogenesis are unclear, however the pathophysiological complications involving the central nervous system most likely include induction of the pro-inflammatory responses, invasion of the blood-brain barrier and the connective tissue and neuronal damage leading to the brain dysfunction.

3.1. Routes of entry into the central nervous system

There are two possible routes of entry into the central nervous system: through the nasal passage or the blood. In humans, *Acanthamoeba* is usually blood-borne, while other free-living amoebae such as *Naegleria* spp. enter via

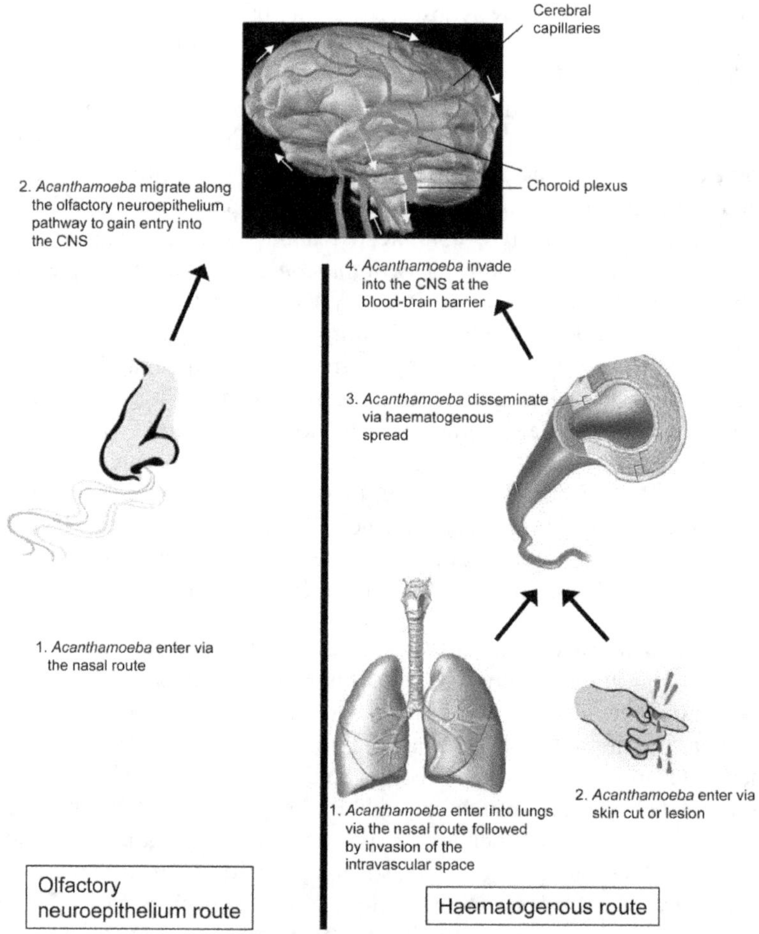

Figure 10. The likely routes of *Acanthamoeba* entry into the central nervous system (CNS). *Acanthamoeba* enter lungs via the nasal route. Next, amoebae traverse the lungs into the bloodstream, followed by haematogenous spread. Finally, amoebae cross the blood-brain barrier and enter into the CNS to produce disease. It is noteworthy that *Acanthamoeba* may bypass the lower respiratory tract and directly enter into the bloodstream via skin lesions. The olfactory neuroepithelium may provide an alternative route of entry into the CNS.

the olfactory route (Cerva, 1967; Culbertson *et al.*, 1959). It is widely accepted that the route of entry for *Acanthamoeba* include the respiratory tract leading to amoebae invasion of the alveolar blood vessels, followed by the haematogenous spread. Skin lesions may provide direct entry into the bloodstream, bypassing the lower respiratory tract (Fig. 10). *Acanthamoeba* entry into the central nervous system most likely occurs through the blood-brain barrier (Martinez, 1985; 1991), which is highly selective in regulating the entry of microbes and/or molecules. The olfactory neuroepithelium route (i.e., invasion of the olfactory part of the nasal epithelium, and migration along nerve fibers, followed by invasion of the olfactory bulb) provides another route of entry into the central nervous system and has been studied in the experimental models (Martinez, 1991; Martinez and Visvesvara, 1997). It is worth noting that the olfactory part of the nasal mucosa in the rat (rodents) is 50% of the total surface area of the nose, while in humans, the olfactory part of the nasal mucosa is only 3 – 8% (Graff and Pollack, 2005; Illum, 2000). The smaller fraction of nasal mucosa associated with the olfactory epithelium in humans can be associated with a lower likelihood of this route leading to human disease. Furthermore, the widespread distribution of lesions in the brain would fit better with the haematogenous spread than the olfactory access. *Acanthamoeba* entry into the central nervous system most likely occurs through the endothelial lining of cerebral capillaries (Martinez *et al.,* 1975; Martinez, 1985; 1991). In support, hematoxylin-eosin-stained sections of the brain tissue of *Acanthamoeba* granulomatous encephalitis patients exhibit large numbers of amoebae in the perivascular space. Pathogens that enter via the choroid plexus route will end up in the cerebrospinal fluid. Affected tissues other than the central nervous system may include subcutaneous tissue, skin, liver, lungs, kidneys, adrenals, pancreas, prostate, lymph nodes and bone marrow, which further suggest haematogenous spread. Also, *in vivo* studies showed that amoebae that were instilled intranasally into mice develop pulmonary disease, followed by neurological signs (Martinez *et al.*, 1975). Pathological examination of these mice showed broncho-pneumonia associated with encephalitis with amoebic trophozoites and cysts in lung and brain, suggesting the involvement of the haematogenous spread (Martinez *et al.*, 1975). Overall, these findings suggested that unlike *Naegleria fowleri* which uses olfactory neuroepithelium route, *Acanthamoeba* invades the central nervous system via the blood-brain barrier. In support, acquired immunodeficiency syndrome (AIDS) patients have been observed who developed cutaneous infection with *Acanthamoeba*, together with the sinus involvement but did not show any evidence of the central nervous system infection (Helton *et al.*, 1993). Moreover, mice infected with amoebae were shown to have amoebae in the brain, lungs, liver, kidneys, heart, and the spleen - affecting all organs of the inoculated mice, further suggesting haematogenous spread as a likely route of amoebae entry into the central nervous system (Górnik and Kuźna-Grygiel 2005). Notably, the cerebrospinal fluid is continuous with the extracellular fluid of the brain parenchyma, and flows unidirectionally from ventricles down to the lumbar region, thus spinal taps of patients may aid in the early diagnosis. Notable, an early diagnosis is critical in the successful treatment of *Acanthamoeba* granulomatous encephalitis (Petry *et al.,* 2006).

The cutaneous and respiratory infections can last for months but the involvement of the central nervous system can result in fatal consequences within days or weeks. *Acanthamoeba* spp. also cause infections of the central nervous system of animals, including gorillas, monkeys, dogs, ovines (sheep-like), bovines, horses, and kangaroos, as well as birds, reptiles, amphibians, fish, and even invertebrates.

3.2. Epidemiology

The epidemiology of *Acanthamoeba* granulomatous encephalitis is rather confusing. The fact that normally this infection is secondary, makes diagnosis difficult and thus contributes to our inability to assess the actual number of *Acanthamoeba* granulomatous encephalitis infections. Perhaps the number of *Acanthamoeba* granulomatous encephalitis cases in human immunodeficiency virus (HIV) / AIDS patients, although not completely accurate, may indicate the real burden of this infection. This has only been made possible by the pioneering work of G. S. Visvesvara (Centers for Disease Control, USA, www.cdc.gov) and the late Dr. A. J. Martinez (University of Pittsburgh School of Medicine, USA). In the USA, there were approximately 350,000 deaths due to HIV/AIDS during 1981 – 1996 with the highest mortality during the mid 90's; 49,000 in 1994 and 50,000 in 1995, which declined to 39,000 in 1996) (Heath *et al.,* 1998; Centers for Disease Control, www.cdc.gov). Over a similar period, the number of *Acanthamoeba* granulomatous encephalitis deaths in HIV/AIDS patients was approximately 55 (Martinez and Visvesvara, 1997). Thus, the approximate rate can be calculated as 1.57 *Acanthamoeba* granulomatous encephalitis deaths per 10,000 HIV/AIDS deaths in the USA, even though the number of *Acanthamoeba* granulomatous encephalitis infections may be much higher in countries with warmer climates, because there will be increased ubiquity of amoebae and/or increased outdoor public activities. The estimated worldwide number of HIV/AIDS patients is a massive 40 – 45 million (as of 2007) and continues to rise sharply. Hypothetically, this figure represents the number of *Acanthamoeba* granulomatous encephalitis susceptible hosts. If this is so, then why are there not a large number of *Acanthamoeba* granulomatous encephalitis infections? There could be several explanations for this. At least for the USA, the number of deaths due to HIV/ AIDS has been declining since late 1990's, i.e., 22,000 deaths in 1997 and 18,000 deaths in 2003 in the USA (Heath *et al.,* 1998; CDC), thus declining the number of *Acanthamoeba* granulomatous encephalitis susceptible hosts. This decline in HIV/AIDS deaths in the USA is attributed to early diagnosis followed by the introduction of novel anti-retroviral therapies, i.e., highly active antiretroviral therapy (HAART), which was first introduced in 1996. As well as improving AIDS symptoms, HAART has protective effects against *Acanthamoeba* and other opportunistic pathogens (Carter *et al.,* 2004; Pozio and Morales, 2005; Seijo Martinez *et al.,* 2000). However, these therapies are not available to the majority of HIV/AIDS patients in less developed or developing countries in other parts of the world. Thus the approximate rate of 1.57 *Acanthamoeba* granulomatous encephalitis infections per 10,000 HIV/AIDS deaths in such countries may provide only a minimum estimate of the burden of *Acanthamoeba* granulomatous encephalitis. The fact that *Acanthamoeba* granulomatous encephalitis cases are not being reported in the developing countries (especially in Africa) is due to the lack of expertise, reporting problems, lack of proper monitoring and the lack of proper health care systems.

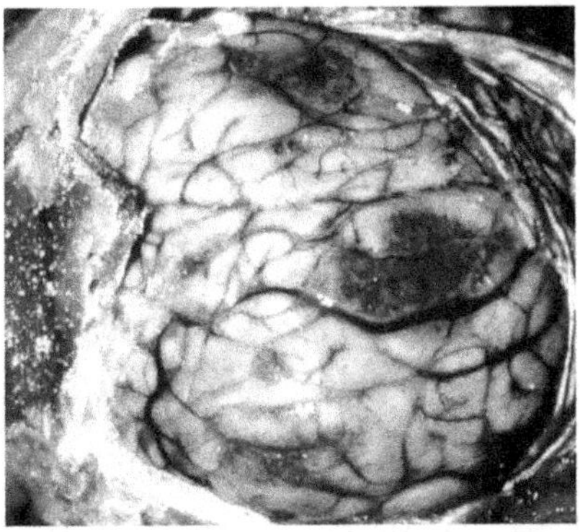

Figure 11. *Acanthamoeba* infected brain exhibiting severity of the disease.

Of interest, there were 5 million new reported cases of HIV/AIDS in 2003 alone (approximately 14,000 infections per day), while 3 million deaths occurred due to HIV/AIDS-related diseases (approximately 8,500 deaths per day), mostly in Africa (even though there hasn't been a single reported case of *Acanthamoeba* granulomatous encephalitis in Africa). And applying 1.57 *Acanthamoeba* granulomatous encephalitis deaths per 10,000 HIV/AIDS deaths, the total number of *Acanthamoeba* granulomatous encephalitis infections in 2003 can be estimated at approximately 471. Although this number is significantly less than the 3 million deaths in total, *Acanthamoeba* granulomatous encephalitis is certainly a contributing factor in AIDS-related deaths, and needs continued attention. In addition, other infections such as diabetes, malignancies, malnutrition, alcoholism, or having a debilitated immune system due to immunosuppressive therapy or other complications may all contribute to *Acanthamoeba* granulomatous encephalitis.

Similar to *Acanthamoeba* keratitis, *Acanthamoeba* of the T4 genotype is the major causative agent of *Acanthamoeba* e granulomatous encephalitis.

3.3. Pathophysiology of *Acanthamoeba* granulomatous encephalitis

The clinical symptoms of *Acanthamoeba* granulomatous encephalitis may resemble viral, bacterial or tuberculosis meningitis: headache, fever, behavioral changes (confusion, irritability, hallucinations, and dizziness), hemiparesis (partial paralysis of one side of body), lethargy (fatigue), visual disturbances, stiff neck, aphasia (inability to comprehend language), ataxia (incoordination of muscle movements), vomiting, nausea, cranial nerve palsies, increased intracranial pressure, focal neurological symptoms, seizures, and death (Table 1). These are due to hemorrhagic necrotizing lesions with severe meningeal irritation and encephalitis (Fig. 11) (Carter *et al.*, 1981; Martinez *et al.*, 1980; Martinez, 1985; 1991). Patients with respiratory infections, skin ulcerations (often a presenting manifestation) or brain abscesses should be strongly suspected for infections due to free-living amoebae. Post mortem examination often shows severe edema and hemorrhagic necrosis in occipital, parietal, temporal, or frontal lobes. The lesions are usually multifocal, located in the basal ganglia, midbrain, thalamus, brainstem, cerebral hemispheres.

The cervical portion of the spinal cord is occasionally affected. Cerebellar tonsillar herniations may be prominent features (Martinez 1991). The characteristic lesion in the central nervous system parenchyma is a necrotizing subacute or chronic granulomatous encephalitis in which trophozoites and cysts are usually found in association with giant cells. It is not known whether this necrotic phase is caused by actively feeding trophozoites or inflammatory processes such as the release of the cytokines. The lesions due to *Acanthamoeba* granulomatous encephalitis are most numerous in the basal ganglia, midbrain, brainstem, and cerebral hemiparesis with characteristic lesions in the central nervous system parenchyma resulting in chronic granulomatous encephalitis. Both the trophozoites and cysts are mainly present in the perivascular spaces of the affected areas. Amoebic trophozoites and cysts may be found piercing the vascular wall (Martinez 1991). At postmortem, involvement of viscera other than the central nervous system may be found suggesting a pre-mortem haematogenous dissemination of trophozoites and cysts. The organs involved may be subcutaneous tissue, and skin, liver, lungs, kidneys, adrenals, pancreas, prostate, lymph nodes, and myometrium (middle layer of the uterine wall consisting of smooth muscle cells and supporting stromal and vascular tissue) (Martinez 1991). To demonstrate the lung involvement, trophozoites and cysts have been found in pulmonary alveoli from *Acanthamoeba* granulomatous encephalitis patients and pneumonitis is a common feature. Although in the immunocompetent individuals, well-developed granulomas form around the organisms, but granuloma formation is weak and/or absent in patients with a severely impaired immune system that is interpreted as impairment of cellular immune response (Martinez *et al.*, 2000). The affected tissues other than the central nervous system may include subcutaneous tissue, skin, liver, lungs, kidneys, adrenals, pancreas, prostate, lymph nodes, and bone marrow. Of note, delta-9-tetrahydrocannabinol, the major psychoactive component of marijuana, exacerbates brain infection by *Acanthamoeba* (Marciano-Cabral *et al.*, 2001).

Apart from brain necropsies or biopsies, amoebae have been isolated from a brain abscess of an *Acanthamoeba* granulomatous encephalitis patient. In this case, a 55-year-old diabetic aboriginal woman presented with a two-week history of fever, altered mental state and convulsions. On the basis of computed tomographic scanning, a diagnosis of cerebral abscess was made. The pus that was drained produced no bacterial growth but, on microscopy, amoebic cysts were observed, which showed positive growth of *Acanthamoeba*. The patient appeared to respond to drainage of the abscess and anti-protozoal therapy. Unfortunately, she developed necrotizing enteritis which led ultimately to her death. Antibiotic sensitivity and pathogenicity testing suggested that *Acanthamoeba* were unusually virulent (Harwood *et al.*, 1988).

Notably, *Acanthamoeba* also cause infections of the central nervous system of animals including gorillas, monkeys, dogs, ovines, bovines, horses, and kangaroos, as well as birds, reptiles, amphibians, fishes, and even invertebrates.

3.4. Diagnosis

3.4.1. Clinical and laboratory diagnosis

Due to the rarity of the disease and complicated symptoms, which are common to other pathogens causing the central nervous system infections, the diagnosis of *Acanthamoeba* granulomatous encephalitis is problematic. The symptoms are similar to other central nervous system pathogens including virus, bacteria and fungi. This makes *Acanthamoeba* granulomatous encephalitis diagnosis challenging and requires very high suspicion and expertise (Table 2). In general, magnetic resonance imaging or computerized tomography of the brain shows ring-enhancing lesions or low-density abnormalities mimicking a single or multiple space-occupying mass in the cerebral cortex. Depending on the location of the lesions, *Acanthamoeba* granulomatous encephalitis exhibits a broad spectrum of neurological signs and symptoms. In majority of cases, lesions are most numerous in the temporal and parietal lobes. However, severely immunocompromised patients may not exhibit such lesions (Martinez, 1985; 1991; Martinez and Visvesvara, 1997). As indicated above, the neurological manifestations of *Acanthamoeba* granulomatous encephalitis may vary and include headache, fever, behavioral changes, hemiparesis, lethargy, stiff neck, agitation, aphasia, ataxia, vomiting, nausea, cranial nerve palsies, increased intracranial pressure (Martinez, 1985; 1991; Martinez and Visvesvara, 1997). For simplicity, the signs of neurological examination can be grouped into four categories as follows:

(i) State of consciousness: this may range from confused to being unconscious.
(ii) Loss of reflex activity: this may be observed on part or whole of body or in the function of organs as well as internal functions.
(iii) Abnormal speech patterns, e.g., aphasia.
(iv) Abnormal motor patterns: imbalance or unsteady gait or seizure.

The later stages of *Acanthamoeba* granulomatous encephalitis involve a complete loss of consciousness, seizures and coma, which are poor prognostic signs, with most patients who reach this level of neurological deficit deteriorating and dying within days. The pathophysiological complications involving the central nervous system most likely include induction of proinflammatory responses, invasion of the blood-brain barrier and neuronal damage leading to brain dysfunction (Benedetto and Auriault, 2002a; 2002b; Khan, 2006; 2007; Mattana *et al.,* 2002; Marciano-Cabral and Cabral, 2003; Marciano-Cabral *et al.,* 2004; Shin *et al.,* 2001; Visvesvara *et al.,* 2007).

The cerebrospinal fluid findings, although not confirmatory of *Acanthamoeba* granulomatous encephalitis but are of value in diagnosing the central nervous system involvement. These include pleocytosis with lymphocytic predominance as an important feature with elevated polymorphonuclear leukocytes (up to 800 cells per μL), increased protein concentrations, decreased glucose concentrations and minimal cloudiness (Marciano-Cabral and Cabral, 2003). However, the cerebrospinal fluid may be devoid of cells in HIV-positive patients. The cerebrospinal fluid is centrifuges at low speed (250 x g) for 10 min to avoid rupture of trophozoites. Trophozoites may be observed in the wet preparations of the cerebrospinal fluid but may be unrecognized due to their resemblance of macrophages. The absence of viral and bacterial pathogens should be strongly suspected of *Acanthamoeba* granulomatous encephalitis. Due to the low density of parasites, the detection of host immune response should be attempted primarily. The demonstration of high levels of *Acanthamoeba*-specific antibodies in patient's serum may provide a useful and straightforward method to further suspect *Acanthamoeba* granulomatous encephalitis. This is performed using indirect immunofluorescence assays. The serial dilutions of the patient's serum are incubated with fixed amoebae-coated slides (preferably T1, T4, T12 isolates as they have been shown frequently to cause *Acanthamoeba* granulomatous encephalitis), followed by incubation with fluorescein isothiocyanate (FITC)-labelled anti-human antibody and visualized under fluorescent microscope. It is important to remember that the levels of anti-*Acanthamoeba* antibodies in normal populations may be in the range of 1:20 to 1:60 (Cerva, 1989; Cursons *et al.,* 1980. However, patients with severely impaired immune system may not develop a high titre, thus other clinical findings should be taken into account for the correct diagnosis. Alternatively, the immunofluorescent staining of infected tissues embedded in paraffin should allow the identification of *Acanthamoeba* in the tissue sections.

The confirmatory evidence comes from direct microscopic observation of amoebae in the cerebrospinal fluid (after centrifugation at low speed) or in the brain biopsy but requires familiarity of morphological characters. Giemsa-Wright, acridine orange or Calcofluor white staining may facilitate morphological-based positive identification of these amoebae. The lack of familiarity with amoebae morphological characteristics may require immunohistochemical studies using antisera made against *Acanthamoeba* to identify the etiologic agent, which should aid in the clinical diagnosis of *Acanthamoeba* granulomatous encephalitis (Willaert and Stevens, 1976). In addition, it is helpful to inoculate few drops of the cerebrospinal fluid and/or brain biopsy for amoebae culturing as described in C2.9. *Acanthamoeba* feeds on bacteria as food source, and depending on the number of amoebae in the specimen, trophozoites can be observed within a few hours (up to 12 h). However in the absence of amoebae, plates should be monitored for up to 7 days (reviewed in Khan, 2006). This method is particularly useful if problems are encountered in differentiating *Acanthamoeba* from monocytes, polymorphonuclear leukocytes and macrophages. As indicated in section C2.9, PCR-based methods have been developed, as well as matrix-assisted laser desorption–ionization time-of-flight mass spectrometry in section 2.9.6, but microscopy and plating-based analysis remain methods of choice.

3.5. Host susceptibility

Acanthamoeba granulomatous encephalitis is a rare disease that occurs mostly in immunocompromised or debilitated patients due to HIV infection, diabetes, immunosuppressive therapy, malignancies, malnutrition, and alcoholism and usually seen as a secondary infection. This is due to the inability of

Acanthamoeba to evade the immune system of immunocompetent individuals. Indeed, the healthy human serum exhibits amoebicidal activities by activating the alternative complement pathway. Of interest, protozoan parasites with the ability to evade the host immune system possess sialic acid on the surface of their plasma membranes, thus blocking the alternative pathway convertase or by expressing special coat or a capsule. For example, the expression of variable surface glycoprotein on the surface of African *Trypanosoma* cover underlying components of the plasma membrane, thus preventing activation of the alternative complement pathway (Ferrante and Allison, 1983). However, the plasma membrane of *Acanthamoeba* lacks sialic acid (Korn and Olivecrona, 1971) or any protective coat or capsule (Bowers and Korn, 1968) and thus amoebae (at least in the trophozoite stage) are exposed to the complement-mediated attack in an antibody-independent pathway (Ferrante and Rowan-Kelly, 1983). In addition, the presence of anti-*Acanthamoeba* antibodies in normal populations provides additional protection against these opportunistic pathogens (Cursons *et al.,* 1980). Overall, the complement pathways and the antibodies together with neutrophils and macrophages show potent amoebilytic activities thus suppressing infection. The conclusion from these findings is clear, in that, a debilitated immune status of the host is usually a pre-requisite in *Acanthamoeba* granulomatous encephalitis, but the core basis of host susceptibility in contracting *Acanthamoeba* granulomatous encephalitis requires further study as it may involve other factors such as host ethnicity (i.e., genetic basis of the host) or the inability of host to induce specific immune response against these pathogens. Interestingly, in a recent study (2001), it was shown that Hispanic individuals were shown to be 14.5 times less likely to develop antibodies against *Acanthamoeba* (T4 genotype) than Caucasians individuals. The precise basis of these variations in the antibody level is not clear, e.g., genetic or lack of exposure of Hispanics to *Acanthamoeba* in the environment. It is unlikely that a lesser seropositivity in Hispanic individuals is due to their lack of exposure to *Acanthamoeba*, compared with Caucasians individuals. At the same time, it is not clear whether lack of anti-*Acanthamoeba* antibody levels in Hispanics makes them more susceptible to *Acanthamoeba* infections. Future studies will identify the precise host factors that play an important role in controlling this fatal infection, and may help develop therapeutic interventions for susceptible hosts.

3.6. Species and organ specificity

Species specificity for *Acanthamoeba* granulomatous encephalitis is not determined. For *in vivo* studies, mice are widely used in the study of *Acanthamoeba* granulomatous encephalitis. For organ specificity, it is shown that when injected in mice, amoebae invaded the central nervous system, but also lungs, kidney, heart, liver, diaphragm and caused significant changes in these organs. The presence of amoebae outside of the central nervous system was found mainly in the early period of infection. Overall, these findings suggested that amoebae were isolated from different organs of dead animals

Table 3. Risk factors associated with *Acanthamoeba* infections

No.	Risk factors associated with *Acanthamoeba* keratitis
1	Handling of contact lenses (CL) with unclean hands
2	Washing CL with home-made saline / tap water
3	CL wear for more than recommended times
4	CL wear during swimming, performing water sport activities or relaxing in a hot tub
5	Washing eyes and/or swimming with corneal trauma – splashing eyes with contaminated water
6	Re-using CL without proper cleaning
7	Incubating CL in disinfectants for less than recommended times
8	Chlorine-based disinfectants are less effective in killing *Acanthamoeba*
No.	**Risk factors associated with *Acanthamoeba* granulomatous encephalitis**
1	HIV/AIDS patients are particularly at risk
2	Individuals with lymphoproliferative disorders, hematologic disorders, diabetes mellitus, pneumonitis, renal failure, liver cirrhosis, rhinitis, pharyngitis, gammalobulinemia, pregnancy, systemic lupus erythematosus, glucose-6-phosphate deficiency, tuberculosis are at risk
3	Alcohol abuse
4	Organ/tissue transplantation with immunosuppressive therapy
5	Excessive use of steroids or antibiotics
6	For all of the above, exposure to contaminated soil/water are potential risk factors
7	For all of the above, activities which may result in skin cut/bruises followed by exposure to contaminated soil/water are potential risk factors

and did not exhibit any specificity in relation to the organs in following infections (Mazur and Jóźwiak, 1993).

3.7. Risk factors

Acanthamoeba granulomatous encephalitis is normally a secondary infection to other primary diseases. Almost all reported cases have been reported in immunocompromised patients due to HIV/AIDS patients, and/or in individuals with lymphoproliferative disorders (refer to several conditions in which lymphocytes are produced in excessive quantities), hematologic disorders (diseases that affect the blood), diabetes mellitus, pneumonitis (inflammation of lung tissues), renal failure, liver cirrhosis or other hepatic diseases, rhinitis (irritation and inflammation of some internal areas of the nose), pharyngitis (inflammation of the pharynx), gamma-globulinaemia (significant changes in the levels of immunoglobulins/antibodies), pregnancy, systemic lupus erythematosus (autoimmune diseases that most often harms the heart, joints, skin, lungs, blood vessels, liver, kidneys, and the nervous system), glucose-6-phosphate dehydrogenase deficiency [glucose-6-phosphate dehydrogenase is an enzyme that catalyzes the oxidation of glucose-6-phosphate to 6-phosphogluconate and in tandem, reduces the oxidized form of nicotinamide adenine dinucleotide phosphate ($NADP^+$) to nicotinamide adenine dinucleotide phosphate (NADPH). NADPH, a required cofactor in many biosynthetic reactions, maintains glutathione in its reduced form. Reduced glutathione acts as a scavenger for dangerous oxidative metabolites in the cell. With the help of the enzyme glutathione peroxidase, reduced glutathione also converts harmful hydrogen peroxide to water. Red blood cells rely heavily upon glucose-6-phosphate dehydrogenase activity because it is the only source of NADPH that protects the cells against oxidative stresses]; tuberculosis or Hodgkin's disease (a type of lymphoma), chronic alcoholism, malnourishment, chronic illness, or debilitation or undergoing radiotherapy. Patients undergoing organ/tissue transplantation with immunosuppressive therapy, steroids and excessive antibiotics are also at risk (Table 3). The risk factors for patients suffering from the above diseases include exposure to contaminated water such as swimming pools, on beaches, or working with garden soil.

3.8. Treatment

For *Acanthamoeba* granulomatous encephalitis, there is no recommended treatment and the majority of cases due to *Acanthamoeba* granulomatous encephalitis are identified, post-mortem. This is due to low sensitivity of *Acanthamoeba* to many anti-amoebic agents but more importantly the inability of these compounds to cross the blood-brain barrier into the central nervous system. Current therapeutic agents include a combination of ketoconazole, fluconazole, sulfadiazine, pentamidine isethionate, amphotericin B, azithromycin, itraconazole, or rifampicin that may be effective against the central nervous system infections due to free-living amoebae, but exhibit severe side effects. Among very few successful outcomes of treatment was a 7-year girl with a single *Acanthamoeba*-induced granulomatous brain tumour, who recovered following total excision of the mass and treatment with ketoconazole. Ketoconazole and rifampicin added to trimethroprim-sulfamethoxazole therapy was used successfully for the treatment of two immunocompetent pediatric patients with the central nervous system infection. In another case, a 40-year old man with *Acanthamoeba* meningitis recovered following treatment with penicillin and chloramphenicol. In this case, *Acanthamoeba* were isolated repeatedly from the cerebrospinal fluid of the patients (Lalitha *et al.*, 1985; Ofori-Kwakye *et al.*, 1986). A successful outcome in a HIV-positive individual with the central nervous system involvement was observed followed by therapy with fluconazole and sulfadiazine. Despite these limited successes, the prognosis for *Acanthamoeba* granulomatous encephalitis remains extremely poor and there is an urgent need to identify novel anti-*Acanthamoeba* drugs. Notably, pentamidine isethionate should be used be caution as some patients demonstrated nephrotoxicity.

Recent studies have suggested that alkylphosphocholine compounds (phospholipid-like molecules that are phosphocholine esters of aliphatic long chain alcohols differing in chain length, unsaturation and position of the cis-double bond and have been shown to possess antimicrobial activities), such as hexadecylphosphocholine (miltifosine), exhibit anti-*Acanthamoeba* properties as well as the ability to cross the blood-brain barrier and may thus have value in the treatment of *Acanthamoeba* granulomatous encephalitis (Kotting *et al.,* 1992; Walochnik *et al.,* 2002). Further studies are needed to determine their precise mode of action against *Acanthamoeba*, and to develop methods of application and more importantly, to assess the success of these compounds *in vivo*. Even with treatment, survivors may develop disability such as hearing loss, vision impairment etc.

4. Cutaneous Acanthamebiasis

Other infections due to *Acanthamoeba* involve nasopharyngeal and, more commonly, cutaneous infections. The cutaneous infections are characterized by nodules and skin ulcerations and demonstrate *Acanthamoeba* trophozoites and cysts. In healthy individuals, these infections are very rare and may be self-limiting. However, in immunocompromised individuals, this may provide a route of entry into the bloodstream, followed by the haematogenous spread to different tissues, which may lead to fatal consequences. The involvement of the central nervous system warrants deaths within weeks (Torno *et al.,* 2000). Histologic examination of cutaneous lesions generally shows foci of necrosis surrounded by inflammatory cells, vasculitis, trophozoites and cysts. However, the histologic appearance of the cutaneous lesions may mimic that of fungi, viruses, mycobacteria, or inflammation due to a foreign body. Both *Acanthamoeba* granulomatous encephalitis and cutaneous infections can occur in combination or independent of each other. The direct demonstration of amoebae in biopsy using Calcofluor white staining, immunofluorescence assay, polymerase chain reaction-based assay, or isolation of *Acanthamoeba* from the clinical specimens using plating assays, provides positive diagnosis as described above. There is no recommended treatment, but topical application of itraconazole, 5-fluorocytosine, pentamidine isethionate, ketoconazole and chlorohexidine gluconate may be of value.

From: *Acanthamoeba*: Biology and Pathogenesis. By: Naveed Khan.

Section D

Pathogenesis

1. Introduction.

Despite advances in the antimicrobial chemotherapy and supportive care, the prognosis for *Acanthamoeba* infections (in particular for *Acanthamoeba* granulomatous encephalitis with a mortality rate of more than 95%) remains poor. In part, this is due to our incomplete understanding of the pathogenesis and pathophysiology of *Acanthamoeba* infections together with difficulties in diagnosis leading to delayed chemotherapy. Another major concern during the course of therapy is the ability of *Acanthamoeba* to transform into dormant cyst forms, which may resist the recommended levels of antimicrobial chemotherapy. In support, it is shown that *Acanthamoeba* cysts develop a relatively impermeable, double-walled structure, consisting of an inner-wall, the endocyst, and an outer-wall, the ectocyst. In addition to the possible drug resistance, cysts may also reactivate following the antimicrobial chemotherapy leading to recurrence of infection.

In view of the devastating nature of *Acanthamoeba* infections and the problems associated with the correct diagnosis and successful therapy, there is a need for better understanding of the pathogenesis and pathophysiology of *Acanthamoeba* infections, which should help develop measures for therapeutic interventions. Much of our understanding of *Acanthamoeba* pathogenesis comes from the availability of vertebrate models to study *Acanthamoeba* infections.

At present, the following models are used to study the pathogenesis of *Acanthamoeba* keratitis and *Acanthamoeba* granulomatous encephalitis.

- *in vivo* models (performing a given experiment using a living organism as a whole, as opposed to a partial or a dead organism);
- *in vitro* models (performing a given experiment in a controlled environment outside of a living organism; for example in a test tube); and,
- *ex vivo* models (performing a given experiment using a living tissue derived from a whole organisms).

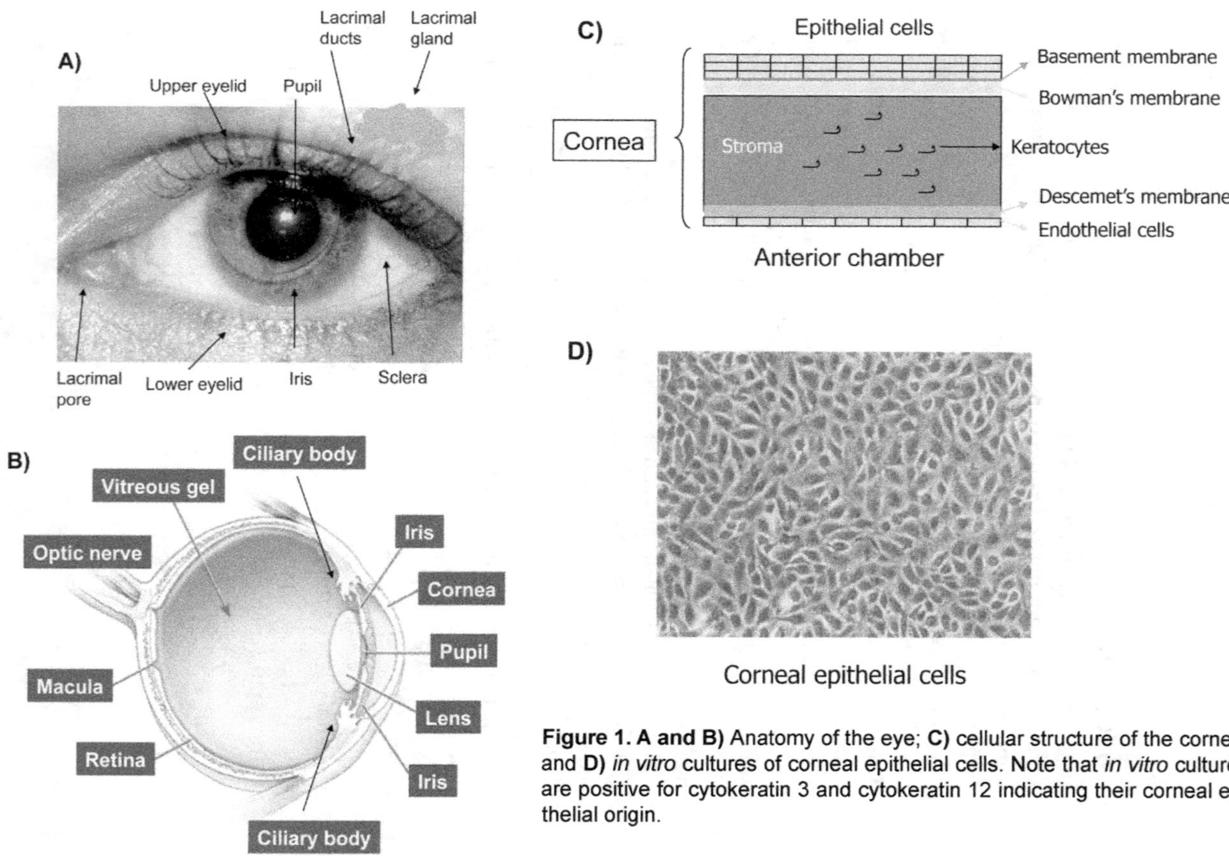

Figure 1. A and B) Anatomy of the eye; **C)** cellular structure of the cornea; and **D)** *in vitro* cultures of corneal epithelial cells. Note that *in vitro* cultures are positive for cytokeratin 3 and cytokeratin 12 indicating their corneal epithelial origin.

2. *Acanthamoeba* keratitis

2.1. *In vivo* models

In vivo models of *Acanthamoeba* keratitis using Yucatan micropigs and Chinese hamster have been developed. Both Chinese hamster and pigs fitted with *Acanthamoeba*-laden soft contact lenses develop corneal infections that clinically and histopathologically mimic the human counterpart. The strong correlation between the clinical and histopathological features of contact lens-induced *Acanthamoeba* keratitis in the Chinese hamster and pig, as well as the anatomical similarity of the pig eye with the human eye, make the porcine model (as well as the Chinese hamster model) a valuable tool for investigations of the pathogenesis and pathophysiology of *Acanthamoeba* keratitis. However, since these animals are expensive, difficult to handle and have ethical issues, there is a need to use alternative models that are equally effective, cheaper and easy to handle.

2.2. *Ex vivo* models

The *ex vivo* models were developed by testing normal, intact corneas (Fig. 1), of 11 mammalian and one avian species. It was interesting to observe species specificity for *Acanthamoeba* keratitis. *Acanthamoeba* trophozoites not only failed to produce cytopathic effects but did not even bind to the corneal epithelium of mice, rats, cotton rats, horses, guinea pigs, cows, chickens, dogs, and rabbits. However, parasites adhered, invaded, and produced severe damage to the human, pig, and the Chinese hamster corneas suggesting their possible use as an *ex vivo* model to study the pathogenesis of *Acanthamoeba* keratitis (Niederkorn *et al*., 1992). But again the associated costs and ethical issues limit their usefulness.

Recently, a novel *ex vivo* model of the human cornea has been developed, which uses live human stroma coated with the human extracellular matrix to mimic the *in vivo* basement membrane (Fig. 2A). Because this is a novel *ex vivo* model, hence it is described in more detail here. Briefly, the human corneo-scleral discs are cut in the region of the Bowman's membrane to separate the epithelium from the stroma. This results in a thinner section of the epithelium (less than 100 μm), and a thicker section of the stroma (more than 500 μm). The epithelial section is transferred to a collagen-coated flask, to be cultured. The stromal section is further cut into 2 – 3 sections of 100 – 200 μm and placed in the depression of a cavity slide, followed by addition of the human extracellular matrix and is left to polymerize for 4 hours in a culture hood. Once polymerized, the human corneal epithelial cells are seeded at a high density (5 x 10^5 cells), onto the scaffold for a further 48 hours. At this density, cells firmly adhere to the human extracellular matrix and colonize the whole surface within three days and thereafter multilayering. At this stage, the construct (stroma + extracellular matrix + epithelial layer) is transferred to a culture insert for the next stage of airlifting to expose its surface. The medium is lowered to just cover the surface and placed on a rocking platform, thus exposing it to the air and submerging it in the medium. This mimics the action of the tear wash of the *in vivo* situation. After more

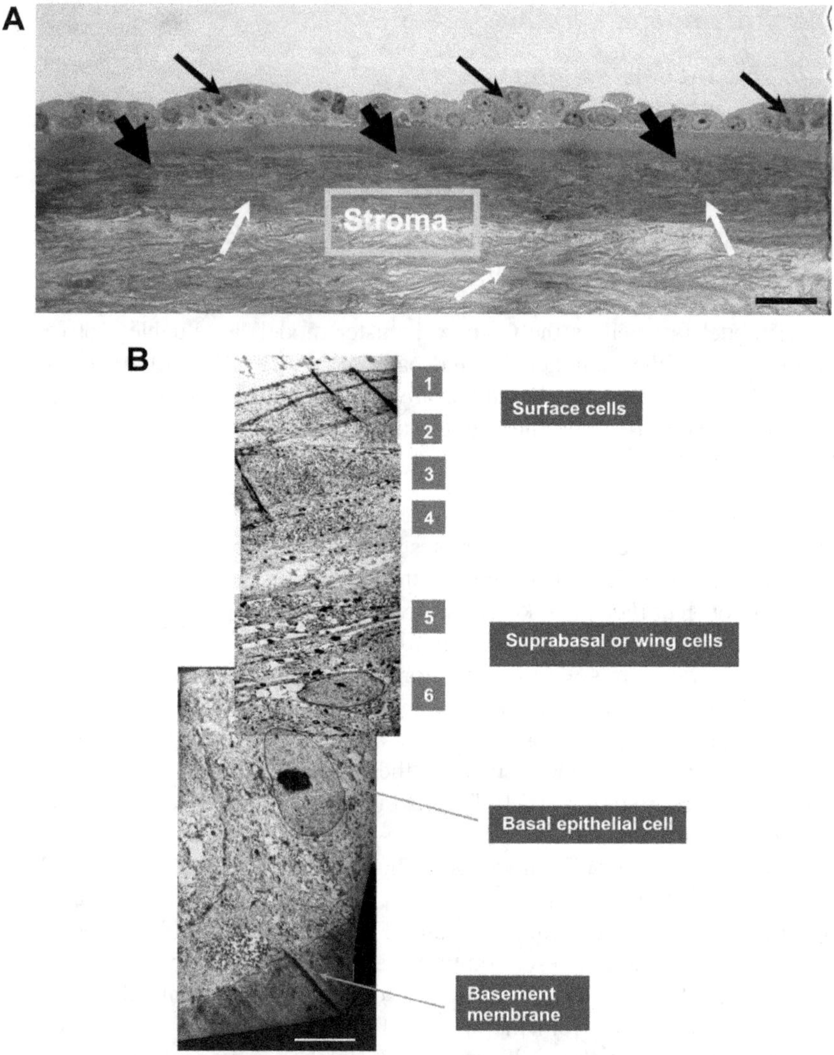

Figure 2. A) Multilayering of epithelial cells in a 14 day-old construct. Black arrows: epithelial cells; Arrowheads: human extracellular matrix (basement membrane); white arrows: fibrillar collagen bundles. Bar = 30μm. **B)** Montage of Transmission Electron Micrograph of a 3 week-old construct. Bar = 2μm.

than 5 days, "cornification" is evident i.e., the surface cells become larger and much flattened and then die. The montage of the transmission electron micrograph in Fig. 2B reveals at least 6 – 7 layers of the epithelial cells on top of a basement membrane. Although not tested, this model should be of potential value to study the pathogenesis of *Acanthamoeba* keratitis as well as of other corneal pathogens, and testing the efficacy of potential antimicrobial compounds.

2.3. *In vitro* models

The availability of *in vitro* models has made it possible to study, at the molecular level, *Acanthamoeba* interactions with the corneal epithelial cells, as well as parasite interactions with the fibroblasts, keratocytes, and the corneal endothelial cells (Fig. 1D). Indeed, the *in vitro* models are shown to be relevant to study *Acanthamoeba* keratitis. This is shown by findings that *Acanthamoeba* isolates pathogenic for animals *in vivo* show severe cytopathic effects *in vitro*, confirming that the *in vitro* models are relevant to study the pathogenesis of *Acanthamoeba* infections (Grillot and Ambroise-Thomas, 1981).

3. *Acanthamoeba* granulomatous encephalitis

3.1. *In vivo* models

At present, the use of mice is the most widely used *in vivo* model to study *Acanthamoeba* granulomatous encephalitis. In this model, *Acanthamoeba* are inoculated intranasally into mice. Because *Acanthamoeba* granulomatous encephalitis is generally limited to individuals with a weaker immune system, this condition can be induced by pretreating mice with corticosteroid (i.e., methylprednisolone) or tetracycline to suppress the host defences. This was shown in a series of experiments where laboratory mice were pretreated with either corticosteroid (i.e., methylprednisolone) or tetracycline to suppress host defences followed by intranasal inoculation of *Acanthamoeba* and the mortality was recorded. The results revealed that the mortality rate for animals receiving methylprednisolone and then *Acanthamoeba* was greater (around 50%) than animals inoculated with *Acanthamoeba* alone (around 10%). Similarly, the mortality rate for animals receiving tetracycline and *Acanthamoeba* was greater (60%) than the animals with *Acanthamoeba* alone (10%). These findings suggested that the host defences, particularly those depending on neutrophils are important in the control of *Acanthamoeba* granulomatous encephalitis. Furthermore, it was shown that when injected in mice, *Acanthamoeba* not only invaded the central nervous system, but also lungs, kidney, heart, liver, diaphragm and produced significant changes in these organs. The presence of *Acanthamoeba* outside of the central nervous system was found mainly in the early period of infection. Overall, these findings showed that amoebae were isolated from different organs of the dead animals and did not exhibit any specificity in relation to the organs following infection, a finding that is consistent with the human infection and confirms the validity of mice as a useful *in vivo* model to study the pathogenesis of *Acanthamoeba* granulomatous encephalitis (Mazur and Jóźwiak, 1993).

Although physiologically relevant, vertebrate models are expensive; not routinely available in many laboratories, require labour-intensive management; and have ethical implications. The remarkable parallels between the innate immune responses of mammals and insects (Scully and Bidocchka, 2006), especially as insects possess a highly selective blood-brain barrier exhibiting functional properties comparable with that of vertebrates (Carlson *et al.*, 2000), and the fact that *Acanthamoeba* granulomatous encephalitis is limited

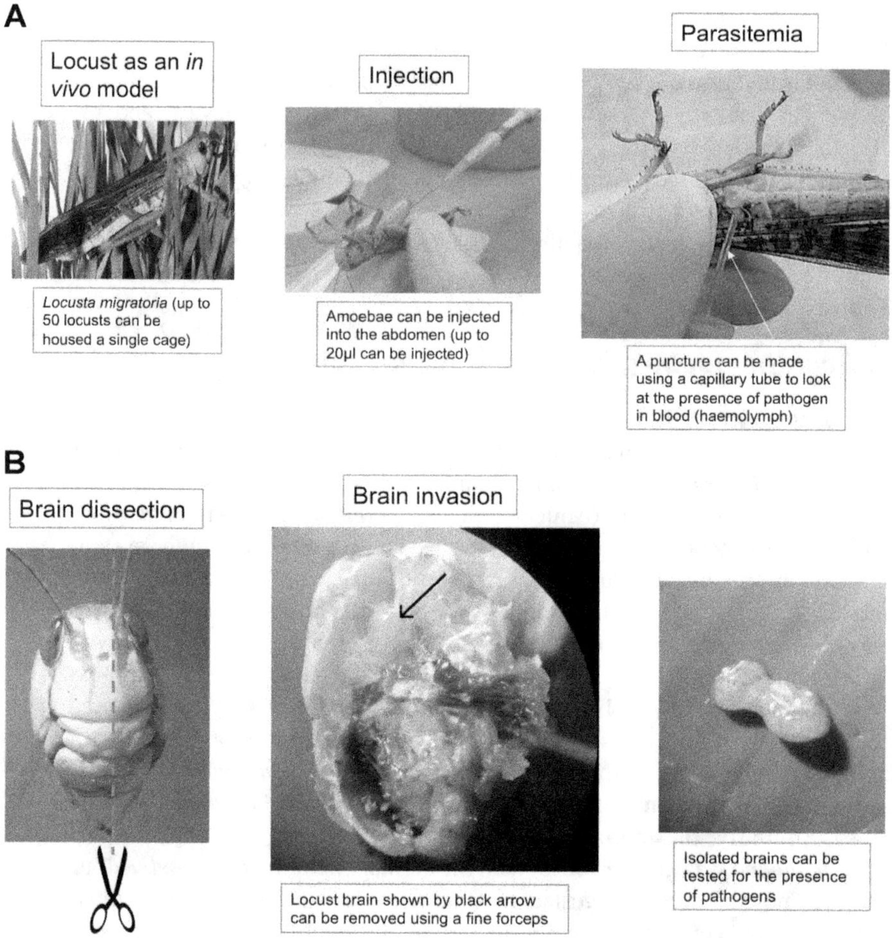

Figure 3. Use of locusts as an *in vivo* model to study *Acanthamoeba* infection. **(A)** Infection can be studied in the blood (i.e., haemolymph) and **(B)** brains can be isolated to study the involvement of the central nervous system.

mostly to immunocompromised populations, suggest that insects could make a useful model in which to study the pathogenesis of *Acanthamoeba* granulomatous encephalitis (Fig. 3). To this end, recent studies have tested the use of *Locusta migratoria*, the African migratory locust as a model to study *Acanthamoeba* granulomatous encephalitis. It is noteworthy that injection of *Acanthamoeba* produced locust death with almost 100% mortality, with as few as 10^3 amoebae, however the length of time required to produce 100% mortality was dependent on the inoculum size. Regardless of the various inocula, *Acanthamoeba* was isolated from the brain lysates of parasite-injected locusts. Thus, it is tempting to speculate that *Acanthamoeba*-mediated locust death is, in part, associated with the parasite's ability to invade the locust central nervous system. It is interesting to note that following injection, *Acanthamoeba* were recovered from the locust haemolymph, as well as other tissues such as fat body and muscle tissues, suggesting that *Acanthamoeba* produced parasitaemia and survived the onslaught of the innate immune defences of locusts, a finding that is consistent with the human form of *Acanthamoeba* granulomatous encephalitis.

Given that the molecular components of the blood-brain barrier in insects such as *Drosophila* are homologous to those of vertebrates (Willott *et al.*, 1993), it is possible that there will be similarities between the mechanisms by which *Acanthamoeba* invades the central nervous system of mammals and that of locusts, and research is underway to determine this. Another interesting finding was that *Acanthamoeba*-injected locusts displayed pronounced weight loss, compared with the control group. A significant weight loss together with reduced faeces production seen in locusts injected with *Acanthamoeba* suggests that this aspect of sickness behaviour may result from reduced sensory input. Previous studies have identified several factors associated with the virulence of *Acanthamoeba* including a mannose-binding protein expressed on the surface membranes of *Acanthamoeba*, serine proteases, phospholipases, and ecto-ATPases (discussed later). By studying the role of these factors *in vivo*, the validity of the locust model will be confirmed further. Although further investigations are needed, these findings suggest that the insect model may offer potential in the study of the human *Acanthamoeba* granulomatous encephalitis and can elicit the full spectrum of host-pathogen interactions, at least at the early stages of research. It is anticipated that the likely existence of additional virulence determinants of the *Acanthamoeba* can be investigated *in vivo* using this insect system, which can later be tested in a vertebrate model.

Axenic cultivation of pathogens over a longer period of time results in the loss of virulence, thus another advantage of studying *Acanthamoeba* granulomatous encephalitis *in vivo* is that parasites regain and/or retain virulence. For example, it is shown that following passage in the mouse brain, *Acanthamoeba* exhibited enhanced virulence, i.e., infected mouse mortality increased from 5% (with axenically cultured *Acanthamoeba*) to 70% (with brain passaged *Acanthamoeba*).

In determining the gene profiles, the mouse brain-passaged *Acanthamoeba* showed upregulations of genes encoding predictive NADH-dehydrogenase, proteasomal ATPase, and GDP-mannose pyrophosphorylase B, suggesting their possible role in the *Acanthamoeba* granulomatous encephalitis (Han *et al.*, 2006).

3.2. *Ex vivo* models

The *ex vivo* model of the human blood-brain barrier has been developed (Fig. 4). This model makes use of astrocytes, pericytes, neurones and the brain microvascular endothelial cells in 24-well Transwells.

3.3. *In vitro* models

The availability of *in vitro* cultures of primary human brain microvascular endothelial cells has been of tremendous value in the identification and characterization of molecular determinants that are required in host-parasite interactions (Stins *et al.*, 1997; Alsam *et al.*, 2003), and which may lead to

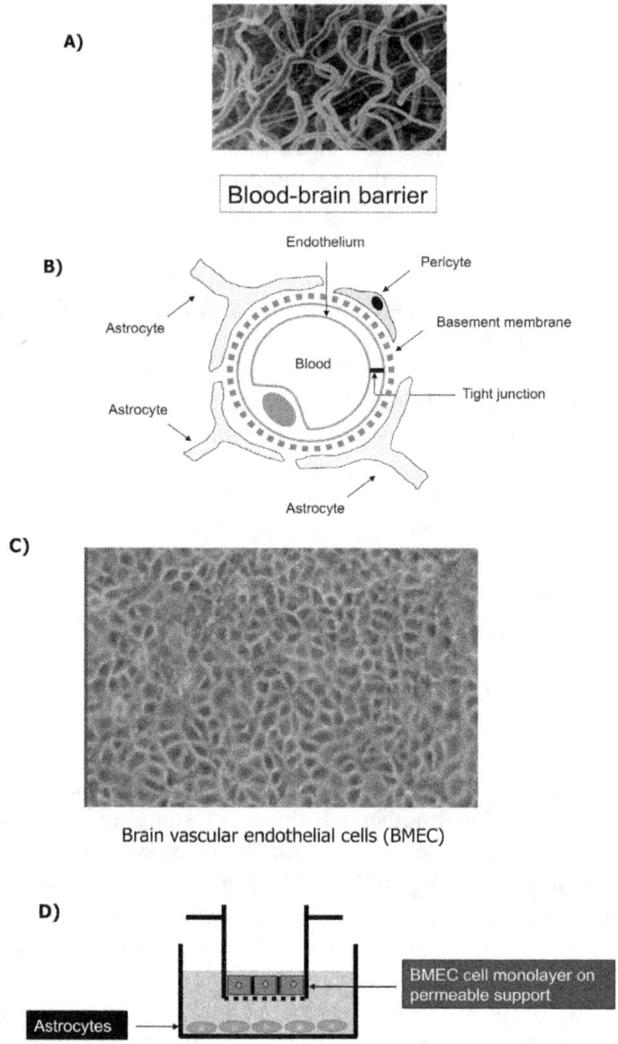

Figure 4. A) and B) Anatomy and the cellular structures of the microvessels of the human brain tissue; **C)** *in vitro* cultures of brain microvascular endothelial cells. Note that *in vitro* cultures are positive for factor VIII, carbonic anhydrase IV, uptake of fluorescently labelled acetylated low density lipoprotein, and gamma-glutamyl transpeptidase indicating their brain endothelial origin; and **D)** *Ex vivo* model of the blood-brain barrier. Transwells can be used to grow brain microvascular endothelial cells into the upper chamber on a permeable support, and astrocytes can be cultured into the bottom chamber.

Acanthamoeba traversal of the blood-brain barrier. The brain microvascular endothelial cells used in our laboratory are positive for brain endothelial markers such as factor VIII, carbonic anhydrase, Ulex Europaeus Agglutinin I, gamma-glutamyl transpeptidase, and took up acetylated low-density lipoprotein (Fig. 4C). In addition, these cells demonstrated the expression of the tight junction proteins and the formation of a polar monolayer.

4. *Acanthamoeba*: An opportunistic pathogen

Acanthamoeba is an opportunistic or incidental or accidental pathogen. This term is generally described for pathogens, in which disease confers no obvious biological advantage for the infectious agent. Such organisms can be further divided into two groups, (i) organisms whose natural habitat is humans but only cause disease in a small minority, and (ii) organisms whose natural habitat is nature, but if they encounter a susceptible host, they produce disease, e.g., *Acanthamoeba*. This is in contrast to obligate pathogens that have to produce disease in human beings in order to continue to survive and propagate, e.g., most viruses for which humans are the only natural host.

4.1. An opportunist with the pathogenic potential

Although *Acanthamoeba* is a free-living amoeba whose natural habitat is nature, however the ability of *Acanthamoeba* to produce infection is not simply due to their mere exposure to the susceptible hosts, but they do possess virulence properties that enable them to produce serious and fatal infections. For example, in a survey of 101 asymptomatic daily wearers of contact lenses storage cases for the presence of microbes, it was found that 81% were contaminated with the bacteria, fungi and protozoa including *Acanthamoeba* (Grey *et al.*, 1995). The occurrence of fungi was higher than protozoa: 24 *versus* 20% respectively, and significantly higher than *Acanthamoeba*, 24 *versus* 8% respectively, even though fungi rarely cause corneal infections in wearers of contact lenses. In another survey microbial contamination of the contact lens care systems of ten people, it was shown that all had *Acanthamoeba* and Gram-negative bacteria (*Pseudomonas* spp. in 6 of them); Gram-positive bacteria (*Bacillus* spp. in 5 of them); and fungi isolated in six care systems. Despite the wide microbial contamination, *Acanthamoeba* keratitis was the most predominant (seven individuals infected with *Acanthamoeba* keratitis), and only one had *Pseudomonas* keratitis, and the remaining two were asymptomatic (Donzis *et al.*, 1989). These findings further suggest that the mere presence of a microbe is not sufficient to produce infection, but that the keratitis-causing microbes such as *Acanthamoeba* possess virulence properties, which enable them to become potential ocular pathogens. And it is the identification of these virulence determinants and a complete understanding of their more of action that should help in the rationale development of therapeutic interventions. Of interest, 75% of individuals in this study used hydrogen peroxide as a disinfectant for their lenses indicating the ability of microbes to resist and survive many available disinfectants for cleaning contact lenses. Indeed, many of the organisms identified as contaminants in this survey exhibited catalase activity, i.e., an enzyme that breaks down hydrogen peroxide to oxygen and water (Gray *et al.*, 1995). Similarly, the ability of *Acanthamoeba* to produce granulomatous encephalitis requires virulence factors such as specific adhesins, production of toxins and the ability to resist immune factors, which as a whole enables this parasite to produce fatal infections.

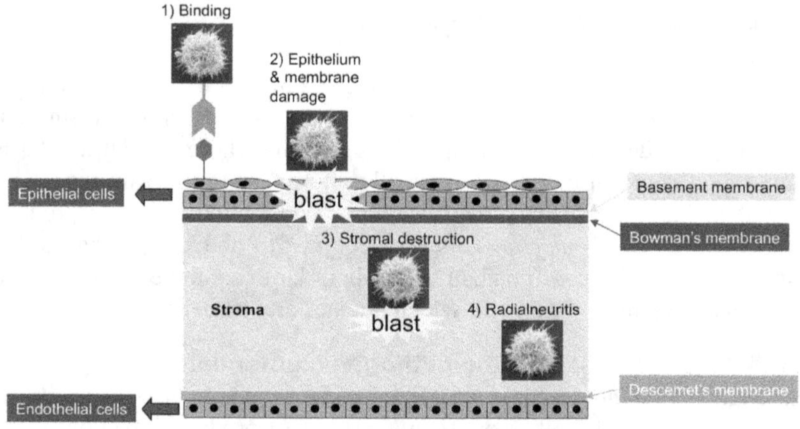

Figure 5. The pathogenic cascade of *Acanthamoeba* keratitis.

5. Crossing of the biological barriers

5.1. Cornea

The pathogenic cascades of *Acanthamoeba* keratitis is depicted in Fig. 5. Using a mannose-binding protein, *Acanthamoeba* adheres to the mannosylated glycoproteins on the corneal epithelium. The mannosylated glycoproteins are upregulated in response to corneal abrasion. The initial binding leads to secondary events such as phagocytosis and toxin production resulting in the host cell death and breachment of the Bowman's membranes, leading to the invasion of the parasite into the collagenous stroma. To this end, it is shown that *Acanthamoeba* proteases contribute to the destruction of the stroma. *Acanthamoeba* trophozoites cluster around corneal nerves, producing radial keratoneuritis and extreme pain. *Acanthamoeba* rarely progresses beyond the corneal endothelium to produce intraocular infections (Fig. 5).

5.2. Traversal of the blood-brain barrier

It is widely accepted that the route of entry for *Acanthamoeba* includes the respiratory tract leading to amoebae invasion of the alveolar blood vessels, followed by the haematogenous spread. However, it is not clear how circulating *Acanthamoeba* breakthrough the blood-brain barrier to gain entry into the central nervous system to produce disease. It is noteworthy that *Acanthamoeba* may bypass the lower respiratory tract and directly enter into the bloodstream via skin lesions. Blood vessels serving the brain can be collectively called as the blood-brain barrier. There are more than 100 billion capillaries in the human brain extending approximately 400 miles with a surface area of $20m^2$ providing an extensive network to fulfill the nutrient demands as well as to maintain homeostasis of the central nervous system (Pardridge, 2002). Two different structures separate the blood from the central nervous system: the blood-brain barrier and the blood-cerebrospinal fluid barrier. The blood-brain barrier includes the endothelial lining of the brain capillaries that are closely associated with pericytes, astrocytes and

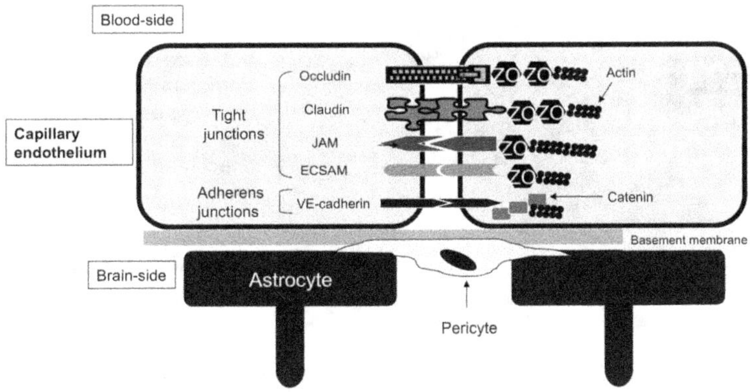

Figure 6. Schematic illustration of the blood-brain barrier at the cerebral capillary endothelium exhibiting tight junctions. The endothelial cells are surrounded by basement membrane, which is ensheathed by astrocytes and pericytes. The tight junctions are composed of integral proteins including occludin, claudin, junctional adhesion molecule (JAM), and endothelial cell-selective adhesion molecule (ECSAM), that interact with their counterparts on the adjacent endothelial cells. Both tight junctions and adherens junctions are composed of multiple proteins. The cytoplasmic tails of these proteins interact with the actin cytoskeleton via a number of accessory proteins including members of zonula-occludens (ZO).

neurones (Fig. 4B). The presence of astrocytes ensheathing the endothelium is limited to the blood-brain barrier. Astrocytes are involved in the induction of permeability barrier properties of the blood-brain barrier. Pericytes also are closely associated with the endothelial cells and are required for capillary maturation, while basement membrane [containing laminin, collagen IV, proteoglycans (especially heparan sulphate), fibronectins, nidogen, and entactin], is important for the blood-brain barrier differentiation. In addition to cellular differences, the blood-brain barrier is characterized by the presence of tight junctions (Fig. 6). The tight junctions are primarily responsible for the barrier function preventing the entry of blood contents into the central nervous system, making it a highly selective barrier compared with the peripheral endothelium (Miller, 1999). For comparison, endothelial cells of the blood-brain barrier exhibit high transendothelial electrical resistance (around 2000 Ω per cm^2) (Pardridge, 1999), while endothelial cells from the human placenta exhibit less than 50 Ω per cm^2 (Jinga *et al.,* 2000). The presence of tight junctions in the blood-brain barrier ensures that even small molecules such as dyes and antibiotics are prevented from entry into the central nervous system by limiting the paracellular route. In contrast, non-brain endothelium is more permissive. However, the small lipophilic molecule substances such as O_2 and CO_2 can diffuse freely across the plasma membrane along their concentration gradient. The nutrients such as glucose, amino acids etc. are transported across the blood-brain barrier via transporters, while larger molecules such as insulin, leptin, iron, transferrin are taken up via receptor-mediated endocytosis to support the neuronal function. The junction complex in the blood-brain barrier comprises of tight junctions

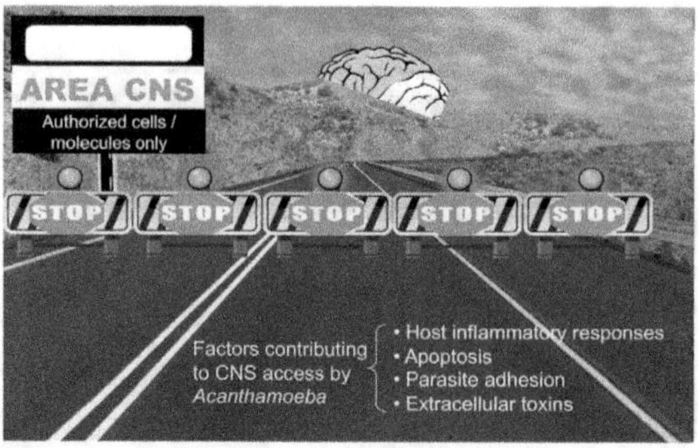

Figure 7. Central nervous system is protected by the presence of the blood-brain barrier that blocks the entry of pathogens/molecules. *Acanthamoeba* uses both direct and indirect virulence factors to traverse this biological barrier. In addition, host inflammatory responses also contribute to blood-brain barrier perturbations.

and adherens junctions. The tight junctions consists of integral membrane proteins (i.e., claudin, occludin, junction adhesion molecule, endothelial cell-selective adhesion molecules), and cytoplasmic proteins [zonula occludens-1 (ZO-1), ZO-2, ZO-3]. Claudins bind to claudins on the adjacent endothelial cells to form the tight junctions. The carboxy terminal of claudins binds to the cytoplasmic zonula occluden proteins. Claudin-5 and claudin-3 are localized in the tight junctions of the brain endothelial cells. Occludin is a phosphoprotein and its cytoplasmic domain is directly associated with zonula proteins. Junctional adhesion molecules and the endothelial cell-selective adhesion molecules also are localized in the tight junctions of the blood-brain barrier. The accessory proteins of the zonula occludens provide structural support and bind to integral membrane proteins of the tight junctions, further linking membrane proteins to the actin cytoskeleton (Fig. 6) (Huber *et al.,* 2001; Kim *et al.,* 2006; Tuomanen, 1996). In contrast, the adherens junctions are composed of membrane protein involving cell-to-cell cadherin homotypic interactions, and cadherin-catenin interactions. Adherens junction proteins include cadherin, actinin and vinculin (analogue of catenin).

The mechanism by which *Acanthamoeba* traverses the blood-brain barrier is complex and is likely to involve both parasite (adhesins, proteases, phospholipases) as well as host determinants [interleukin-beta, interleukin-alpha, tumor necrosis factor- alpha, interferon-gamma, host cell apoptosis] (Fig. 7). The overall outcome is increased permeability and/or apoptosis of the brain endothelial cells, which promote the blood-brain barrier disruptions leading to the parasite invasion of the central nervous system. The understanding of the cross-talk between parasite-host interactions, as well as between the blood-brain barrier and the central nervous system in disease will provide insights to explain the neuropathogenesis and may help develop novel therapeutic interventions. A model of *Acanthamoeba* granulomatous encephalitis is depicted in Fig. 8.

6. Direct virulence factor

Overall, the pathogenesis of *Acanthamoeba* infections is highly complex and involves several determinants working in concert to produce disease. *Acanthamoeba*-mediated human cell death most likely occurs in two phases: the induction of apoptosis during the early stages of infection, followed by an independent and rapid necrosis during the later stages of the infection,

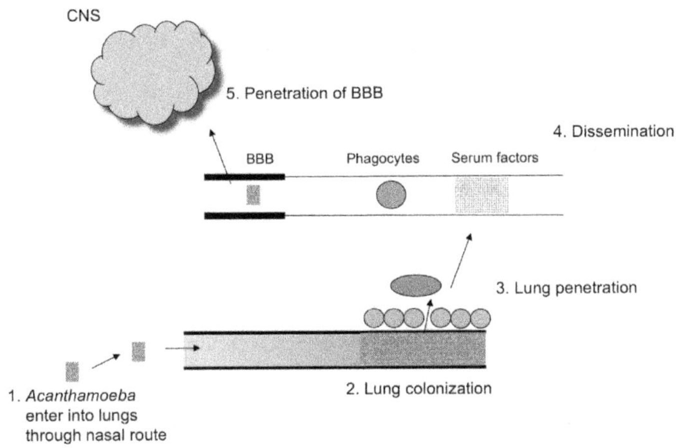

Figure 8. The model of *Acanthamoeba* granulomatous encephalitis. *Acanthamoeba* are thought to enter lungs via the nasal route. Next, amoebae traverse the lungs into the bloodstream, followed by haematogenous spread. Finally, *Acanthamoeba* cross the blood-brain barrier and enter into the central nervous system (CNS) to produce disease. It is noteworthy that *Acanthamoeba* may bypass the lower respiratory tract and directly enter into the bloodstream via skin lesions. The olfactory neuroepithelium may provide an alternative route of entry into the CNS.

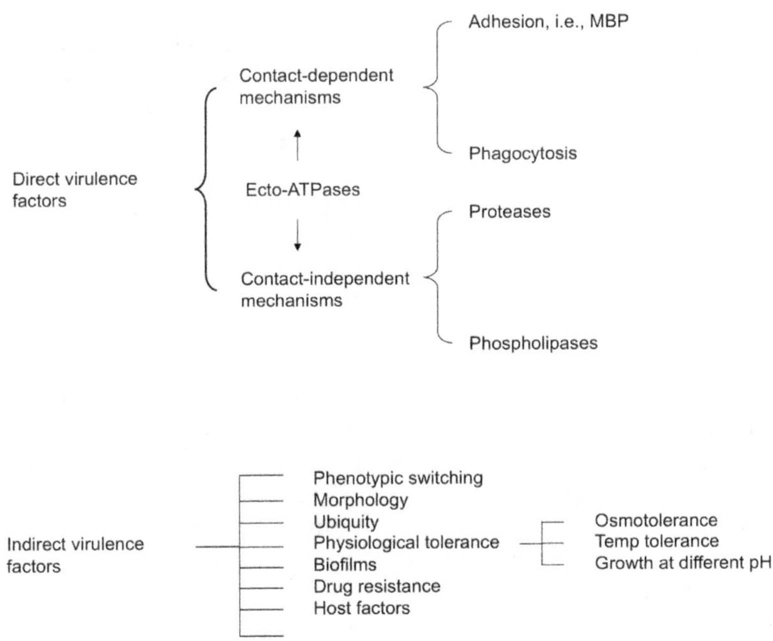

Figure 9. Direct and indirect virulence factors that contribute to *Acanthamoeba* infections.

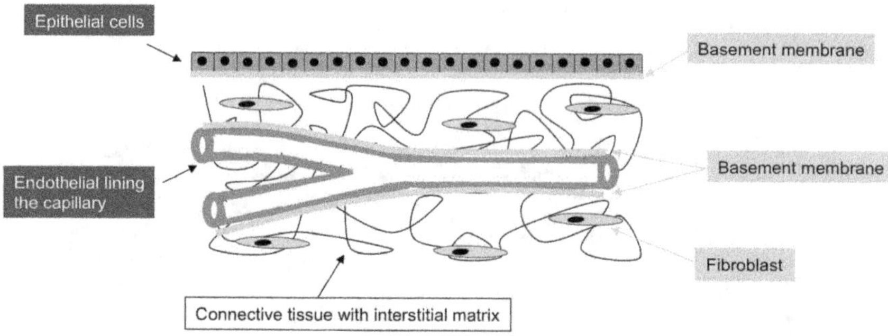

Figure 10. Extracellular matrix in relation to epithelium, endothelium and the connective tissue.

mediated by phagocytosis, proteases etc. For simplicity, in the following account these factors are described separately as contact-dependent and contact-independent mechanisms (Fig. 9).

7. Contact-dependent mechanisms

The ability of the *Acanthamoeba* to bind to the host is a critical first step in the pathogenesis of *Acanthamoeba* infections.

7.1. *Acanthamoeba* binds to the extracellular matrix proteins

The extracellular matrix is the extracellular part of the tissue, which provides structural support to cells (Fig. 10). The extracellular matrix is the defining feature of the connective tissue. The extracellular matrix includes the interstitial matrix and the basement membrane. Interstitial matrix is present between various cells (i.e., in the intercellular spaces) and composed of polysaccharides and fibrous proteins. Basement membranes are sheet-like depositions of the extracellular matrix on which various epithelial cells rest. It is shown that *Acanthamoeba* binds to various components of the extracellular matrix including collagen type IV, laminin, and fibronectin. *Acanthamoeba* binding to various components of the extracellular matrix is mediated by mannose-binding protein. This is shown by findings that the exogenous free alpha-methyl-mannopyranoside inhibited *Acanthamoeba* binding to collagen IV, laminin, and fibronectin in a concentration-dependent manner. The calcium enhances the adhesion of *Acanthamoeba* to the extracellular matrix proteins. However, it is not clear whether the calcium acts intracellularly or at the cell surface. In addition, the laminin is a major mannosylated glycoprotein constituent of the extracellular matrix and the basement membrane of the host cells and recent studies have shown that *Acanthamoeba* binds to the laminin using a laminin-binding protein (Gordon *et al.*, 1993; Wang *et al.*, 1994).

7.2. Adhesion to the host cells

Acanthamoeba has been shown to bind to the host cells (Fig. 11). A mannose-binding protein expressed on the surface of *Acanthamoeba* has been identified, which is important in parasite binding to the host cells. Indeed,

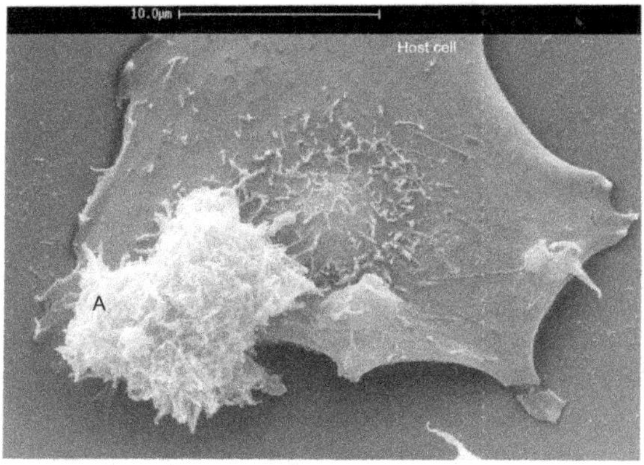

Figure 11. Binding of *Acanthamoeba* to host cells is mediated by acanthopodia. *Acanthamoeba castellanii* belonging to the T4 genotype were incubated with host cells, followed by several washes and observed under scanning electron microscope. Note that parasites were able to exhibit binding to the host cells and binding was mediated by acanthopodia. A is *Acanthamoeba*.

it is shown that *Acanthamoeba* adhesion to various host cells is mediated by the mannose-binding protein. These included human corneal epithelial cells, human brain microvascular endothelial cells, pig corneal epithelium, Chinese hamster corneal epithelium, human corneal fibroblasts, rat microglial cells, monkey kidney cells, neurones etc. These findings suggested that the mannose-binding protein provides initial attachment to the host cells and play a role in the pathogenesis of *Acanthamoeba* keratitis, as well as *Acanthamoeba* granulomatous encephalitis. It is not clear whether there are other more specific mechanisms of *Acanthamoeba* binding to the host cells, but at least the initial binding seems to be dependent on the expression of mannose-binding protein and its binding to mannose-containing glycoproteins on the surface of the host cell. Indeed, *Acanthamoeba* will bind to mannose-containing glycolipids, mannose-coated tissue culture plates and the adherence of *Acanthamoeba* even to inert surfaces is mediated by mannose-binding protein and adherence can be inhibited by the presence of exogenous free mannose. The significance of mannose-binding protein is further shown by the observation that it is expressed only during the infective trophozoite stage of *Acanthamoeba* and dormant cysts lack the expression of mannose-binding protein, which might explain the inability of cysts to bind to the host cells. Later studies identified a mannose-binding protein gene in *Acanthamoeba*, containing 6 exons and 5 introns that spans 3.6 kbp. The 2.5 kbp cDNA codes for an 833 amino acids precursor protein with a signal sequence (residues 1 – 21 aa), an *N*-terminal extracellular domain (residues 22 – 733 aa) with five *N*- and three *O*-glycosylation sites, a transmembrane domain (residues 734 – 755 aa), and a *C*-terminal intracellular domain (residues 756 – 833 aa). Notably, *Acanthamoeba* mannose-binding protein is itself a mannose-containing glycoprotein. This was shown by findings that *N*-glycanase treatment to remove *N*-linked oligosaccharides shifted the subunit molecular mass of the mannose-binding protein from 130 kDa to 110 kDa. The hexosamine analysis revealed that mannose-binding protein of *Acanthamoeba* lacks detectable levels of *N*-acetyl-galactosamine, suggesting the absence of *O*-linked oligosaccharides. The functional role of mannose-binding protein was demonstrated with findings that the oral immunization

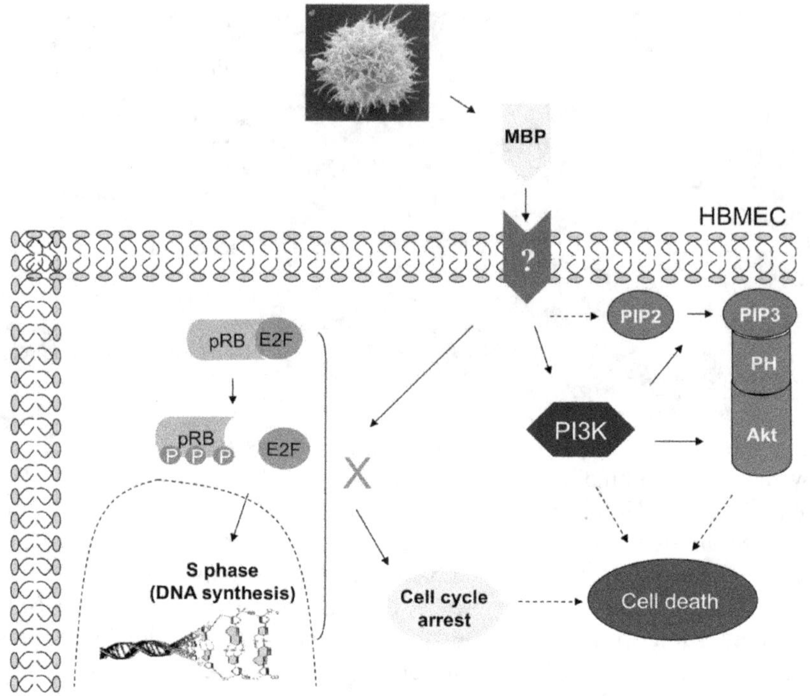

Figure 12. Host intracellular signaling in response to *Acanthamoeba*. Note that *Acanthamoeba* induces cell cycle arrest in the host cells by i) altering expression of genes as well as by ii) modulating protein retinoblastoma (pRb) phosphorylations. In addition, *Acanthamoeba* have also been shown to induce host cell death via phosphatidylinositol 3-kinase (PI3K). By secreting proteases, amoebae disrupt tight junctions by targeting zonula-1 and occludin proteins. MBP is mannose-binding protein; E2F is a transcription factor that controls cell proliferation through regulating the expression of essential genes required for cell cycle progression; PIP2 is phosphatidylinositol-4,5-bisphosphate; PIP3 is phosphatidylinositol-3,4,5-trisphosphate; Akt (protein kinase B)-PH domain, a serine/threonine kinase is a critical enzyme in signal transduction pathways involved in cell proliferation, apoptosis, angiogenesis, and diabetes.

with recombinant mannose-binding protein ameliorates *Acanthamoeba* keratitis in the Chinese hamster animal model. Beside these studies, an understanding of the precise events in mannose-binding protein binding to mannose-containing glycoprotein and/or additional *Acanthamoeba* adhesins secondary to mannose-binding protein should be a focus for future studies. Of interest in this context is a recently identified laminin-binding protein from *Acanthamoeba* with a predicted molecular mass of 28.2 kDa. Laminin is a major mannosylated glycoprotein constituent of the extracellular matrix and the basement membrane of the host cells. In addition, a >207 kDa adhesin in soluble *Acanthamoeba* membrane preparations is identified, which requires further investigation (Morton *et al.,* 1991; van Klink *et al.,* 1992; 1993); Badenoch *et al.,* 1994; Kennett *et al.*, 1999; Shin *et al.*, 2001; Alsam *et al.*, 2003; Sissons *et al.*, 2004a; Panjwani *et al.*, 1992; Yang *et al.*, 1997; Dudley *et al.*, 2005; Garate *et al.*, 2005; 2006; Hong *et al.*, 2004; Imbert-Bouyer *et al.*, 2004).

7.3. Host intracellular signaling in response to *Acanthamoeba*

The initial binding of *Acanthamoeba* to the surface of the host cells interferes with the host intracellular signaling pathways. For example, it is well-established that *Acanthamoeba* induces apoptosis in the host cells. Apoptosis or programmed cell death, is known to be dependent on the host cell's own signaling pathways. One of the primary changes in response to *Acanthamoeba* is the alterations in the levels of calcium. The increase in cytosolic levels of calcium is dependent on the transmembrane influx of the extracellular calcium. Among other roles, the alterations in the levels of intracellular calcium are that it exerts effects on the cytoskeletal structure, induce morphological changes, or alter the permeability of the plasma membrane, finally leading to target cell death. An understanding of the complex intracellular signaling pathways is crucial to identify targets for therapeutic interventions. There are more than 10,000 signaling molecules in a single host cell at any one time, so the identification of key molecules and how they interact in response to *Acanthamoeba*, leading to a functional outcome (e.g., apoptosis) is clearly a challenge. However, it most likely involves events both at the transcriptional-level and at the post-translational-level. It is well-established that proteins that regulate cell fate require tyrosine phosphorylations as well as serine/threonine phosphorylations for intracellular signaling. To this end, recent studies have shown that *Acanthamoeba* upregulates or downregulates the expression of a number of genes important for regulating the cell cycle. Overall, it is shown that *Acanthamoeba* upregulates the expression of genes such as GADD45A and p130 Rb, associated with the cell cycle arrest, as well as inhibiting the expression of other genes, such as those for cyclins F, G1 and cyclin dependent kinase-6 that encode proteins important for the cell cycle progression. The overall response of these events is shown to be the arrest of the host cell cycle (Fig. 12). This is further supported by the dephosphorylation of the retinoblastoma protein (pRb). In the unphosphorylated form, pRb remains bound to the E2F transcription factors (required for DNA synthesis) in the cytoplasm and inhibits E2F translocation into the nucleus. However, when phosphorylated by cyclin-dependent kinases, pRb undergoes a conformational change resulting in E2F-pRb complex dissociation. The released E2F translocates into the nucleus and initiates the DNA synthesis for the S-phase (Fig. 12). Thus pRb is a potent inhibitor of G1/S cell cycle progression. Recent studies showed that *Acanthamoeba* inhibits pRb phosphorylations in the human corneal epithelial cells as well as in the human brain microvascular endothelial cells indicating that the *Acanthamoeba* induces cell cycle arrest in the host cells. Other studies have shown that *Acanthamoeba* induces apoptosis in the host cells, but whether cell cycle arrest and apoptosis are independent of each other, or whether the cell cycle arrest is a primary event that leads subsequently to apoptosis, remains to be determined. Of note, recent studies have shown that *Acanthamoeba*-mediated host cell death is dependent on the activation of phosphatidylinositol 3-kinase (Fig. 12). This was shown using LY294002, a specific phosphatidylinositol 3-kinase inhibitor, which blocked *Acanthamoeba*-mediated host cell death.

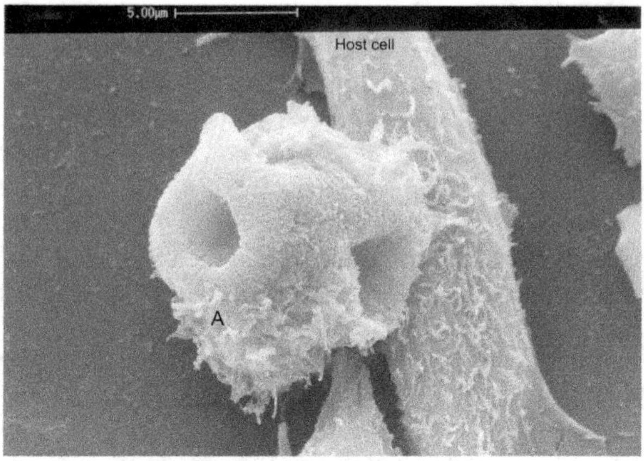

Figure 13. *Acanthamoeba* exhibiting amoebastomes (mouth-like openings). *Acanthamoeba* incubated with the host cells exhibited the presence of amoebastomes within 30 minutes of incubation. These structures are known be involved in the phagocytosis of *Acanthamoeba*. A is *Acanthamoeba*.

These findings were further confirmed using host cells expressing mutant p85, i.e., a regulatory subunit of phosphatidylinositol 3-kinase (i.e., host cells expressing dominant negative phosphatidylinositol 3-kinase), and these cells were significantly less susceptible to *Acanthamoeba*-mediated damage. Phosphatidylinositol 3-kinase has been known traditionally to be important for the cell survival pathways, so these results are not surprising. For example, it is shown that interferon-alpha induces phosphatidylinositol 3-kinase-mediated apoptosis in myeloma cells without the Akt phosphorylations (Akt is a downstream effector of phosphatidylinositol 3-kinase). It was further shown that downstream effectors of phosphatidylinositol 3-kinase-mediated apoptosis involves activation of proapoptotic molecules, Bak and Bax, loss of mitochondrial membrane potential and release of cytochrome *c*: all well-known mediators of the apoptosis. Similar mechanisms may exist in *Acanthamoeba*-mediated host cell death (Taylor *et al.,* 1995; Alizadeh *et al.,* 1994; Mattana *et al.*, 1995; Sissons *et al.,* 2004a; 2005a; Thyrell *et al.,* 2004).

7.4. Phagocytosis

Adhesion of *Acanthamoeba* to the host cells leads to secondary processes such as phagocytosis or secretion of toxins. The ecological significance of *Acanthamoeba* phagocytosis is probably in the uptake of the food particles such as bacteria, plasmids, or fungal cells. However, the ability of *Acanthamoeba* to form food cups or amoebastomes during incubations with the host cells suggests they have a role in the pathogenesis of *Acanthamoeba* (Fig. 13) (Diaz *et al.*, 1991; Khan, 2001; Pettit *et al.,* 1996). Phagocytosis has been shown in a number of protozoan parasites including *Balamuthia mandrillaris* and *Trichomonas vaginalis* in which its been shown to ingest different mammalian cells (Heat, 1981; Brasseur *et al.*, 1982; Gonzalez-Robels *et al.*, 1995; Rendon-Maldonado *et al.*, 1998).

Phagocytosis is an actin-dependent process involving the polymerization of monomeric G-actin into filamentous F-actin. To this end, it is shown that cytochalasin D (a toxin that blocks actin polymerization) inhibits *Acanthamoeba*-mediated host cell death, confirming that actin-mediated cytoskeletal rearrangements play an important role in the pathogenesis of *Acanthamoeba* (Niederkorn *et al.*, 1999; Taylor *et al.*, 1995).

7.4.1. Phagosome formation and fusion with the lysosome

The particle ingestion by *Acanthamoeba* is rapid. Within 40 sec, bound particles can be surrounded by pseudopods, brought into the cytoplasm, and released as phagosome into the cytoplasmic stream. Under electron microscope, the phagosome appears as a flask-like invagination of the surface. Separation from the surface occurs by fragmentation of the attenuated neck of the invagination. The separated phagosome membrane has a 3 to 4-fold greater density of intramembrane particles than the plasma membrane from which it derives. This change is evident within 15 min of ingestion. Phagosome membrane has a higher protein to phospholipids ratio and a higher glycosphingolipid content than the plasma membrane. These phagosomes are subsequently fused with lysosomes to form phago-lysosomes. Phago-lysosomes show multiple small vesiculations of characteristic morphology. The fusion of phago-lysosomes has been shown during incubation of phago-lysosome-containing homogenates of *Acanthamoeba*, suggesting that *Acanthamoeba* homogenates system is a useful model of the fusion process *in vivo*. The rate of phago-lysosome fusion in *Acanthamoeba* homogenates may be regulated by cyclic nucleotides, with enhancements of the fusion rate by the cyclic AMP and the inhibition of the rate by cyclic GMP (Oates and Touster, 1978; Bowers, 1980).

During phagocytosis, cells ingest and re-circulate a quantity of membrane equivalent to the entire cell surface in a period of minutes. The food vacuoles (i.e., phagosomes) are fused with lysosomes, acquiring acid phosphatase and glycosidases. Interestingly, lysosomal enzymes, acid phosphatase and glycosidases are sequestered into the early phagosomes and not to be distributed either by small vesicle transport or phagosome-phagosome fusion to those formed later. Based on this, it is likely that the newly internalized surface membrane in phagosomes is not rapidly randomized with internal pools, but is recycled to the surface as a function of the digestive process (Hohman and Bowers, 1993). However, it is not known whether there is a mixing of the surface and the lysosomal membrane components after fusion or how membranes are returned to the cell surface. Although it is accepted that a major portion of the membranes returns to the surface via small vesicles, however the actual mechanisms may be more complex than a simple re-insertion of the unaltered plasma membrane by way of small vesicles or defecation vacuoles deriving from the phagosome. In an attempt to measure the surface areas of the three membrane compartments, cell surface membrane, the vacuole membrane (*Acanthamoeba* digestive system) and associated membrane of the small vesicles, it was revealed that the total membrane area

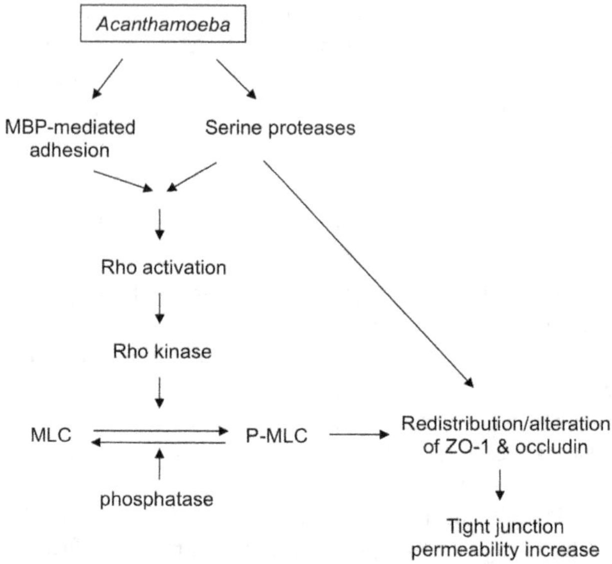

Figure 14. Diagrammatic representation of Rho activation leading to the blood-brain barrier permeability changes. RhoA activation induces myosin light-chain phosphorylation (P-MLC) via Rho kinase causing redistribution/alteration of tight junction proteins, zonula occludens-1 (ZO-1) and occludin, resulting in the elevation of barrier permeability.

in the 3 compartments was 3.2 μm^2 per μm^3 of the protoplasmic volume. In pinocytosing cell, 32% of the membrane was in cell surface membrane, 25% in the vacuole membrane and 43% in the associated membrane of the small vesicles. Interestingly, the vacuole compartment occupies approximately 20% of the total cell volume, and the small vesicle, approximately 3% (Bowers, 1980; Bowers *et al.*, 1981).

Phagocytosis in *Acanthamoeba* is associated with the oxidative metabolism. This is shown by findings that during uptake of latex beads, cyanide-insensitive oxygen consumption increased nearly two-fold. Cell fractionation studies revealed a localization for a portion of the NADPH oxidase activity in the phago-lysosomes. In addition, the phago-lysosome membrane was enriched about two-fold in a *b*-type cytochrome. These results show that oxidative metabolism in *Acanthamoeba* has some striking similarities to the respiratory burst oxidase of neutrophils (Brooks and Schneider, 1985). In addition, phagocytosis in *Acanthamoeba* is associated with extracellular release of AMP, a respiratory burst which may be necessary for the generation of the reactive oxygen metabolites during subsequent digestion, and with enhanced degradation of phosphatidylinositol 4,5-biphosphate and phosphatidylinositol 1,4,5-triphosphate. Previous studies demonstrated that binding and internalization of yeast particles can be inhibited by exogenous mannose in a concentration-dependent manner. However, mannose did not inhibit the uptake of latex beads suggesting the specificity for food particles, and the role of mannose-binding protein in *Acanthamoeba* phagocytosis.

Furthermore the binding and internalization of yeast by *Acanthamoeba* stimulated degradation of phosphatidylinositol 4,5-biphosphate to inositol 1,4,5-triphosphate (IP3). The IP3 is a secondary messenger molecule used in the signal transduction in biological cells. These findings led to the conclusion that the *Acanthamoeba* mannose-binding protein stimulates the degradation of phosphatidylinositol 4,5-biphosphate to IP3 as an initial event in the phagocytosis (Dawidowicz, 1990; 1990a).

The ability of *Acanthamoeba* to phagocytose is an intracellular signaling-dependent process. Genistein (a protein tyrosine kinase inhibitor) inhibits while sodium orthovanadate (protein tyrosine phosphatase inhibitor) enhances *Acanthamoeba* phagocytosis, indicating that a tyrosine kinase-induced actin polymerization signal is important in *Acanthamoeba* phagocytosis (Alsam *et al.,* 2005a). Rho GTPases are the major regulators of the actin cytoskeleton, which link external signals to the cytoskeleton (Mackay and Hall, 1998). As their name indicates, Rho GTPases bind and hydrolyze GTP and, in the process, stimulate pathways that induce specific cytoskeletal rearrangements resulting in distinct phenotypes (Mounier *et al.,* 1999). There are three well-studied pathways including: the RhoA pathway, leading to stress fibre formation, Rac1 activation triggering lamellipodia formation; and Cdc42 activation, promoting filopodia formation (Mackay and Hall, 1998). Recent studies have shown that the Rho kinase inhibitor, Y27632, blocks partially *Acanthamoeba* phagocytosis. Y27632 blocks stress fibre formation by inhibiting myosin light chain phosphorylation and cofilin phosphorylations but independent of profilin pathway (Fig. 14). Overall, these findings suggested that, in addition to the RhoA pathway, Rac1 and Cdc42 pathways may also be involved in *Acanthamoeba* phagocytosis. This is further supported with the finding that LY294002, a specific inhibitor of phosphatidylinositol 3-kinase, inhibits *Acanthamoeba* phagocytosis. Phosphatidylinositol 3-kinase is shown to be involved in Rac1-dependent lamellipodia formation (Wennström *et al.,* 1994) and Cdc42-dependent cytoskeletal changes (Jimenez *et al.,* 2000). Future research into the identification of additional molecules/pathways; how various intracellular signaling pathways interact; and/or whether they are independent of each other, will enhance our understanding of *Acanthamoeba* phagocytosis, and should be of value in the development of therapeutic interventions. As described above, mannose but not other saccharides, inhibits *Acanthamoeba* phagocytosis indicating that *Acanthamoeba* phagocytosis is a receptor-dependent process, and *Acanthamoeba* adhesin (i.e., mannose-binding protein), is involved in its ability to phagocytose (Allen and Dawidowicz, 1990; Alsam *et al.,* 2005a). Overall, these findings suggest that the mannose-binding protein is crucial in the *Acanthamoeba* binding to the target cells as well as its phagocytotic processes.

7.4.2. Lysosomal enzymes

The phosphomannosyl targeting system of eukaryotic cells have 3 major components:

1. the lysosomal acid hydrolases
2. the enzymes which generate the phosphomannosyl recognition marker and
3. the receptors that recognise this marker and shuttle the acid hydrolases to lysosomes.

With a few exceptions, all lysosomal hydrolases are thought to express a common protein domain that is recognized by *N*-acetyl-glucosamine-1-phosphotransferase, thereby allowing the selective phosphorylation of mannose residues on this class of glycoproteins.

It is shown that *Acanthamoeba* membranes contain high levels of *N*-acetyl-glucosamine-1-phosphotransferase activity, i.e., greater than 100 times compared with human fibroblasts or rat liver. Like rat enzyme, the *N*-acetyl-glucosamine-1-phosphotransferase (molecular weight of enzyme is 250 kDa) of *Acanthamoeba* phosphorylates the lysosomal enzymes, cathepsin D and uteroferrin (Ketcham and Kornfeld, 1992; 1992a; Lang *et al.*, 1986). *N*-acetyl-glucosamine-1-phosphotransferase transfers *N*-acetyl-glucosamine-1-phosphate from UDP-*N*-acetyl-glucosamine to the 6-position of terminal mannose residues on lysosomal enzymes. Then, *N*-acetyl-glucosamine-1-phosphodiester α-*N*-acteyl-glucosaminidase cleaves off the *N*-acetyl-glucosamine residue to expose the resulting Man-6-P recognition marker. The phosphorylated lysosomal enzymes bind with high affinity to Man-6-P receptors and are subsequently targeted to lysosomes. Thus, the initial and discriminating enzyme in the Man-6-P-dependent sorting system is *N*-acetyl-glucosamine-1-phosphotransferase. Furthermore, the *N*-acetyl-glucosamine-1-phosphotransferase has a requirement for the presence of at least one Manα1,2→Man sequence on the acceptor glycoprotein (Couso *et al.*, 1986). However, *Acanthamoeba* extracts do not contain detectable levels of *N*-acetyl-glucosamine-1-phosphodiester α-*N*-acetyl-glucosaminidase, the second enzyme in the biosynthetic pathway for the mannose-6-phosphate recognition marker, suggesting that *Acanthamoeba* does not utilize the phosphomannosyl sorting pathway despite expression of very high levels of *N*-acetyl-glucosamine-1-phosphotransferase (Ketcham and Kornfeld, 1992).

7.5. Ecto-ATPases

Ecto-ATPases are glycoproteins present in the plasma membranes that have their active sites facing the external medium rather than the cytoplasm. Ecto-ATPases hydrolyze the extracellular ATP and other nucleoside triphosphates (Sissons *et al.*, 2004b). The resultant ADP can have toxic effects on the host cells. For example, it has been shown that ADP released by *Acanthamoeba* binds to $P2y_2$ purinergic receptors on the host cells, causing an increase in the intracellular calcium, inducing caspase-3 activation and finally resulting in apoptosis (Mattana *et al.*, 2002). A P2 receptor antagonist, suramin, inhibits *Acanthamoeba*-mediated host cell death (Mattana *et al.*, 2002; Sissons *et al.*, 2004b), suggesting that ecto-ATPases play an important role in *Acanthamoeba* pathogenesis in a contact-independent mechanism. Furthermore, clinical isolates exhibit higher ecto-ATPase activities compared with the environmental isolates. The ecto-ATPase activity is significantly increased in the presence of exogenous mannose but not by other sugars. However, the mannose-mediated enhanced ecto-ATPase activity is not observed in weak- and/or non-pathogenic *Acanthamoeba*. Also, the weak- and/or non-pathogenic *Acanthamoeba* do not bind to the host cells (Alsam *et al.,* 2003). Taken together, these findings suggest that the engagement of *Acanthamoeba* adhesin (i.e., mannose-binding protein) enhances ecto-ATPase activities and thus ecto-ATPases may also play a role in the *Acanthamoeba* pathogenesis but this time in a contact-dependent mechanism. How do the mannose-binding protein is associated with ecto-ATPase activities is not clear. Of interest, several ecto-ATPases of approximate molecular weights of 62, 100, 218, 272 and more than 300 kDa have been described in *Acanthamoeba* (Sissons *et al.,* 2004b). The differences in ecto-ATPases have been attributed to strain, species / genotype differences. Future research will elucidate their function in *Acanthamoeba* biology, and investigate their precise role in the contact-dependent and the contact-independent mechanisms of *Acanthamoeba* pathogenesis, and their usefulness as diagnostic targets for the genotype differentiation.

7.6. Neuraminidase activity

The trophozoites and cysts of *Acanthamoeba* exhibit neuraminidase activity, which is membrane-associated, as well as released into the culture medium at the start of the exponential growth phase. The enzyme activity is optimal at pH 5 and at temperatures of 25 – 30°C. The live parasites release sialic acid from the human cells. Therefore, the neuraminidase of *Acanthamoeba* could be relevant in the colonization of the parasite, and also important in producing damage of the sialic acid-rich corneal epithelium and in the alterations of glycolipids associated with meningoencephalitis. Interestingly, the neuraminidases of *Trypanosoma cruzi* and *Acanthamoeba* are immunologically related as demonstrated by antibodies against neuraminidase of *Trypanosoma cruzi*, which reacted with *Acanthamoeba* parasites using immunofluorescence, immunoblotting and enzyme-linked immunosorbent assays (Pellegrin *et al.*, 1991; 1992).

7.7. Superoxide dismutase

The enzyme, superoxide dismutase catalyzes the dismutation of the superoxide into oxygen and hydrogen peroxide. As such, it is an important antioxidant defence in nearly all cells exposed to oxygen. The superoxide is one of the main reactive oxygen species in the cell and as such, the superoxide dismutase plays an important antioxidant role.

Two superoxide dismutases have been identified in *Acanthamoeba*: an iron superoxide dismutase (approximate molecular weight of 50 kDa) and a copper-zinc superoxide dismutase (approximate molecular weight of 38 kDa). These enzymes occur as cytoplasmic and detergent-extractable fractions. These enzymes may be potential virulence factors of *Acanthamoeba* by acting both as antioxidants and anti-inflammatory agents. These enzymes may provide additional targets for chemotherapy and immuno-diagnosis of *Acanthamoeba* infections (Cho *et al.*, 2000).

7.8. *Acanthamoeba* activates plasminogen

Acanthamoeba display plasminogen activator activity by catalyzing the cleavage of host plasminogen to form plasmin, which can activate host proteolytic enzymes, such as pro-matrix metalloproteinases. Once activated, the matrix metalloproteinases degrade the basement membranes and the components of the extracellular matrix such as type I and type II collagens, fibronectins and laminin. Thus the matrix metalloproteinases are involved in the tissue remodeling. The pathogenic *Acanthamoeba* show positive chemotactic response to the endothelial extracts (van Klink *et al.*, 1992).

8. Contact-independent mechanisms

The primary *Acanthamoeba* and the host cell interactions lead to secondary events such as interference with the host intracellular signaling pathways, phagocytosis and toxin productions, thus ultimately leading to the host cell death. Factors that are secreted and/or released by *Acanthamoeba* and take part in *Acanthamoeba* infection are indicated below (Fig. 9).

8.1. Hydrolytic enzymes

To produce damage to the host cell and/or tissue migration, the majority of pathogens rely upon their ability to produce hydrolytic enzymes. These enzymes may be constitutive that are required for routine cellular functions or inducible, which are produced under specific conditions, for example upon contact with the target cells. These enzymes can have devastating effects on the host cells by causing membrane dysfunction or physical disruptions. Cell membranes are made of proteins and lipids, and *Acanthamoeba* is known to produce several types of hydrolytic enzymes: proteases, which hydrolyze peptide bonds; phospholipases, which hydrolyze phospholipids; glycosidases and glycosyl transferases that are responsible for the breakdown and re-assembly of carbohydrates by the hydrolysis and formation of the glycosidic bonds.

8.1.1. Elastase

Because penetration of the mucosal surfaces by *Acanthamoeba* most likely depends on the capacity to the damage tissue, it is not surprising that *Acanthamoeba* is known to produce elastase. Elastases are known to degrade a range of connective tissue proteins such as elastin, an elastic fiber, fibrinogen, collagen, and proteoglycans, which together determined the mechanical properties of the connective tissue. Furthermore, tissues altered by prior elastase treatment are more susceptible to oxygen radical attack, suggesting their involvement in the pathogenesis and pathophysiology of *Acanthamoeba* infections (Ferrante and Bates, 1988).

8.1.2. Proteases

Proteases are degradative enzymes which catalyze the total hydrolysis of proteins. Proteases exhibit a variety of complex physiological functions.

They play an important role in many pathological processes such as protein catabolism, blood coagulation, inflammation, tumour growth, the release of hormones and the transport of secretory proteins across membranes. In general, extracellular proteases catalyze the hydrolysis of large proteins into smaller molecules for subsequent absorption by the cell, whereas intracellular proteases play a role in the regulation of the cell function. The microbial extracellular proteases are primarily involved in keeping the cells alive by providing them with the necessary amino acid pool as nutrition. Advances in analytical techniques have demonstrated that proteases can also conduct highly specific and selective modifications of proteins such as activation of zymogenic forms of enzymes by limited proteolysis. Since proteases are physiological necessary for living organisms, they are found in a wide diversity of sources such as plants, animals and microorganisms. Proteases are subdivided into two major groups i.e., exopeptidases and endopeptidases depending on their site of action. Exopeptidases cleave the peptide bonds proximal to the amino- or carboxy-terminal of the substrate, whereas endopeptidases cleave peptide bonds distant from the termini of the substrate. Based on the functional group at the action site, proteases are classified into six major groups i.e., serine, aspartic, cysteine, metalloproteases, threonine and glutamic acid proteases. Proteases play an important role in the tissue invasion, migration and the host pathology of other parasites. The protease activity has been reported in a number of other protozoa including, *Entamoeba, Giardia, Trypanosoma,* and *Leishmania*. The amount of proteolytic enzymes produced by *Entamoeba histolytica* has been correlated with the virulence showing the importance of proteases. Proteases are now well-known virulence factors in the majority of viral, bacterial, protozoan and multicellular pathogens.

All *Acanthamoeba* isolates tested to date exhibit proteolytic activities. Interestingly, both clinical and non-clinical isolates of *Acanthamoeba* exhibit protease activities, but larger amounts are observed with the former. This suggests that the principal physiological role of *Acanthamoeba* proteases is to degrade the substrate for feeding purposes. This is consistent with the fact that *Acanthamoeba* is primarily a free-living environmental organism and its role in the human / animal infections is secondary or opportunistic. At the same time, the excretory and secretory products from trophozoites of *Acanthamoeba* induced damage to the collagen shields in an *in vitro* and an *in vivo* model (He *et al*., 1990). In addition, the pathogenic *Acanthamoeba* has been shown to exhibit increased extracellular protease activities. The link between pathogenicity and the increased levels of extracellular proteases suggest that pathogenic *Acanthamoeba* utilize proteases to facilitate invasion of the host. So far *Acanthamoeba* is known to produce serine, cysteine and metalloproteases, as described below.

The presence of a 35 kDa serine protease in *Acanthamoeba* isolates was demonstrated (some isolates most likely belonged to the T4 genotype). Other studies using T4 isolates have identified the extracellular serine proteases of approximate molecular weights of 33, 36, 49, and 66 kDa, 107 kDa, 55, 97 and 230 kDa, but other studies showed the presence of 27, 47, 60, 75,

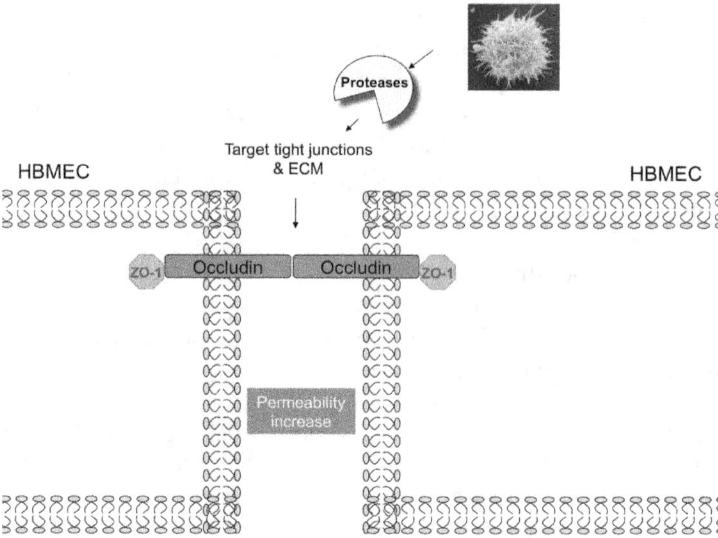

Figure 15. *Acanthamoeba* proteases target tight junctions proteins and the extracellular matrix to induce perturbation of the blood-brain barrier.

100, >110 kDa serine proteases. Furthermore, a 40 kDa serine protease was demonstrated in *Acanthamoeba* belonging to the T4 genotype. This protease was shown to activate plasminogen, whose physiological function is to degrade the extracellular matrix components. This serine protease was not inhibited by amiloride which is a strong inhibitor of urokinase-type plasminogen activator. Additionally, the enzyme is not inhibited by plasminogen activator inhibitor-1 which is the primary physiological inhibitor of both urokinase and tissue-type plasminogen activator. It does not cross-react with antibodies specific for the human urokinase or tissue-type plasminogen activator. This protease activates plasminogen from several mammalian species, including humans, cows and pigs. This diversity of serine proteases within a single genotype of *Acanthamoeba* may be due to strain differences, differences in their virulence, culture under diverse conditions or differences in assay methods.

Additionally a 33 kDa serine protease from *Acanthamoeba healyi* (belonging to the T12 genotype) was identified, which demonstrated degradation of type I and IV collagen and fibronectin, which are main components of the extracellular matrix, as well as fibrinogen, IgG, IgA, albumin, and hemoglobin. Other studies have identified serine proteases with approximate molecular weights of 42 kDa and 12 KD that degrade immunoglobulins, protease inhibitors and interleukin-1, but their genotypes are not known (Mitra *et al.*, 1995; Alfieri *et al.*, 2000; Hadas and Mazur 1993; Mitro *et al.*, 1994; Khan *et al.*, 2000; Cao *et al.*, 1998; Cho *et al.*, 2000; Na *et al.*, 2001; 2002; Kong *et al.*, 2000; Kim *et al.*, 2006).

In addition, a 133 kDa serine protease, called MIP133 is identified as a crucial component of the pathogenic cascade of *Acanthamoeba* pathogenesis.

The MIP133 serine protease is shown to induce the apoptosis of keratocytes, iris ciliary body cells, the retinal pigment epithelial cells, the corneal epithelial cells and the corneal endothelial cells.

In addition, *Acanthamoeba* serine proteases have been shown to be involved in destruction of the tight junction proteins of the blood-brain barrier, including, occludin and zonula-occludens-1, resulting in increased permeability of the blood-brain barrier (Fig. 15).

These properties of serine proteases most likely facilitate *Acanthamoeba* invasion of the corneal stroma and the blood-brain barrier leading to secondary reactions such as edema, necrosis and the inflammatory responses. A direct functional role of serine proteases in *Acanthamoeba* infections is indicated by the observations that the intrastromal injections of *Acanthamoeba* conditioned medium produces corneal lesions *in vivo*, similar to those observed in *Acanthamoeba* keratitis patients, and this effect is inhibited by phenylmethylsulfonyl fluorise (PMSF), a serine protease inhibitor (He *et al.,* 1990; Na *et al.,* 2001). In addition, the chemically synthesized small interfering RNA (siRNA) that is highly specific and efficient in silencing the catalytic domain of the extracellular serine proteases of *Acanthamoeba*, reduced protease activity and parasite-mediated host cell cytotoxicity. These results support that the extracellular serine proteases are directly involved in the pathogenesis and virulence of *Acanthamoeba* (Lorenzo-Morales *et al.*, 2005).

Among cysteine proteases, although data are limited, a 65 kDa extracellular cysteine protease is reported in *Acanthamoeba* isolates belonging to the T4 genotype as well as 43, 70 and 130 kDa cysteine proteases. It is worth noting that in these studies 65 kDa and 70 kDa may refer to the same protease as the difference may have been due to variation in the protease mobility in the substrate gels. In addition, a 24 kDa cysteine protease has been identified from *Acanthamoeba* isolate belonging to the T12 genotype that is most likely intracellular. However, the physiological roles of cysteine proteases in the biology and pathogenesis of *Acanthamoeba* remain to be identified (Alfieri *et al.*, 2000; Hadas and Mazur, 1993; Hong *et al.,* 2002).

In addition to serine and cysteine proteases, there is evidence for metalloprotease activity in *Acanthamoeba*. For example, a 80 kDa metalloprotease was identified in the co-cultures of *Acanthamoeba* and the host cells, but its origin (whether *Acanthamoeba* or the host cells) was not established. Later studies identified a 150 kDa extracellular metalloprotease from *Acanthamoeba*, isolated from an *Acanthamoeba* encephalitis patient belonging to the T1 genotype). This metalloprotease exhibited properties of the extracellular matrix degradation as evidenced by its activity against collagen I and III (major components of the collagenous extracellular matrix), elastin (elastic fibrils of the extracellular matrix), plasminogen (involved in proteolytic degradation of the extracellular matrix), as well as degradation of casein, gelatin, and hemoglobin, suggesting a role both in *Acanthamoeba* encephalitis and *Acanthamoeba* keratitis infections.

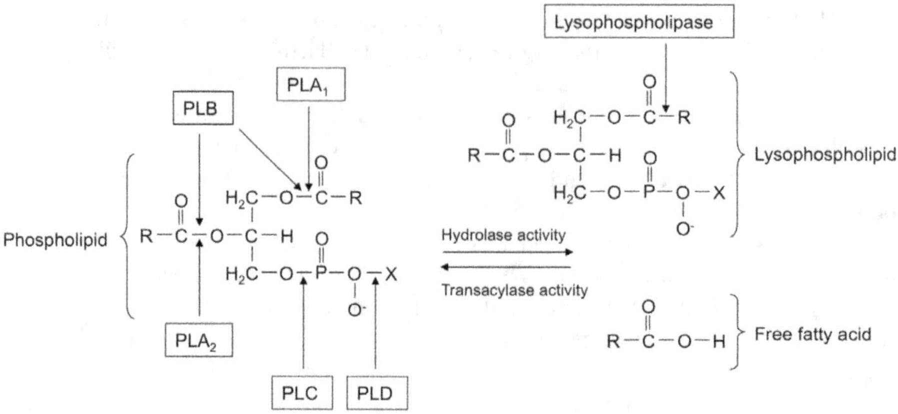

Figure 16. Phospholipases and their sites of action.

Recently, the complete sequence of a type-1 metacaspase from *Acanthamoeba* is reported, comprising of 478 amino acids. The metacaspase was recovered from an expression library using sera specific for membrane components implicated in stimulating encystation. A central domain of 155 amino acid residues contains the cysteine / histidine catalytic dyad and is the most conserved region containing at least 30 amino acid identities in all metacaspases. The *Acanthamoeba* metacaspase has the most proline-rich *N*-terminus so far reported in type-1 metacaspases with over 40 proline residues in the first 150 residues. Alanine-proline-proline is present 11 times. Phylogeny studies using only the conserved proteolytic domains or the complete sequences show identical branching patterns, differing only in the rates of change (Trzyna *et al.*, 2006).

Overall, these studies have shown that *Acanthamoeba* exhibits diverse proteases and elastases, which could play important roles in *Acanthamoeba* infections. However, their precise modes of action at the molecular-level are only beginning to emerge. That some of the above proteases are secreted only by the clinical isolates may indicate their role as potent virulence factors and/ or diagnostic targets. Future studies in the role of proteases as vaccine targets, search for novel inhibitors by screening of chemical libraries, or rational development of drugs based on structural studies should enhance our ability to targets these important pathogens (Alfieri *et al.*, 2000; Mitro *et al.*, 1994; Cao *et al.*, 1998; Alsam *et al.*, 2005b; Sissons *et al.*, 2005b; Ferrante and Bates, 1988).

8.1.3. Phospholipases

Phospholipases are a diverse group of enzymes that hydrolyze the ester linkage in glycerophospholipids and can cause membrane dysfunction. The five major phospholipases are A1, A2, B, C, D, and each has the ability to cleave a specific ester bond in the substrate of the target membrane (Fig. 16). All phospholipases are present in multiple forms.

The fact that there is a large turnover of the plasma membrane in *Acanthamoeba* during the phagocytosis and pinocytosis processes indicates that there is large controlled local degradation of phospholipids leading to an instability of the membrane phospholipid bilayer, which would then reform after the acylation of the lysophospholipid. To this end, all of the enzymes that are needed for such as cycle are present in the plasma membrane of *Acanthamoeba* including, phospholipase A2, acyl CoA:lysolecithin acyltransferase, and acyl CoA synthetase. In addition, phospholipase A1 and lysophospholipase are present in the plasma membrane of *Acanthamoeba*. The plasma membrane lysophospholipase may also serve to protect the cell from the lytic effect of lysophospholipids either of exogenous or endogenous origin. Also, the plasma membranes have the enzymatic capability of modulating the fatty acyl composition of phospholipids by de-acylation and acylation.

Our knowledge of phospholipases in the virulence of *Acanthamoeba* is fragmented, however several studies have shown that the pathogenic *Acanthamoeba* that exhibit cytopathic effects on mammalian cells *in vitro* liberate more phospholipase suggesting their possible involvement in *Acanthamoeba* infections. Because phospholipases cleave phospholipids, it is reasonable to suspect that they play a role in the membrane disruptions, penetration of the host cells, and the cell lysis, however it remains to be determined. Other actions of phospholipases may involve interference with intracellular signaling pathways. Phospholipases generate lipids and lipid-derived products that act as second messengers.

Overall, these studies suggest that *Acanthamoeba* phospholipases and/or lysophospholipases may play a role in causing host cell damage or affect other cellular functions such as induction of the inflammatory responses thus facilitate the virulence of *Acanthamoeba*, however, this remains to be fully established. More studies are needed to identify and characterize *Acanthamoeba* phospholipases and to determine their potential role in the development of therapeutic intervention. This is not a novel concept: earlier studies have shown that the phospholipase C from *Clostridium perfringens* induces protection against *Clostridium perfringens*-mediated gas gangrene. In addition, the targeting of phospholipases using synthetic inhibitor compounds has been shown to have the potential to prevent *Candida* infections. Antibodies produced against *Acanthamoeba* phospholipases may also be of potential value in the development of sensitive and specific diagnostic assays as well as of therapeutic value. In addition, acid phosphatases are demonstrated in *Acanthamoeba* (Im 1990; Cursons *et al.*, 1978; Hax *et al.*, 1974; Victoria and Korn, 1975a; 1975b; Visvesvara and Balamuth, 1975; Dennis *et al.*, 1991; Serhan *et al.*, 1996; Oishi *et al.*, 1988; Bryant and Stevens, 1996; Bunting *et al.*, 1997; Robertson and Lands, 1964; Hanel *et al.*, 1995; Kameyama *et al.*, 1975).

8.1.4. Glycosidases (also called glycoside hydrolases)

Glycoside hydrolases catalyze the hydrolysis of the glycosidic linkage to generate two smaller sugars. Glycoside hydrolases are ubiquitous in nature and are involved in the degradation of biomass such as cellulose and in a variety

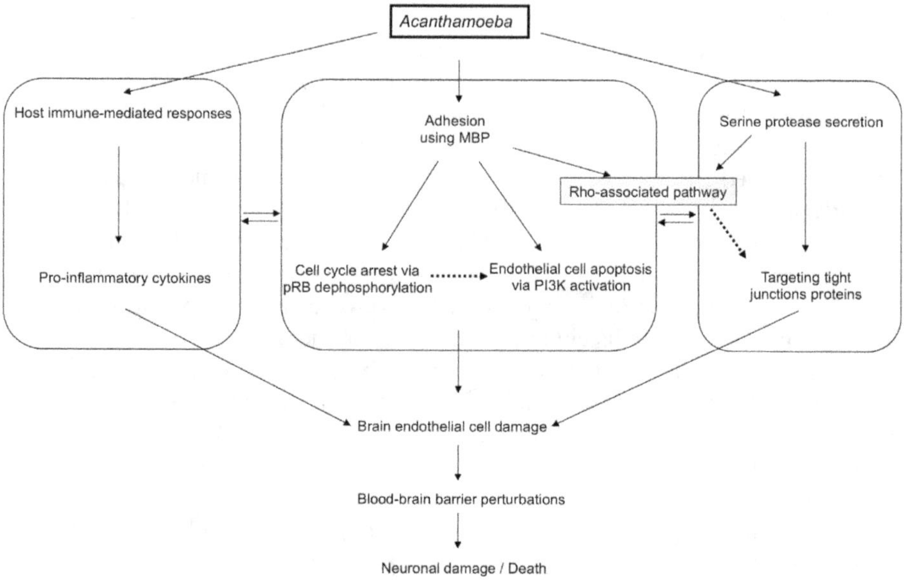

Figure 17. Schematic illustrates involvement of *Acanthamoeba* mannose-binding protein (MBP), serine proteases, retinoblastoma protein (pRB), phosphatidylinositol 3-kinase (PI3K), Rho activation, tight junction proteins leading to host cell damage, a pre-requisite in *Acanthamoeba* translocation of the biological barriers. In addition, host inflammatory responses may also contribute to blood-brain barrier perturbations.

of cellular functions. Together with glycosyltransferases, glycosidases form the major catalytic machinery for the synthesis and breakage of glycosidic bonds.

Acanthamoeba is shown to has a variety of glycosidases activities including β-glycosidase, α-glucosidase, β-galactosidase, β-*N*-acetyl-glucosaminidase, β-*N*-acetyl-galactosaminidase, α-mannosidase. *Acanthamoeba* extracts are also shown to mediate an enzymatic lysis of cell walls from several species of bacteria including *Micrococcus lysodeikticus*, *Micrococcus roseus*, *Streptococcus faecalis*, *Bacillus megaterium*, *Sarcina lutea*, *Micrococcus radiodurans* and limited activity against *Bacillus subtilis*, *Bacillus cereus* but has no effects on *Acanthamoeba* cyst walls or chitin (from lobster shells). An exhaustive digestion of *Micrococcus lysodeikticus* cell walls releases free *N*-acetyl-glucosamine, *N*-acetyl-muramic acid, glycine, alanine, glutamic acid, and lysine suggesting that *Acanthamoeba* possesses both endo- and exo-hexosaminidases and β-*N*-acetyl-hexosaminidases. Since *Acanthamoeba* is known to utilize maltose, cellobiose, sucrose, or lactose, some of the glycosidases indicated above may suggest the utilization of these disaccharides (Rosenthal *et al.*, 1969; Im, 1990). Overall *Acanthamoeba* uses both contact-dependent as well as contact-independent mechanisms to produce host cell perturbations (Fig. 17).

9. Indirect virulence factor

The ability of *Acanthamoeba* to produce human diseases is a multifactorial process and is, amongst other factors, dependent on its ability to survive outside its mammalian host for various times and under diverse environmental conditions (high osmolarity, varying temperatures, food deprivation, and resistance to chemotherapeutic drugs). The ability of *Acanthamoeba* to overcome such conditions can be considered as contributory factors towards disease and are indicated as indirect virulence factors (Fig. 9).

9.1. Morphology

The infective forms of *Acanthamoeba* or trophozoites do not have a distinct morphology. However, they do possess spine-like structures known as acanthopodia on their surface, which may play a role in the pathogenesis of *Acanthamoeba* infections by modulating binding of pathogenic *Acanthamoeba* to the host cells. It would not be surprising if it was found that the mannose-binding protein, which is involved in the binding of *Acanthamoeba* to the host cells is localized on the acanthopodia. In addition, their amoeboid motion resembles that of macrophages/neutrophils and it may be possible that *Acanthamoeba* use similar mechanisms to traverse the biological barriers such as the blood-brain barrier.

9.2. Temperature tolerance, osmotolerance and growth at different pH

Upon contact with the tear film and the corneal epithelial cells, *Acanthamoeba* is exposed to high osmolarity (due to salinity in tears), high temperatures as well as alterations in pH. For successful transmission, *Acanthamoeba* must withstand these burdens and exhibit growth. Pathogenic *Acanthamoeba* seems undisturbed by the types of stresses they must deal with, when entering their hosts compared with the weak pathogens. In addition, the pathogenic *Acanthamoeba* showed relatively high background levels of heat shock proteins (i.e., HSP60 and HSP70) compared with the weak pathogens. These studies suggest that the higher levels of the heat shock proteins in *Acanthamoeba* may indicate their involvement in (i) tolerance induction to the hosts' stressors and/or (ii) in species' virulence.

It is not surprising that growth at high temperature and high osmolarity are the hallmarks of the pathogenic *Acanthamoeba*. These studies have shown that the ability of *Acanthamoeba* to grow at high temperature and high osmolarity correlate with the pathogenicity of *Acanthamoeba* isolates, and may provide a good indicator of the pathogenic potential of a given isolate.

However, the precise mechanisms by which the pathogenic *Acanthamoeba* adapt to higher temperatures and maintain their metabolic activities remain entirely unknown. Interestingly, temperature tolerance studies in *Candida neoformans* have identified the calcium-dependent protein phosphatase calcineurin as a requirement for its growth at 37°C (Odom *et al.*, 1997a; 1997b). Furthermore, strains of *Candida neoformans* strains in which the calcineurin gene has been disrupted are avirulent in a model of cryptococcal meningitis *in vivo*. These studies might serve as a basis for research into determining the physiological properties of *Acanthamoeba* (De Jonckheere, 1983; Khan *et al.*, 2001; 2002; Walochnik *et al.*, 2000).

Pathogenic *Acanthamoeba* can grow at pH ranging from 4 – 12, which gives it the potential to colonize several niches. For example, the ability of *Candida albicans* to grow at diverse pH is crucial for its virulence and two pH-regulating genes, *PHR1* (expressed at neutral and basic pH) and *PHR2* (expressed at acidic pH) have been identified. The deletion of *PHR2* results in the loss of virulence, while deletion of the *PHR1* results in the reduced virulence in a systemic model. The clinical significance of the ability of *Acanthamoeba* to exhibit growth at different pH remains to be determined (Khan *et al.*, unpublished data; Davis *et al.*, 2000; De Bernardis *et al.*, 1998; Pérez-Serrano *et al.*, 2000).

9.3. Phenotypic switching

Phenotypic switching in *Acanthamoeba* is described as an ability to differentiate into a morphologically distinct dormant cyst form or a vegetative trophozoite form. This is a reversible change dependent on environmental conditions. Cysts are resistant to various antimicrobial agents (discussed in section F) and adverse conditions such as extremes in temperatures, pH, osmolarity, desiccation and cysts can be airborne: all of which presents a major problem in chemotherapy because their persistence may lead to recurrence of the disease. Furthermore, *Acanthamoeba* cysts can survive for several years while maintaining their pathogenicity (Mazur and Zozwiak 1989; Mazur *et al.*, 1995; Byers, 1979; Cordingley *et al.*, 1996; Turner *et al.*, 2000; Weisman, 1976).

These characteristics suggest that the primary functions of cysts lie in withstanding adverse conditions and in the spread of amoebae throughout the environment. In addition, this may represent the ability of *Acanthamoeba* to alternate expression of surface proteins/glycoproteins, in response to changing environments and/or immune surveillance. Overall, phenotypic switching represents a major factor in the transmission of *Acanthamoeba* infections, however, the underlying molecular mechanisms in these processes remain to be elucidated. At present it is not clear whether *Acanthamoeba* shows antigenic variations, and their possible involvement in phenotypic switching, which should be investigated in future studies.

9.4. Chemotaxis

Chemotaxis (chemo meaning chemicals and taxis is movement) is the phenomenon in which cells direct their movements according to certain chemicals in their environment. This is important as *Acanthamoeba* finds food by moving towards the highest concentration of food molecules, or to flee from poisons. Chemotaxis is called positive if movement is in the direction of a higher concentration of the chemical in question, and negative if the direction is opposite.

Acanthamoeba exhibits chemosensory responses as observed by their response to a variety of bacterial products or potential bacterial products by moving actively toward the attractant. *Acanthamoeba* responds to the chemotactic peptide formyl-methionyl-leucyl-phenylalanine, formyl-methionyl-leucyl-phenylalanine benzylamide, lipopolysaccharide, and lipid A. In addition, significant responses to cyclic AMP, lipoteichoic acid, and *N*-acetyl-glucosamine were also found. Interestingly, chemotactic peptide antagonists, mannose, mannosylated bovine serum albumin, and *N*-acetyl-muramic acid all yielded non-significant responses. Pretreatment of *Acanthamoeba* with chemotactic peptides, bacterial products, and bacteria reduced the directional response to attractants. *Acanthamoeba* that had been grown in the presence of bacteria appeared more responsive to chemotactic peptides. Treatment of *Acanthamoeba* with trypsin reduced the response of cells to chemotactic peptides, though sensitivity was restored within a couple of hours. This suggests that *Acanthamoeba* membrane may have receptors, sensitive to these bacterial substances, which are different from the mannose-binding protein involved in binding to the host cells to produce cytotoxicity or involved in binding to bacteria during phagocytosis. The rate of movement was relatively constant (*ca.* 0.40 μm per sec), indicating that the locomotor response to these signals is a taxis, or possibly a klinokinesis, but not an orthokinesis (Schuster and Levandowsky, 1996).

9.5. Ubiquity

Acanthamoeba has been found in diverse environments, from drinking water to distilled water wash bottles, so it is not surprising that human encounter and interact regularly with these organisms, as is evidenced by the finding that in some regions, most (up to 100% of the population in some areas) individuals tested possess *Acanthamoeba* antibodies. This clearly suggests that these are one of the most ubiquitous protozoa and often come in contact with the humans. This provides *Acanthamoeba* with a wider access to approach the limited available susceptible hosts.

9.6. Biofilms

Biofilms are known to play an important role in the pathogenesis of *Acanthamoeba* keratitis. Biofilms are microbially-derived sessile communities, which can be formed in aqueous environments as well as on any materials and medical devices including intravenous catheters, contact lenses, scleral buckles, suture material, and intraocular lenses. In the instance of contact lenses, biofilms are formed through contamination of the storage case. Once established, biofilms provide attractive niches for *Acanthamoeba*, by fulfilling their nutritional requirements as well as providing resistance to disinfectants. In addition, this allows higher binding of *Acanthamoeba* to contact lenses. For example, it is shown that *Acanthamoeba* exhibits significant higher binding to used and *Pseudomonas* biofilm-coated hydrogel lenses compared with the unworn contact lenses. In addition, the abundant nutrient provided by the biofilm encourages transformation of *Acanthamoeba* into the vegetative, infective trophozoite form, and it is important to remember that binding of *Acanthamoeba* to the human corneal epithelial cells most likely occurs during the trophozoite stage as cysts exhibit no and/or minimal binding. Overall, these findings suggest that biofilms play an important role in

Acanthamoeba keratitis in wearers of contact lenses and perhaps preventing biofilm formation is an important preventative strategy (Zegans *et al.*, 2002; Beattie *et al.*, 2003; Dudley *et al.*, 2005; Garate *et al.*, 2006).

9.7. Effect of cholesterol (or sterol biosynthesis) on *Acanthamoeba* virulence

When repeatedly passaged through *p*-chlorophenoxyisobutyrate (CPIB) (an inhibitor of sterol biosynthesis), it led to the gradual loss of invasiveness, inhibition of multiplication, lowering of the sterol and phospholipid contents and phospholipase A activity of the trophozoites. In contrast, passage through cholesterol caused gradual induction of invasiveness, stimulation of multiplication, increase in sterol and phospholipid contents and phospholipase A activity of trophozoites (Misra *et al.*, 1986).

9.8. Host factors

The factors that enable *Acanthamoeba* to produce disease are not limited solely to the parasite, but most likely involve the host determinants. Evidence for this comes from recent studies in the UK, Japan and New Zealand, which suggest that the storage cases of contact lenses of 400 – 800 per 10,000 asymptomatic wearers are contaminated with *Acanthamoeba*. This number is remarkably high compared with the incidence of *Acanthamoeba* keratitis in wearers of contact lenses, which is around 0.01 – 1.49 per 10,000. These findings suggest that factors such as host susceptibility, tissue specificity, tear factors, sIgA, corneal trauma, as well as the environmental factors such as osmolarity may be important in initiating *Acanthamoeba* infections. In addition, malnutrition, mental stress, age, metabolic factors, and other primary diseases may play a role in the pathogenesis of *Acanthamoeba* infections.
However, the extent to which such host factors contribute to the outcome of *Acanthamoeba* infections is unclear because host factors are more complex and difficult to study than those of the parasite. For example, in bacterial infections such as *Salmonella*, the genetic constitution of the host determines susceptibility. These studies with *Salmonella* were possible only because transgenic animals were available (Devonshire *et al.*, 1993; Gray *et al.*, 1995; Larkin *et al.*, 1990; Watanabe *et al.*, 1994; Fleiszig *et al.*, 1994; Harrington and Hormaeche, 1986).

Overall, it can be concluded that *Acanthamoeba* traversal of the biological barriers such as the blood-brain barrier, is a complex process that involves both parasite as well as host factors.

From: *Acanthamoeba*: Biology and Pathogenesis. By: Naveed Khan.

Section E

Immune Response

1. Immunology

Once in the human body, *Acanthamoeba* is encountered by a highly professional defence system. The fact that *Acanthamoeba* granulomatous encephalitis normally occurs in immunocompromised people suggests that immune system plays an important role in *Acanthamoeba* infections. Here, I have briefly described the human immune system, followed by mechanisms used by *Acanthamoeba* to evade human immune system. Traditionally, the human defence system has been divided into two components (Fig. 1);

1. Non-specific / constitutive / innate immune system that includes: skin that acts as a physical barrier, neutrophils that police the bloodstream and attack any foreign invaders, complement and cytokines that direct the activities of neutrophils, and natural killer cells that kill virus-infected cells.
2. Specific / inducible / acquired immune system that includes: antibodies (produced by B-lymphocytes) and T-lymphocytes.

These defence systems do not operate independently but communicate with each other to build an effective defence against the invading organism.

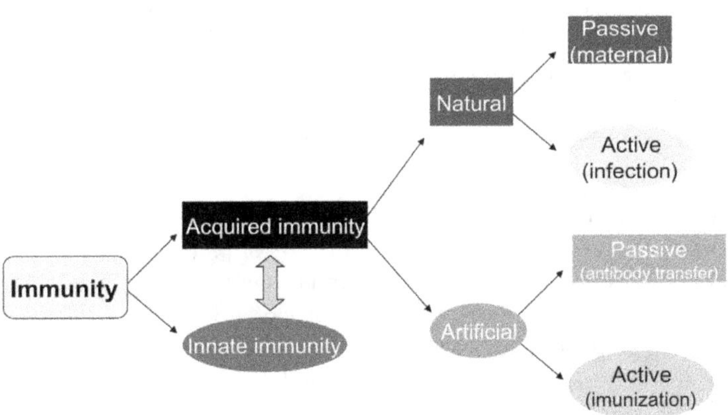

Figure 1. An overview of human immune system.

1.1. Non-specific immune system

1.1.1. Skin

The outer layers of the skin are dry, thick layers of keratinized cells that provide a physical barrier to prevent entry of pathogens. Natural openings in the skin such as pores, hair follicles, and sweat glands are protected by secretion of toxic chemicals such as;

- Fatty acids (produced by sebaceous glands in the skin and possess anti-microbial properties), and
- Lysozyme (present in mucosal secretions and breaks down cell walls).

Both fatty acids and lysozyme work in combination to prevent colonization by potential pathogens. Generally, pathogens can only breach this barrier through burns, wounds, insect bites, animal bites and human behaviour (sharing syringes, smoking etc). Underneath the skin, is skin-associated lymphoid tissue that connects the non-specific / innate immune defences with the specific / acquired immune defences. In skin-associated lymphoid tissue, it is the resident macrophages that attack microbial pathogens and initiate specific immune defences.

1.1.2. Mucosa

Within the human body, various systems are exposed to the environment including the following:

- Respiratory tract: this tract is lined with cilia that beat in such a way to propel particles towards the throat, where they can be expelled by coughing and swallowing and excretion.
- Intestinal tract: hydrochloric acid secreted by stomach is lethal or inhibitory to many (but not all) microorganisms.
- Genito-urinary-tract: commensal bacteria in the vagina produce lactic acid resulting in a low pH which is lethal or inhibitory to many (but not all) microorganisms.

These tracts are covered in the mucosal membranes. In addition, cells in the mucosal membranes are protected by thick layer of the mucus. The mucus is sticky and slimy that can trap microorganisms so they can be expelled and do not invade body. The mucus consists of proteins and polysaccharides that prevent microbes from reaching the epithelial cells. The mucus also contains:

1. Lactoferrin: iron-binding protein that deprives microbes of iron.
2. Lysozyme: an enzyme that digests cell walls of pathogens, by disrupting the *N*-acetyl-muramic acid – *N*-acetyl-glucosamine linkages. It is present in tears, saliva, sweat, and other mucosal secretions.
3. Defensins: small proteins which interfere with pathogen's intracellular signaling pathways and metabolism, and may produce holes in the microbial membranes by a charge or voltage-dependent mechanism.
4. Cathelicidins: antimicrobial peptides that were originally discovered as insect defence peptides.

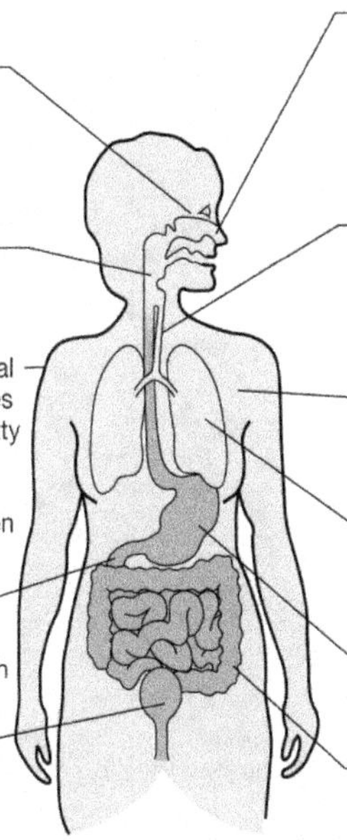

Figure 2. Some of the human innate defences.

5. Collectins: proteins that bind to carbohydrates on microbial surfaces and promote their elimination.

Mucin is constantly shed and replaced, together with any trapped microbes. In addition, the infection may be limited by the short life of the intestinal epithelial cells, which are shed approximately every 30 hours, together with any microbial pathogen. The skin and mucosal defences are highly effective in preventing pathogens from invading into the body and can exclude more than 99% microbes that we encounter. Similar to the skin, mucosal membranes are associated with the specific immune defence by mucosa-associated lymphoid tissue. For the intestinal tract, it is called the gastrointestinal-associated lymphoid tissue. In the gastrointestinal-associated lymphoid tissue, the resident macrophages (similar function to macrophages of the skin-associated lymphoid tissue) initiate specific defences.

1.1.3. Normal flora

In addition to the above, the skin and mucus membranes of the body are home to a variety of microbes that constitute the normal flora and include virus, bacteria, protozoa and fungi. The normal flora plays an important role in protecting the body by competing with potential pathogens, a situation known as microbial antagonism. The normal flora achieves this by the followings:

(i) consuming the available nutrients, making them unavailable to pathogens;
(ii) stimulating the body's defences; and
(iii) occupying binding sites required by pathogens (Fig. 2).

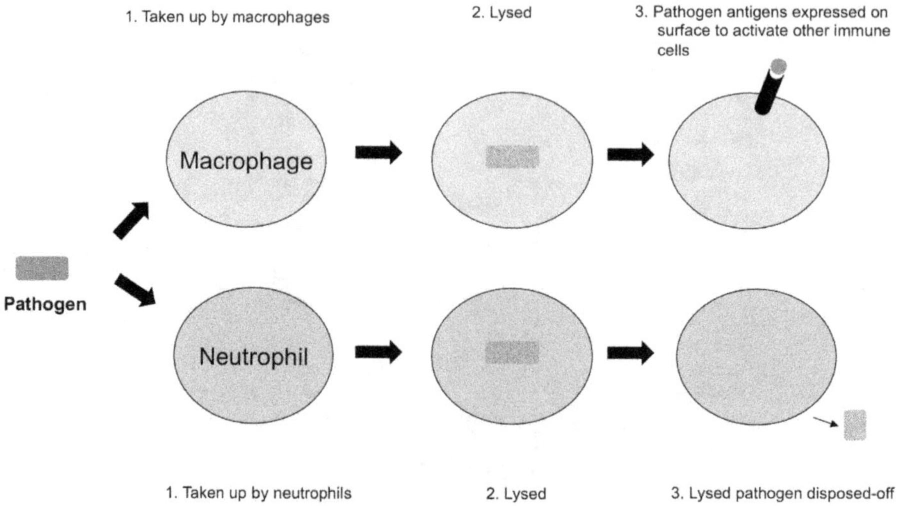

Figure 3. Uptake of pathogens by phagocytes leading to their degradation.

1.1.4. Recruitment of phagocytes and their mode of killing

The violation of initial defences may lead to the inflammation. The inflammation is a tissue reaction to infection or injury characterized by redness (increased blood flow), swelling (increased extravascular fluid and phagocyte infiltration), heat (increased blood flow and pyrogens, which are fever-inducing agents), pain (local tissue destruction and irritation of the sensory nerve) and the loss of function. The inflammation leads to recruitment of phagocytes to the site of infection, which leads to ingestion and destruction of microbes (Fig. 3). The major phagocytes at the site of infection are;

- Polymorphonuclear leukocytes: can be further classified as neutrophils, eosinophils, and basophils.
- Macrophages: can be further classified as alveolar macrophages, microglial cells, Kupffer cells, and Langerhans cells.

As the name suggests, the polymorphonuclear leukocytes possess a nucleus with multi-lobed structure that may appear to be multi-nuclei when viewed in cross-sections. Among polymorphonuclear leukocytes, neutrophils are the key cells in the innate immune defences (Fig. 4). Neutrophils are primarily found in the bloodstream. Other major phagocytic cells are initially monocytes, which leave the blood and mature into the phagocytic macrophages (Fig. 4). Macrophages may be named differently depending on their location. For example, wandering macrophages in the blood target microbes in the blood as well as leave the blood by squeezing through cells, lining the capillaries and target microbes in tissues (similar to neutrophils). Other macrophages do not wander. These include alveolar macrophages of the lungs, microglial cells of the central nervous system, and Kupffer cells of the liver, Langerhans cells of the skin / mucosa / cornea, where they await microbial invaders. Despite various differences in these macrophages, for

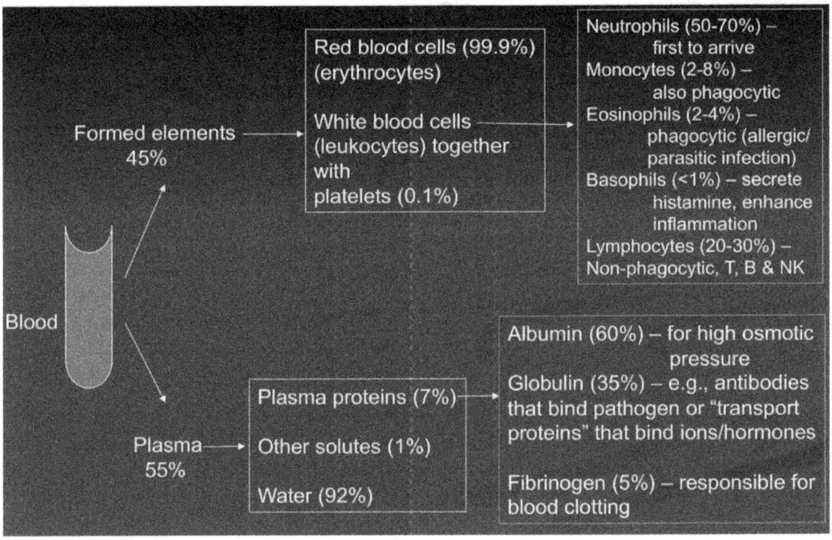

Figure 4. Components of blood.

simplicity, I have referred to all as "macrophages" throughout this chapter. A general function of the macrophage is to phagocytoze microbes within a specific organ, which is similar to neutrophils. Although both neutrophils and macrophages kill invading microbes, a key difference between them is that macrophages not only kill microbes but stimulate the activation of other immune cells to respond to the pathogen.

The steps of ingestion and killing by phagocytic cells (such as neutrophils) are as follows:

1. Attachment of the microbe, which involves the formation of pseudopodia.
2. Uptake of the microbe in a vacuole "phagosome".
3. Killing of the microbe in a lysosomal-dependent and lysosomal-independent mechanism.

   Lysosomal-independent mechanism:

   Here the phagosome does not fuse with the lysosome. There are two pathways within this mechanisms as follows:

   (i) generation of oxygen radicals that can damage membranes, proteins, and DNA thus killing the microbe.

   (ii) production of nitric oxide that deprives microbes of essential growth element by binding to iron and is highly toxic.

   Lysosomal-dependent mechanism:

   This is the most usual mechanism of microbial killing. Here, the phagosome does fuse with the lysosome (containing antimicrobial chemicals and enzymes) to form phago-lysosome. The lysosome contains myeloperoxidase enzyme that is normally inactive but following fusion with the phagosome, this enzyme becomes activated and produces hypochlorite and chlorine (from hydrogen peroxide and

Table 1. Examples of major cytokine families.

Family	Member (examples)	Comments
Interleukin (IL)	IL-1 to IL-32	Secreting by different cells and function differently (e.g., IL-1 is involved in fever, inflammation, T-cell activation, macrophage activation; IL-2 is involved in proliferation etc).
Interferon (IFN)	IFN-α, IFN-β, IFN-γ	Leukocyte IFN (inhibit viral replication) Fibroblast IFN (inhibit viral replication) Lymphocyte IFN (many functions including macrophage activation, induction of MHC-I & II)
Tumor necrosis factor (TNF)	TNF-α TNF-β	Secreted by monocytes and other cells (fever, activates macrophages and endothelial cells) Secreted by T-cells (activity similar to TNF-α)
Colony stimulating factor (CSF)	G-CSF, M-CSF, GM-CSF,	Secreted by leukocytes, endothelial cells (helps differentiate bone marrow cells into particular cell type, e.g., neutrophils, macrophages)
Chemokine	MCP, Eotaxin	Controls the migration of cells between and within tissues
Growth factor	TGF, IGF	Anti-inflammatory effects

chloride), which are highly toxic to the invading microbes. In addition, the high acidity of the phago-lysosome is microbicidal. Furthermore, various enzyme systems result in the production of superoxide, hydrogen peroxide, singlet oxygen, hydroxyl radical. All of these chemical species have microbicidal activities and are called, oxygen-dependent killing mechanism. Notably, tissue macrophages do not have myeloperoxidase enzyme, but use catalase to produce hydrogen peroxide, which is use to generate chemical species.

4. The overall result is killing and digestion of the ingested microbe. Often death of the neutrophils also occurs and if in large numbers, the dead neutrophils forms pus.
5. Finally release of digested products to the outside.

1.1.5. Phagocytic call for help (production of cytokines)

When cells of the innate immune system encounter pathogen, they can respond by targeting them directly by phagocytosis. At the same time, they secrete proteins known as cytokines, which are used as smoke signal to recruit other immune cells. To this end, an important family of receptors, Toll-like receptors (13 TLRs are known in humans) on the surface of cells of the immune defences have been identified that can recognize pathogen-associated molecules such as lipopolysaccharide, lipoproteins etc. This binding results in the production of cytokines which activate other immune cells. Cytokines are large number of proteins that are usually less than 20 kDa, and serve a hormone-like function in allowing cells to communicate with each other. However, hormones act in an endocrine manner, i.e., produced in one organ and travel through the bloodstream to another organ. Instead, cytokines are produced and act locally in a paracrine manner (hormones produced by cells in a tissue bind to receptors on other cells in the immediate vicinity) and autocrine manner (hormone produced by cells bind to receptors on the same cells). Thus the role of cytokines is to enable cells to communicate with each other locally. They are produced primarily by the endothelial cells (which form the blood vessels) and phagocytes. There are many cytokines that are divided into various families (Table 1). Cytokine producing cells express CD14 receptor on their surface, which binds to the specific microbial components. This leads to the release of cytokines. Basically, cytokines function as smoke-signals, as they attract phagocytes to the site of infection and link non-specific and the specific immune systems. The major cytokines are interleukins, colony-stimulating factors, interferons, tumor-necrosis factors, chemokines and growth factors. Cytokines control many aspects of the cell behaviour both in normal physiology as well as play a crucial role in the infection. Fig. 5 provides a summary of the innate / non-specific immune responses.

Innate / non-specific defences →

- Barriers → Skin and mucous membranes, prevent pathogen entry
- Phagocytes → Macrophages, neutrophils, eosinophils ingest & destroy pathogens
- Extracellular killing → Eosinophils and NK kill pathogens without phagocytozing them
- Complement → Attract phagocytes, and attack pathogen
- Interferons → Increase resistance of cells, in particular to viral infection
- Defensins → Interfere with membranes
- Inflammation → Increased blood flow, capillary permeability, and migration of leukocytes into infected areas
- Fever → Inhibits pathogen, accelerates repairs, mobilize defences

Figure 5. Two pathways by which complement is activated, resulting in the formation of the membrane attack complex (MAC) in the host cell membranes.

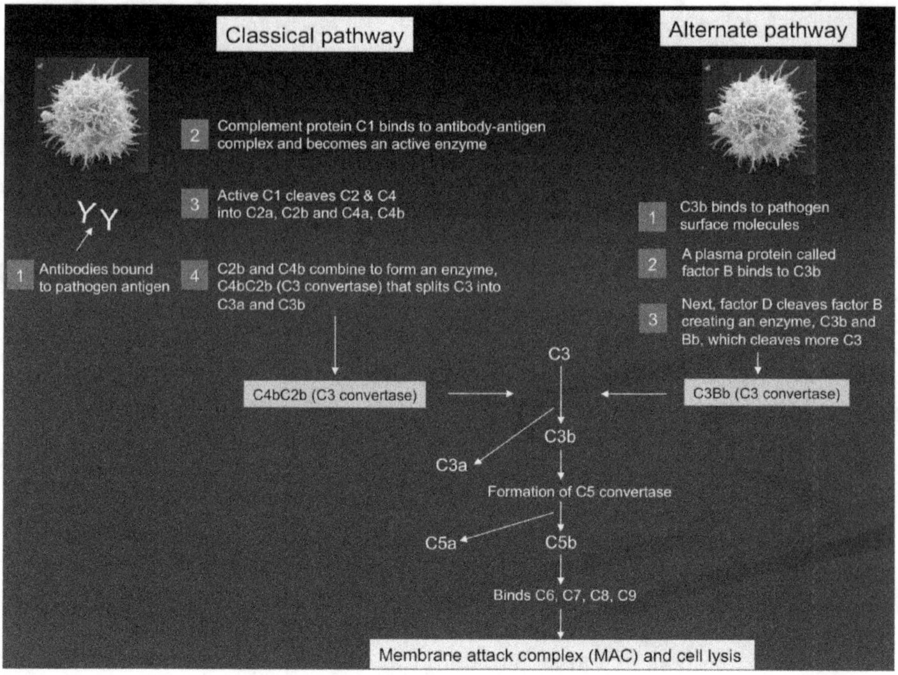

Figure 6. Summary of components of the innate / non-specific defences.

In addition, the plasma (acellular part of the blood) also contains: (i) lactoferrin, an iron binding protein that deprives pathogens of iron; (ii) lysozyme, that digest cell walls of pathogens; and (iii) defensins, which interfere with pathogen's intracellular signaling pathways and metabolism, and may produce holes in the pathogen membranes.

1.1.6. Complement

In addition to phagocytes, the plasma contains a potent complement system. Complement is a set of proteins produced in part by the liver that circulate in the blood and are activated sequentially by the proteolytic cleavage, a process called the complement activation. Complement attaches to invading microbe and directly cause damage to the microbial cell through a process known as the membrane attack complexes or help phagocytes to eliminate microbes from the body (Fig. 6). There are nine complement proteins, C1 – C9. There are two pathways of complement activation as follows:

1. The classical pathway which is initiated by the antibody-antigen complexes and,
2. The alternative pathway which is triggered by the microbial surface molecules such as cell membranes.

The activation causes complement proteins to be proteolytically cleaved. The cleaved products are indicated by small fragment 'a' or large fragment 'b'. In the classical pathway, antibodies specifically bind to the microbial antigens. A complement protein, C1, binds to the antibody-antigen complex and becomes an active enzyme. Active C1 splits several molecules of C2 and C4. The fragments of C2 and C4 combine to form an enzyme, C4bC2b (C3 convertase), that splits C3 into C3a and C3b. The C3b combines with the remaining fragments of C2 and C4 to form an enzyme that cleaves C5 into C5a and C5b. The C5b combines with C6, C7, C8 and several molecules of C9 to form a membrane attack complex. A membrane attack complex drills a circular hole in the pathogen's membrane leading to hypotonic lysis of the cell. In contrast, the alternate pathway occurs independently of antibodies. It begins with the cleavage of C3 into C3a and C3b. This occurs naturally in the plasma but at a slow rate and C3b is further degraded. However, in the presence of a microbe, C3b binds to the microbial surface, to a protein called factor B. Another plasma protein, factor D, cleaves factor B creating an enzyme composed of C3b and Bb, which continues with the complement pathway and membrane attack complex. The ultimate effect of complement is damage of the microbe by creating pores (i.e., membrane attack complex) or their opsonization (pathogen uptake by phagocytes).

In addition, there is a lectin pathway, which is very similar to classical pathway, except for the very first step. Here, instead of antibody binding to an antigen on a pathogen, mannose-binding protein (MBL; a protein that is always present in the serum) binds to mannose residues on the surface of the pathogen. These mannose residues may be components of glycoproteins or polysaccharides. When the MBP has bound to the pathogen, a protease called MBP-associated serine protease bind to the MBP. This complex cleaves C4 into C4a and C4b and C2 into C2a and C2b in the same way as for the classical pathway. Once C4b is generated, it binds to the cell surface and the remaining sequence for complement is the same as for the classical pathway.

1.2. Specific immune system

This system involves cells that respond to the invading pathogen in a specific fashion. These include CD8 cytotoxic T-cells, CD4 T-helper cells, B-cells (produce antibodies) and macrophages (Table 2). As described above, macrophages are phagocytes that are mostly found in tissues. Macrophages are first produced by the stem cells as monocytes, which circulate in the bloodstream. On attraction to a site of action or entry into a tissue they differentiate into macrophages. The invading pathogens are taken up by macrophages and killed (like in neutrophils) but instead of releasing the digested products to the outside, some microbial products are loaded onto a complex of proteins called the major histocompatibility complex proteins and presented on the cell surface. Hence, these cells are called antigen-presenting cells (APC). There are two major types of the major histocompatibility complexes, known as (i) the major histocompatibility complex class-I (MHC class-I), and (ii) the major histocompatibility complex class-II (MHC class-II). Of note, all cells express MHC class-I molecules, but MHC class-II are confined to the immune cells such as macrophages, B-cells, dendritic cells etc. Why certain microbial products are selected to be expressed on the antigen-presenting cells and not other microbial products, remains incompletely understood. Finally, the antigen expressed on the major histocompatibility complex is

Table 2. Cells involved in immune defence against microbes

Eosinophils and neutrophils (PMN)	Phagocytic cells
Basophils	Non-phagocytic cells, that release inflammatory chemicals
Monocyte	Immature macrophage – limited killing but produce cytokine
Macrophage	Phagocytic cell that kills, and present antigens to other immune cells
Dendritic cell	Phagocytic cell, active in presenting antigens to other immune cell
NK cell	Kills cells infected with cytoplasmic pathogens
B cells	When activated, produce antibodies
CD8 T cells	Cytotoxic T cells, kills cells containing cytosolic bacteria
CD4 T cells	Helper cells for B cells and activates macrophages
Fibroblasts	Act in repair after infection
Platelets	Produce clotting, which act as barrier to pathogen dissemination and blood loss

recognized by T-cells (either CD4 T-helper cells or CD8 cytotoxic T-cells). In general, the extracellular microbes / antigens are loaded onto MHC class-II complex and expressed on the surface of the antigen presenting cell, while the intracellular pathogens / antigens (e.g., virus that are obligate intracellular pathogens) are loaded onto MHC class-I complex and expressed on the surface of the antigen presenting cells. Notably, if microbial antigens are loaded onto the MHC class-I, they will be recognized by CD8 cytotoxic T-cells. In contrast, antigen-MHC class-II complex will be recognized by the CD4 T-helper cells (Fig. 7). The activation of CD4 T-helper cells or CD8 cytotoxic T-cells results in the generation of T-cells within 24 – 72 hours. Because of the time course, this response is known as delayed type hypersensitivity (DTH) responses. Since, these responses do not involve antibody, thus also are called cell-mediated responses or cell-mediated immunity.

1. In the case of interaction between the CD4 T-helper cell and the macrophage (i.e., antigen presenting cell), CD4 T-helper cells are activated and produce cytokines, such as interferon-gamma, interleukin-2, and interleukin-2, which activate macrophages and stimulate B-cells to produce antibodies.
2. In the case of interaction between CD4 T-helper cell and the B-cell, CD4 T-helper cells are activated as above, which in turn stimulates B-cells to differentiate into antibody-producing cells, that are called the plasma cells.

CD4 T-helper cells are capable of influencing a variety of the immune cells. In order to be effective, CD4 T-helper cells must determine which cytokines will allow the immune system to be most useful or beneficial for the host. Upon interaction with antigen presenting cells, CD4 T-helper cells (T_h0) develop into effector CD4 T-helper cells by differentiating into two major subtypes of cells known as T_h1 (Type 1) and T_h2 (Type 2) cells, respectively (Fig. 7).

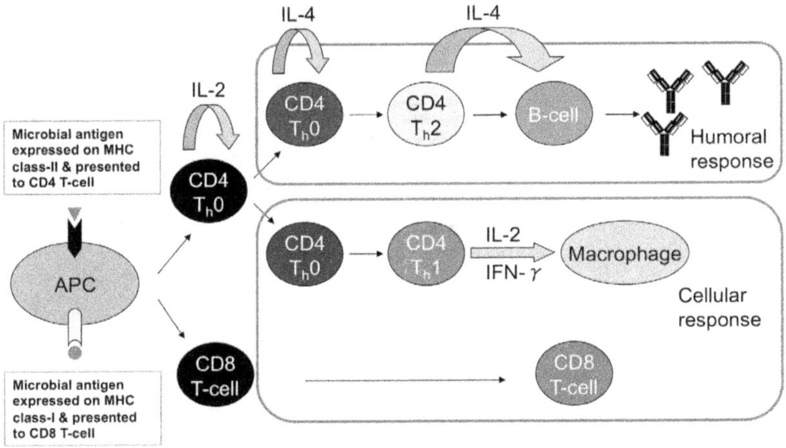

Figure 7. Activation of CD4 T-helper cells (T_h0) into Type 1 (T_h1) and Type 2 (T_h2). Microbes or antigens are taken up by antigen presenting cells (APC), processed and presented to lymphocytes resulting in their activation. IFN-γ is interferon-gamma; IL-2 is interleukin-2; and IL-4 is interleukin-4.

As described above, antigen-MHC class-II recognition is generally used in response to the extracellular microbial pathogens. In contrast, if the microbial pathogen is intracellular, their products are loaded onto the MHC class-I proteins. The MHC class-I complex displays proteins on almost all cells. Finally, the antigen-MHC class-I complex is recognized by the CD8 cytotoxic T-cells. The CD8 cytotoxic T-cells kill the host cells displaying antigen-MHC class-I complex, thus killing any intracellular microbe within it. This is particularly useful against viral infections.

1.2.1. Antibodies

Antibodies are proteins called immunoglobulins (Igs). They are made of two heavy and two light chains, which create two functionally important areas (Fig. 8). The region that recognizes and binds antigen is called the Fab region, while the Fc region binds to phagocytes (phagocytes have an Fc receptor). The result of an antibody binding to a specific antigen is:

- Enhanced phagocytosis by providing a binding site (Fc) for phagocytes.
- Microbes trapped in the mucin.
- Complement activation.
- Neutralization of toxins.
- Inactivation of proteins.
- Inhibition of binding of toxins/microbes to the host cells.

This ultimately results in the neutralization and removal of microbes and/or their toxins from the host. The antibodies present in the serum are IgG (70 – 75% of total antibodies) and IgM (10% of the total antibodies), while IgA (15% of the total antibodies) are present in secretions such as milk, tears, saliva and other mucosal secretions, and are called secretory IgA (sIgA). In addition, IgE and IgD are present in the serum and are involved in the allergic reactions.

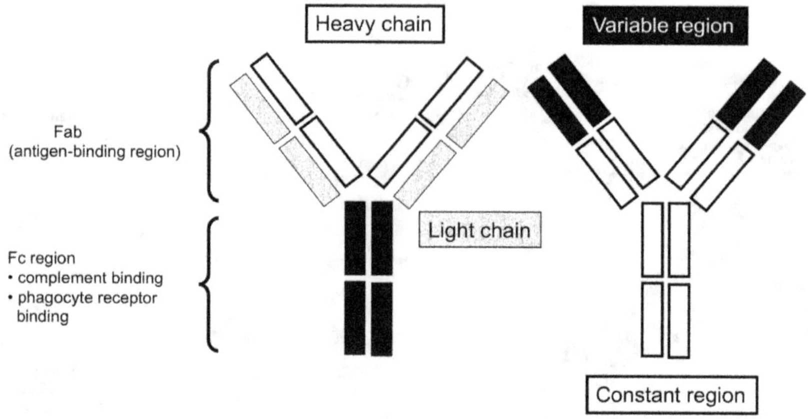

Figure 8. Structure of an antibody.

2. What are the basic types of parasite immune evasion strategies?

Upon invasion into the human body, generally pathogenic protozoa are recognized as foreign and have to cope with the highly professional host immune system. The involvement of the immune system in protozoan infections is further shown with the finding that the majority of parasitic infections are limited to babies/children or individuals with less developed (impaired) immune systems and these infections may be self-limiting in adults with a fully developed immune system. To this end, parasite-specific antibodies in the presence of phagocytes provide a highly effective defence system. For example, the role of antibodies in malaria is demonstrated with the finding that passive transfer of antibodies from immune individuals to those suffering from acute malaria results in the reduction of parasitemia. In addition, B-cell deficiency severely impairs the host's ability to clear malaria. Antibodies specifically bind to antigens on microbial surface with their Fab region, and binds to phagocytes (neutrophils/macrophages) using their Fc region. Thus, phagocytes in the presence of specific antibodies can effectively clear parasites from the human body.

Therefore, parasites must develop mechanisms to evade the host immune system. This is essential for some parasites to buy sufficient time to complete development (reproduction) and in particular for parasites that cause chronic infections, lasting months to years. There are several strategies used by the pathogenic protozoa to overcome the host's immune response as described below (Fig. 9).

2.1. Non-induction of immune response

Some protozoa pathogens may target organs and/or tissues that have a limited immune response, thus not detected by the immune system. For example, *Acanthamoeba* keratitis is limited to the cornea, which is avascular and as such has limited immune system.

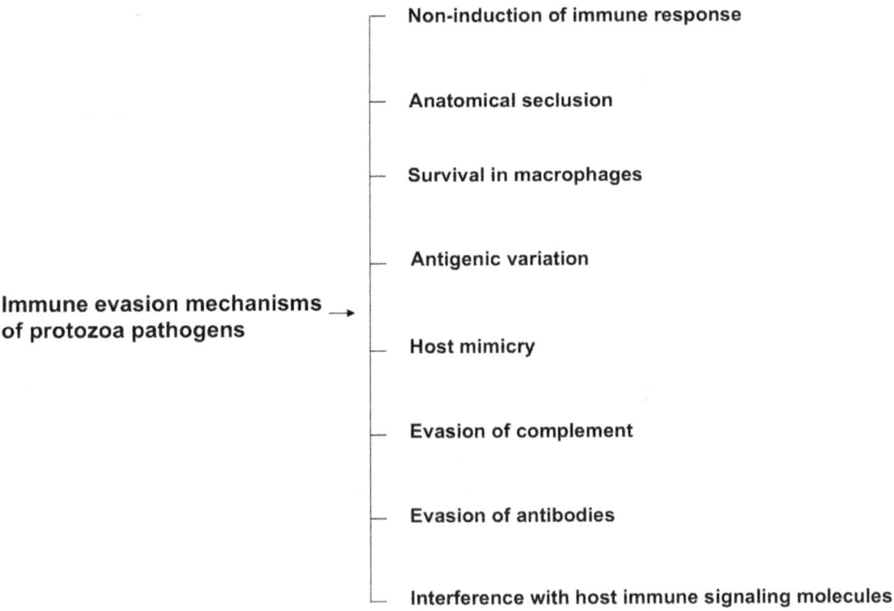

Figure 9. An overview of parasite immune evasion mechanisms.

2.2. Anatomical seclusion

The parasite can invade non-immune cells, replicate intracellularly and hide from the immune cells, thus avoiding an overwhelming immune response. For example, some stages of *Plasmodium* spp. live inside the liver cells, while other stages, during its life cycle live inside erythrocytes, which do not have a nucleus, thus are not recognized by the CD8 cytotoxic T-cells or natural killer cells. Other benefits of anatomical seclusion are to avoid being exposed to antimicrobials.

2.3. Mechanisms of survival in macrophages

Even if taken up by the immune cells, some parasites have the ability to survive the onslaught. For example, *Leishmania* spp. are taken up by macrophages but evade macrophage proteolytic processes by (i) the induction of cytokines such as tumor growth factor-beta, that inhibit or deactivate macrophages, (ii) by inhibiting phagolysosome formation, (iii) by producing anti-oxidases to counter products of the macrophage's oxidative burst, or (iv) by inhibiting inducible nitric oxide synthase expression or activity.

2.4. Antigenic variation

Parasites can evade the immune response by changing antigens that are expressed on their surfaces, every few generations. This is achieved by random activation or rearrangement of genes that code for surface antigens. Thus antibodies produced against the former antigens become ineffective. This is best studied in the African trypanosomes. *Trypanosoma brucei* continuously changes its surface glycoprotein coat called "variable surface glycoprotein

(VSG)" to stay one step ahead of the immune system that mounts an antibody response to that particular VSG. The host's antibodies facilitate neutralization and killing of approximately 99% of original *Trypanosoma* population. However, during this time a few of the *Trypanosoma* have shed their coat, i.e., switched VSG, and covered themselves with a new antigenically distinct VSG coat. These distinct *Trypanosoma* give rise to a new population expressing the new VSG coat. The immune system again responds to this proliferated population by producing a new set of antibodies that successfully kill 99% of the *Trypanosoma* population. But once again, VSG switching among a small portion of *Trypanosoma* renders them undetectable, and parasite successfully evade the host immune response and this cycle continues.

2.5. Host mimicry

Some parasites use molecular mimicry to escape the host immunosurveillance. This can be achieved by expressing proteins/glycoproteins on the parasite surface that mimic the host proteins/glycoproteins or by adsorbing host components on their surfaces. The ultimate purpose is to be recognized as host cells and thus avoid an overwhelming immune response.

2.6. Evasion of the complement cascade

Complement is a set of proteins that circulate in the blood and are activated (i) through antibody binding to the microbial antigens or directly by bacterial surface leading to the formation of membrane attack complex (forming pores in the microbial membranes) or opsonizes pathogen (helps pathogen uptake by phagocytes). In some protozoa such as in *Trypanosoma*, the conversion of parasites from epimastigotes to infectious trypomastigotes coincides with the acquisition of resistance to lysis by the alternative complement pathway. *Leishmania* spp. resists complement by shedding the membrane attack complex and by inactivation of some membrane attack complex components (C3, C5, C9) by phosphorylation.

2.7. Evasion of antibodies

At the mucosal surfaces, protozoa have to cope with potent secretory immunoglobulins (i.e., sIgA) and then again in the bloodstream, IgG and other antibodies are highly effective in trapping microbes and neutralizing toxins. The pathogenic protozoa have been shown to evade antibodies by antigenic variation (indicated above), and by inactivating antibodies. For example, *Entamoeba histolytica* is known to evade the humoral immunity by degradation of antibodies by secretion of proteases.

2.8. Interference with the host immune signaling molecules

Some protozoan pathogens interfere with the host immune signaling molecules thus affecting the host ability to build an effective immune response. For example, *Leishmania* spp. reduces the production of interferon-gamma by the host which is an important cytokine in the activation of macrophages, i.e., the immune cell important in clearing protozoa from the host as well as building an effective immune response. In addition, *Leishmania* spp.

suppresses the transcription of the interleukin-12 gene, by interfering with the protein kinase C, protein tyrosine kinases, Janus kinases (JAK), and Signal Transducers and Activators of Transcription (STAT), JAK/Stat pathway, thus blocking the CD4 T-helper cells Type 1 (Thl) response. *Entamoeba histolytica* induces interleukin-4 and interleukin-10 and modulates CD4 T-helper cells Type 1 (T_h1) response. *Trypanosoma brucei* increases the levels of interferon-gamma and decreases interleukin-2 and interleukin-2 receptor, which affects T-cell responses. *Trypanosoma cruzi* blocks interleukin-12 and tumor necrosis factor-alpha by interfering with mitogen-activated protein kinase re-phosphorylation. *Toxoplasma* interferes with nuclear factor-kappa B (NF-κB) translocation, Stat-1 translocation and mitogen-activated protein kinase re-phosphorylation resulting in blockage of effector molecules, i.e., interleukin-12, tumor necrosis factor-alpha and MHC class-II complex.

3. Immune response and *Acanthamoeba* infections

Despite the ubiquitous presence of these organisms in diverse environments, the number of infections due to *Acanthamoeba* has remained very low. This is due to the complex nature of *Acanthamoeba* infections that depends on the virulence properties of the parasite, the host susceptibility (e.g., immuno-compromised people or contact-lens wearers are more susceptible), and environmental conditions (personal hygiene, exposure to the contaminated water/soil). For *Acanthamoeba* granulomatous encephalitis, in normal circumstances, a competent immune system may be sufficient to control these pathogens. But due to the complexity of the immune system, the precise factors which contribute to the host resistance and the associated mechanisms remain unclear. By contrast, *Acanthamoeba* keratitis can occur in the normal individuals, although the oral immunization can prevent this infection. This suggests that the immune system still plays an important role in this condition. Although there are obvious similarities in the immune response to *Acanthamoeba* granulomatous encephalitis and to *Acanthamoeba* keratitis, they are described separately for simplicity.

3.1. *Acanthamoeba* keratitis and the immune response

The involvement of the immune response in *Acanthamoeba* keratitis is demonstrated with the findings that steroid administration during the course of *Acanthamoeba* keratitis exacerbates the infection. Thus, the steroid administration needs to be withdrawn extremely slowly to avoid the recurrence of immunological corneal reaction. These findings suggested that the immune response plays a key role in *Acanthamoeba* keratitis (Horsburgh *et al.*, 1991). The presence of macrophages in corneas exposed to the parasite-laden contact lenses prevents the development of full-blown *Acanthamoeba* keratitis *in vivo*, suggesting the involvement of the immune system (Van Klink *et al.*, 1993).

Because the normal cornea is avascular, many primary host defences are provided by the eyelids and secreted tear film. The tear fluid produced by the lacrimal system, together with constant eye lid movement, provides the first line of defence in *Acanthamoeba* keratitis. The tear film contains lysozyme, lactoferrin, beta-lysins, sIgA, prostoglandins, and other compounds with antimicrobial and immunological properties. Of the three distinct layers in the tear film, (i.e., the oil layer, the aqueous layer and the mucous layer), the aqueous layer is the source of compounds with antimicrobial properties such as non-specific antimicrobial lysozyme, lactoferrin and the specific sIgA. A role for sIgA in *Acanthamoeba* keratitis is indicated by findings that *Acanthamoeba* keratitis patients show decreased levels of sIgA, and specific anti-Acanthamoebic sIgA inhibits *Acanthamoeba* binding to the corneal epithelial cells, suggesting that sIgA normally plays an important role in alleviating this infection. However sIgA may not be the only determinant. Factors present in tears, in addition to sIgA are shown to partially inhibit *Acanthamoeba*-mediated binding to and cytotoxicity of human corneal epithelial cells, as shown by using tears that were depleted of sIgA antibodies. Although, the non-sIgA factor can be inactivated by proteinase K, that suggests its proteinaceous nature, but the correct identity remains to be determined. Tears also contain complement that is composed of serum-borne molecules in a cascade-like manner. *Acanthamoeba* is shown to directly activate the complement system via the alternative pathway. However, the pathogenic amoebae are resistant to complement-mediated lysis due to expression of complement regulatory proteins including decay accelerating factor. In addition, complement regulatory protein are expressed on corneal cells and are present in tears, thus activation of the complement system has little or no effect on the resistance of *Acanthamoeba* keratitis (Alizadeh *et al.*, 2001; Leher *et al.*, 1999; Walochnik *et al.*, 2001; Nassif, 1996; Qu and Leher, 1998).

3.1.1. From the contact lens to the cornea

The mechanisms of how binding is inhibited are unclear, but could involve interference with the *Acanthamoeba* adhesins and the associated pathways. Overall, the tear film in conjunction with the blink reflex is highly effective in blocking microbial access to the corneal epithelial cells and expelling *Acanthamoeba* from the surface of the eye and carries it to the conjunctiva. Although the cornea is protected by only limited immune mechanisms, the conjunctiva is highly vascular with lymphoid tissue and contains mainly sIgA-producing plasma cells, T-lymphocytes, natural killer cells and macrophages, which are highly effective in targeting *Acanthamoeba* and stimulating acquired immune responses. For example, conjunctival macrophage depletion exacerbates *Acanthamoeba* keratitis symptoms *in vivo*, and increases the rate of infection to 100%. Thus in order to proceed with infection, *Acanthamoeba* must remain at the corneal surface and invade into the stromal tissue. Of interest, the tear film in wearers of contact lenses differs in terms of volume and make up, and it seems that the use of extended contact lens wear alters the levels of inflammatory mediators in tears and contributes to increased inflammation. Other major changes are exclusion of atmospheric oxygen and thinning of the basal tear film. It has been proposed that the hypoxia is responsible for the metabolic compromise of the cornea and also leads to less resistance to microbial infection. This indicates the cornea is a stressed environment in the presence of a contact lens, and this affects components of the tear film making invasion by the *Acanthamoeba* more likely. In addition, the highly virulent strains of *Acanthamoeba* evade these primary defences to traverse the cornea and invade into the stroma. Once in the stroma, *Acanthamoeba* secrete proteolytic enzymes causing the stromal degradation leading to phagocyte infiltration, which modulate B-cells and T-cells activity to clear *Acanthamoeba*. The macrophages provide defence against *Acanthamoeba* keratitis directly by clearing *Acanthamoeba* and by inducing an inflammatory response, in particular secretion of the macrophage inflammatory protein-2. This results in the recruitment of other immune cells such as neutrophils, which are highly potent in destroying *Acanthamoeba* in a myeloperoxidase-dependent manner. Overall, these studies suggest that upon entry into the eye, *Acanthamoeba* has to deal with normal tear film containing non-specific antimicrobial compounds together with the sweeping action of the eye lids as well as specific immunoglobulins and cell-mediated immunity as described below (Thakur and Willcox, 2000; van Klink *et al.*, 1996; Weisman and Mondino, 2002; Hurt *et al.*, 2001; 2003).

3.1.2. Effect of contact lens wear for extended times on the innate immunity of the ocular surface

The extended wear of the contact lens is one of the predisposition factors in contracting *Acanthamoeba* keratitis. The earlier studies suggested "improper contact-lens hygiene" as a key factor, in that people who wear contact lenses are more susceptible to get *Acanthamoeba* keratitis. However, the improper contact-lens hygiene can not explain the phenomenon that *Acanthamoeba* was found in normal individuals' lens-cases. So there might be other factors associated with the *Acanthamoeba* keratitis. Recently, more attention has been paid on the influence of extended wear contact lenses on the innate immunity of the ocular surface. It has been proven that in contact lens wearers the reactivity of neutrophils and the concentration of certain inflammatory mediators were significantly altered compared with that in non-contact lens wearers. Moreover, other studies showed the important contributions of the innate immunity on occurrence and development of *Acanthamoeba* keratitis. With the contribution of extended wearing contact lens on the immunity and the relation between the innate immunity and *Acanthamoeba*, it is suggested that the impaired innate immunity of the ocular surface may be a key bridge between the extended wear contact lens and *Acanthamoeba* keratitis. With the impaired innate immunity caused by extended contact-lens wearing, the *Acanthamoeba* trophozoites and cysts could not be easily killed, therefore *Acanthamoeba* keratitis has been shown to occur and aggravated. Thus an understanding of the immunological mechanisms of extended contact lens wearing on the *Acanthamoeba* keratitis may give more contributions on the research of the disease, and facilitate the production of contact lens with much higher biocompatibility (Li and Sun, 2008).

3.1.3. Antimicrobial peptides

Antimicrobial peptides are cationic host defence peptides with microbicidal and cell-signaling properties. They show promise as potential therapeutic agents. A beta-defensin antimicrobial peptide gene was isolated from the ocular surface, and its expression was characterized in the presence of the ocular inflammation and/or the infection. It was shown that the beta-defensin antimicrobial peptide, DEFB-109 is constitutively expressed. The gene showed significantly decreased expression in *Acanthamoeba* infection. The discovery of DEFB-109 on the ocular surface enhances our knowledge of the profile of antimicrobial peptide at this important mucosal surface. The fact that the expression of DEFB-109 is significantly reduced in both inflammatory and infective ocular surface disease reflects not only an intimate balance between this host defence gene and microbes but indicates a role other than purely microbicidal. Future studies should dissect mechanisms behind the reduced gene expression of an antimicrobial peptide in disease states to be uncovered (Abedin *et al.*, 2008).

3.1.4. Acquired immunity

3.1.4.1. Immunization using whole *Acanthamoeba* antigens

The recurrence of *Acanthamoeba* keratitis is common, suggesting that the corneal infection alone does not induce protective immunity. In a series of experiments, it was determined whether ocular infection with *Acanthamoeba* trophozoites would elicit protection against re-infection. Animals were infected with parasite-laden soft contact lenses to abraded corneal surfaces and, as expected developed *Acanthamoeba* keratitis. Animals were subsequently challenged with parasite-laden soft contact lenses, and again they developed *Acanthamoeba* infection. These findings indicated that ocular infection alone failed to stimulate and/or provided effective immunity to subsequent ocular challenge with infectious parasites. In support, it is shown that the corneal infection with *Acanthamoeba* failed to induce either the primary cell-mediated immune responses (i.e., delayed-type hypersensitivity) or the serum IgG antibody against the parasite antigens response. The inability of corneal infections to induce primary cell-mediated immune responses is perhaps due to the absence of resident antigen-presenting cells in the central cornea. This was further confirmed with findings that the induction of macrophages migration into the central cornea, prior to infection with *Acanthamoeba* promoted the development of the parasite-specific cell-mediated immune responses. Overall, the findings indicate that in the absence of resident antigen-presenting cells, the corneal infection with *Acanthamoeba* fails to stimulate primary cell-mediated or the humoral immunity. Thus it is not surprising that the evaluation of the histopathologically sections of *Acanthamoeba* keratitis biopsies from humans and the experimental animals reveals the absence of lymphocytes. The serum IgG antibody and the cell-mediated immune response can be generated by systemic immunization with *Acanthamoeba* antigens, however they fail to protect against the subsequent ocular challenge with *Acanthamoeba* trophozoites, suggesting that the T-cell-mediated immunity and the serum antibody, following systemic injections of parasite antigens, do not directly protect against *Acanthamoeba* keratitis.

Next, it was examined whether injection of parasite antigens either intramuscularly, subconjunctivally, or by both routes would induce immunity against *Acanthamoeba* keratitis. Following injections, animals were challenged with the parasite-laden soft contact lenses and observed clinically for the appearance of *Acanthamoeba* keratitis. It was observed that the intramuscular injection of *Acanthamoeba* antigens failed to protect against the ocular infection even though hosts developed high titers of IgG antibodies and displayed lymphocyte blastogenic responses to parasite antigens. By contrast, 50% of the host immunized subconjunctivally were protected against corneal disease, and 100% of the animals immunized by a combination of intramuscular and subconjunctival administration of the parasite antigens were completely protected against two separate ocular challenges with the infectious parasites. Protection did not correlate with either IgG antibody titers or blastogenic potentials of the peripheral blood lymphocytes. Thus, administration of the parasite antigen via the subconjunctival route can protect against *Acanthamoeba* keratitis, suggesting the involvement of the acquired immunity. Moreover, the intramuscular immunization with parasite antigens elicited robust delayed-type hypersensitivity (the cell-mediated response) and IgG antibody responses (Alizadeh *et al.*, 1995; Van Klink *et al.*, 1997).

Later studies revealed that animals immunized orally with *Acanthamoeba* antigens mixed with the cholera toxin, showed significantly lower infection rates than did the control groups, which included cholera toxin alone (21.4% *versus* 72.6% respectively). The mucosal specimens including tears and stool samples showed the presence of parasite-specific sIgA in animals, orally immunized with *Acanthamoeba* antigens mixed with the cholera toxin, but not in the control animals. The mucosal anti-*Acanthamoeba* sIgA profoundly inhibited (more than 75%) binding of parasites to the corneal epithelial cells, *in vitro*. These studies suggested that the oral immunization with *Acanthamoeba* antigens mixed with the cholera toxin induces the production of parasite-specific sIgA in mucosal secretions and perhaps other immune responses, thus prevent the corneal infection. Notably, *Acanthamoeba*-specific sIgA antibody does not affect the viability of *Acanthamoeba* but prevents the early stages of infection, i.e., the adherence of trophozoites to the corneal epithelial cells (Leher *et al.*, 1999).

3.1.4.2. Immunization using specific *Acanthamoeba* antigens

The mannose-binding protein is shown to be a critical adhesin required for *Acanthamoeba* interactions with the host cells. Oral immunization using recombinant mannose-binding protein ameliorates *Acanthamoeba* keratitis and that this protection is associated with the presence of elevated levels of anti-mannose-binding protein sIgA in the tear fluid of the immunized animals.

Among other molecules, it was determined whether the mannose-induced cytopathic protein (a serine protease of approximately molecular weight of 133 kDa, known as MIP-133) and *Acanthamoeba* plasminogen activator (aPA) play key roles in the pathogenesis of *Acanthamoeba* keratitis. Chinese hamsters were orally immunized with four weekly doses of aPA or MIP-133 conjugated with the cholera toxin. The animals were immunized before infection to determine the prophylactic effect of oral immunization. The therapeutic effect of the oral immunization with aPA and MIP-133 was determined after the corneal infection had been established. The animals were then infected via *Acanthamoeba*-laden contact lenses. The results revealed that the oral immunization with the MIP-133 before and after infection with *Acanthamoeba* significantly reduced the severity of the corneal infection which includes infiltration and ulceration ($P<0.05$) and shortened the duration of the disease. The immunization with aPA alone did not significantly affect the course of disease the ($P>0.05$). These findings suggested that once trophozoites invade the cornea, MIP-133 production is necessary to initiate the corneal disease and plays an important role in the subsequent steps of the pathogenic cascade of *Acanthamoeba* keratitis (Alezadeh *et al.*, 2007).

With the exception of the mucosal antibody, elements of the adaptive immune system fail to prevent infection or contribute to its resolution in experimental animals.

3.2. Antibody levels in *Acanthamoeba* patients

The prevalence of serum IgG and tear sIgA against *Acanthamoeba* in the healthy subjects and patients with *Acanthamoeba* keratitis was tested. All healthy subjects and patients with *Acanthamoeba* keratitis had detectable serum IgG antibodies against *Acanthamoeba*. However, patients with *Acanthamoeba* keratitis had significantly higher anti-*Acanthamoeba* IgG antibody titers than the healthy subjects. In contrast, *Acanthamoeba*-specific tear sIgA was significantly lower in patients with *Acanthamoeba* keratitis in comparison with the healthy subjects. The total serum immunoglobulins did not differ significantly between the healthy subjects and patients with *Acanthamoeba* keratitis.

The sera revealed two predominant bands at 29 kDa and at 47 kDa, respectively. Interestingly, compared to the control sera, the sera of *Acanthamoeba* keratitis patients showed rather weak immuno-reactivities and they lacked the 29 kDa and the 47 kDa band in the sIgA immunoblot against the pathogenic strain. Overall these findings suggested that perhaps a low level of anti-*Acanthamoeba* sIgA antibody in the tears appears to be associated with *Acanthamoeba* keratitis and/or the immunological predisposition might also be of importance in *Acanthamoeba* keratitis (Alizadeh *et al.*, 2001; Walochnik *et al.*, 2001).

3.3. Can cysts evade immune responses?

For the majority of pathogenic protozoa, immune evasion strategies include the modulation of the host immune response, prevention of the complement-mediated lysis, degradation of antibody, and the development of cysts. *Acanthamoeba* cysts are not eliminated from the corneas of the human subjects or experimentally infected animals. The persistence of *Acanthamoeba* cysts in the cornea indicates that either the cysts escape the immunological elimination or are not recognized by the host's immunological elements. The aim of this study was to determine the immunogenicity and antigenicity of the *Acanthamoeba* cyst. The mice were immunized intraperitoneally and serum anti-*Acanthamoeba* IgG was measured by enzyme-linked immunosorbent assay. The delayed type hypersensitivity response to *Acanthamoeba* cyst and trophozoite antigens were used to determine the cell-mediated immune responses against *Acanthamoeba* cysts. *Acanthamoeba* cysts were both immunogenic and antigenic, producing anti-*Acanthamoeba* serum IgG, T-lymphocyte proliferation, and delayed type hypersensitivity responses. These results indicate that *Acanthamoeba* cysts are recognized by the immune system. The persistence of the organism in the human cornea means that these adaptive immune responses fail to kill *Acanthamoeba* cysts (McClellan *et al.*, 2002). Overall, it is suggested that the low numbers of cysts in infected corneas combined with their location in an immunologically privileged site contribute to their failure to stimulate the adaptive immune response. This is supported with findings that both macrophages and neutrophils are not chemotactically attracted to intact cysts suggesting the ability of *Acanthamoeba* cysts to lie dormant in the conjunctiva and the stroma without being detected and eliminated by the immune system.

In addition to the development of cysts, it is shown that *Acanthamoeba* serine proteases are known to degrade sIgA *in vitro*. Furthermore, it is shown that the pathogenic *Acanthamoeba* can resist the complement-mediated lysis *in vitro*.

3.4. Macrophages *in vivo*

The role of conjunctival macrophages in *Acanthamoeba* keratitis infection is studied in a Chinese hamster model of this disease. Initially, a selective elimination of macrophages was achieved by repeated subconjunctival injection of liposomes containing dichloromethylene diphosphonate in a Chinese hamster model of *Acanthamoeba* keratitis. The incidence of infection in normal animals was approximately 60% but rose to 100% on day 4 in animals treated with liposomes containing dichloromethylene diphosphonate. Moreover, the clinical appearance of the keratitis in animals treated with liposomes containing dichloromethylene diphosphonate was much more severe than in animals treated with liposomes containing phosphate-buffered saline alone, at all time points. There was also a major change in the chronicity of keratitis, with an earlier onset and a prolonged and chronic course in the animals treated with liposomes containing dichloromethylene diphosphonate.

Furthermore, the conjunctival macrophages were activated via subconjunctival injection of liposomes containing supernatants from concanavalin A-stimulated spleen cell cultures resulted in rapid resolution of the corneal infection in the Chinese hamster model of this disease suggesting that activated macrophages eliminate *Acanthamoeba in vivo*. Approximately 80%

of animals treated with the phosphate buffered saline-containing liposomes demonstrated evidence of the corneal disease at day 14, compared with 10% incidence of infection in the concanavalin A-treated group. Moreover, at all time points examined, the clinical appearance of the keratitis in animals treated with liposomes containing concanavalin A supernatant was significantly reduced compared to the group treated with liposomes containing phosphate buffered saline ($P<0.05$) (Van Klink *et al.*, 1996).

3.5. Neutrophils *in vivo*

The role of neutrophils in *Acanthamoeba* keratitis was determined *in vivo*. The inhibition of neutrophils recruitment was achieved by subconjunctival injection with an antibody against macrophage inflammatory protein-2, a powerful chemoattractant for neutrophils which is secreted by the cornea. In other experiments, neutrophils were depleted by the intraperitoneal injection of anti-Chinese hamster neutrophil antibody. The inhibition of neutrophils to the cornea resulted in an earlier onset and more severe infection, compared with the control group. Anti-macrophage inflammatory protein-2 antibody treatment produced an almost 35% reduction of myeloperoxidase activity in the cornea 6 days post-infection, while levels of the endogenous macrophage inflammatory protein-2 secretion increased significantly. The recruitment of neutrophils into the cornea via the intrastromal injections of recombinant macrophage inflammatory protein-2 generated an initially intense inflammation that resulted in the rapid resolution of the corneal infection. The profound exacerbation of *Acanthamoeba* keratitis seen when neutrophil migration was inhibited, combined with the rapid clearing of the disease in the presence of increased neutrophils, strongly suggests that neutrophils play an important role in combating *Acanthamoeba* infections in the cornea (Hurt *et al.*, 2001).

3.6. Macrophages and neutrophils *in vitro*

Despite aggressive treatment regimen, infection recurrence does occur. This is due to the ability of parasite to transform into a resistant cyst form, which can lie dormant for long period of time. Even with the extensive damage control therapeutic measures such as the corneal transplant, *Acanthamoeba* cysts may not be eliminated from the conjunctiva and the stroma of humans and can excyst, resulting in infection of the corneal transplant. In one study, it was determined whether elements of the innate immune apparatus, neutrophils and macrophages, were capable of detecting and eliminating *Acanthamoeba* cysts and the mechanism by which they kill the cysts were examined. Results showed that neither macrophages nor neutrophils were attracted chemotactically to the intact cysts, yet both were attracted to the lysed cysts. Both macrophages and neutrophils were capable of killing significant numbers of cysts, yet neutrophils were 3-fold more efficient than macrophages. The activation of macrophages with lipopolysaccharide and interferon-gamma did not increase their cytolytic ability. The conditioned medium isolated from macrophages did not lyse the cysts; however, prevention of phagocytosis by cytochalasin D inhibited 100% of macrophage-mediated killing of cysts.

The conditioned medium from neutrophils did kill significant numbers of the cysts, and this killing was blocked by quercetin, a potent inhibitor of the myeloperoxidase. These results indicate that neither macrophages nor neutrophils are chemoattracted to intact cysts, yet both are capable of killing the cysts. Macrophages killed the cysts by phagocytosis, whereas neutrophils killed cysts through the secretion of myeloperoxidase (Hurt *et al.*, 2002).

In contrast, macrophages stimulated with interferon-gamma supernatants killed significant numbers of the trophozoites *in vitro* ($P<0.05$). The killing was inhibited by cytochalasin D, but not by L-N6-1-iminoethyl-L-lysine dihydrochloride (L-NIL), which is a selective inhibitor of the inducible nitric oxide synthase (Alizadeh *et al.*, 2007).

3.7. *Acanthamoeba* granulomatous encephalitis and the immune response

It is accepted that *Acanthamoeba* granulomatous encephalitis develops into a full blown disease with a combined action of the pro-inflammatory cytokines and by destruction of the host cells by *Acanthamoeba*. The immunization with *Acanthamoeba* antigens using intranasal, intraperitoneal, intravenous or oral routes of administration protected animals from *Acanthamoeba* granulomatous encephalitis. Of note, *Acanthamoeba* antigens were prepared using several methods including, *Acanthamoeba* conditioned medium, whole cell sonicate, freezed-thawed extract, or live parasites. It was shown that *Acanthamoeba* lysates prepared by sonication provided the best protection. Also noteworthy is that protection was specific, in that immunization with one species of *Acanthamoeba* did not protect mice against challenge with other species suggesting the expression of varied antigens / toxins.

An overview of the immune response during the course of the infection in an *in vivo* model of *Acanthamoeba* granulomatous encephalitis infection is described below: The mice infected with 3×10^3 trophozoites showed mortality rate of 17%, and 34% in the mice infected with 10^4 trophozoites and 65% with 10^5 trophozoites. All animals showed cell-mediated immune responses (i.e., delayed type hypersensitivity) responses with the level increasing on day 7, and declining on day 14 post-infection. Interestingly, the cytotoxic activity of the natural killer cells was activated on day 1 post-infection, and declined to the level of control group on day 2, in all experimental groups. On day 5 post-infection, the natural killer cell cytotoxicity was significantly suppressed as compared with the control group. The serum antibody titers of the infected mice increased after day 7, but there was no statistical difference between the three infected groups. Overall, there was a significant difference in the cell-mediated immune responses between the infected and the control mice. It is considered that the cell-mediated immune responses should be involved in the murine model infected with *Acanthamoeba* (Kim *et al.*, 1990).

3.7.1. Complement

In support of the complement activation in response to *Acanthamoeba*, previous studies have shown that the normal human serum exhibited the complement-mediated lysis of *Acanthamoeba* (Ferrante and Rowan-Kelly, 1983; Ferrante, 1991; Stewart *et al.,* 1992). This is perhaps due to the fact that the plasma membranes of *Acanthamoeba* lack sialic acid (Korn and Olivecrona, 1971) or any protective coat or capsule (Bowers and Korn, 1968) and thus *Acanthamoeba* are exposed to the complement-mediated killing (Ferrante and Rowan-Kelly, 1983). However, recent studies have shown that a sub-population of the virulent strains of *Acanthamoeba* are resistant to the complement-mediated lysis, perhaps due to the expression of complement regulatory proteins including decay accelerating factor on the trophozoite cell membranes that may disable complement cascade and renders the parasite resistant to complement-mediated lysis (Sissons *et al.,* 2005c; Toney and Marciano-Cabral, 1998). *In vitro* studies have demonstrated that *Acanthamoeba* activates the alternative complement pathway, which is independent of the antibody activation. However, in immunocompetent individuals, antibodies and the complement components promote recognition of *Acanthamoeba* by neutrophils, macrophages and probably lymphocytes. Other studies have shown that *Acanthamoeba* binds to the C1q that would block binding of C1q to C1, the first component of the classical complement cascade. This would result in the inhibition of the classical complement pathway.

But the complement pathway in the presence of phagocytes is highly effective in clearing *Acanthamoeba* (Stewart *et al.,* 1994). Overall, the complement pathways and antibodies together with neutrophils and macrophages show potent lytic activities against *Acanthamoeba* thus suppressing the infection by clearing *Acanthamoeba* (Toney and Marciano-Cabral, 1998; Marciano-Cabral and Toney, 1998). It is important to indicate that although macrophages and/or neutrophils from the naïve animals are able to destroy *Acanthamoeba*, but cells from the immune animals exhibit significantly increased lytic activities against *Acanthamoeba*. The macrophage-mediated killing is contact-dependent and is inhibited with cytochalasin D, i.e., an actin polymerization inhibitor (van Klink *et al.,* 1997). This is further confirmed with the findings that the conditioned medium obtained from macrophages after treatment with lipopolysaccharide and interferon-gamma had no effects on *Acanthamoeba* (Marciano-Cabral and Toney, 1998). Similarly, the neutrophils exhibit potent amoebicidal effects. The neutrophils-mediated killing is significantly increased in the presence of anti-Acanthamoebic antibodies (Stewart *et al.,* 1994). These interactions also stimulate secretion of pro-inflammatory cytokines including interleukin-1-beta, interleukin-6 and tumor necrosis factor-alpha (Toney and Marciano-Cabral, 1998; Marciano-Cabral and Toney, 1998). Other studies in mice have shown significant increased natural killer cells activities in *Acanthamoeba*-infected animals suggesting that natural killer cells may also play a role in the protective immunity (Kim *et al.,* 1993). The fact that up to 100% individuals of a given population may possess anti-*Acanthamoeba* antibodies suggests that immunocompetent individuals exhibit *Acanthamoeba*-specific T-cell responses (in particular CD4 T-helper cells) (Cursons *et al.,* 1980; Tanaka *et al.,*

1994). Based on these findings, it is clear that the debilitated immune status of the host is usually a pre-requisite in *Acanthamoeba* encephalitis but the core basis of the host susceptibility in contracting *Acanthamoeba* encephalitis are not fully understood and may also involve the host ethnic origin (i.e., genetic basis of the host) or the inability of the host to induce specific immune response against these pathogens. Of interest, both *Acanthamoeba* cysts and trophozoites are immunogenic, and immune sera from animals infected with *Acanthamoeba* trophozoites cross-react with cysts suggesting that some of the antigenic epitopes are retained during the encystation process (McClellan *et al.*, 2002). Future studies will identify the precise host factors, which play an important role in controlling this fatal infection and may help develop therapeutic interventions in the susceptible hosts.

3.7.2. Cytokines

In the central nervous system, the cytokine-primed microglial cells (the resident macrophages of the brain) play a pivotal role in the host defence against parasite infections. The purified recombinant cytokines including tumor necrosis factor-alpha, interleukin-1-alpha, and interleukin-1-beta, alone or in combination, were not cytolytic for *Acanthamoeba* trophozoites. The conditioned medium obtained from the macrophage cultures has neither trophistatic nor trophicidal for *Acanthamoeba* spp. Furthermore, the conditioned medium obtained from the activated macrophage cultures after treatment with lipopolysaccharide and interferon-gamma was neither trophistatic nor trophicidal for *Acanthamoeba* spp.

However, the co-cultures of *Acanthamoeba* and the activated macrophages (by prior treatment with the recombinant interferon-gamma and lipopolysaccharide) triggered, in a dose-dependent manner, trophistatic activity and the release of the endogenous interleukin-1alpha, interleukin-1-beta, interleukin-6 and tumor necrosis factor-alpha. More than 88 or 95% of trophistatic activity was abrogated by anti- interleukin-6 or tumor necrosis factor-alpha neutralizing antibodies, suggesting that these endogenously produced cytokines may participate in the antimicrobial capacity. Consistent with these findings, the priming of macrophages with recombinant interferon-gamma and lipopolysaccharide elicited the release of pro-inflammatory interleukin-1-alpha, interleukin-1-beta, interleukin-6-and tumor necrosis factor-alpha. Since L-canavanine affected amebistatic activity only during the priming process but not during the infection process (co-culture), nitric oxide-dependent pathway appears to be not the sole anti-parasitic mechanism involved in this function. These data suggest that the activation of macrophages upregulate the release of the endogenous tumor necrosis factor-alpha, interleukin-6 and interleukin-1-beta by these cells, which could trigger antimicrobial activity against *Acanthamoeba* infection in the brain (Benedetto *et al.*, 2003).

However, the effects of cytokine and cytokine-primed microglia (macrophages) on *Acanthamoeba* may be varied depending on the virulence of the parasites. For example, it is shown that *Acanthamoeba* activated macrophages (microglial cells) were capable of phagocytozing weakly virulent *Acanthamoeba*. In contrast, the virulent *Acanthamoeba* showed cytopathic effects on microglia and destroyed microglial cells (Marciano-Cabral *et al.*, 2004; Shin *et al.*, 2001).

3.7.3. T-lymphocytes

The T-cell responses to *Acanthamoeba* antigens in healthy human subjects were examined. It was interesting to note that the T-cell clones derived from the asymptomatic healthy individuals proliferate in response to *Acanthamoeba* antigens. The T-cell clones were categorized as being CD4 T-helper cell Type 1 (T_h1) involved in cell-mediated immune response (i.e., delayed-type hypersensitivity) reactions in that they produced interferon-gamma in response to amoebic antigens. These findings led to the conclusions that the low incidence of symptomatic disease in humans might be due to acquired protective immunity induced by T-cell responses.

3.74. Evasion of cytokines

A secretory protease from the pathogenic *Acanthamoeba* was shown to degrade interleukin-1-alpha and interleukin-1-beta. Its activity was not inhibited by the endogenous protease inhibitors, such as alpha 2-macroglobulin, alpha 1-trypsin inhibitor, and alpha 2-anti-plasmin. The degradation of host's defense-oriented or regulatory proteins by the *Acanthamoeba* protease suggested that the enzyme might be an important virulence factor in the pathogenesis of *Acanthamoeba* infection (Na *et al.*, 2002; Marciano-Cabral and Toney, 1998; Benedetto and Auriault, 2001).

3.7.5. Evasion of antibodies

A secretory protease from the pathogenic *Acanthamoeba* was shown to degrade secretory sIgA, IgG, and IgM. Its activity was not inhibited by endogenous protease inhibitors, such as alpha 2-macroglobulin, alpha 1-trypsin inhibitor, and alpha 2-anti-plasmin. The degradation of host's defence-oriented or regulatory proteins by the *Acanthamoeba* protease suggested that the enzyme might be an important virulence factor in the pathogenesis of *Acanthamoeba* infection (Na *et al.*, 2002).

3.7.6. Evasion of complement

In a test to evaluate the susceptibility of *Acanthamoeba* species to complement lysis, it was shown that *Acanthamoeba* depleted hemolytic complement activity from the normal human serum, yet were resistant to its lytic effects. The examination of the microtiter plates containing *Acanthamoeba* incubated in the normal human serum demonstrated formation of a pellet in the wells. The pellet formation was not observed when *Acanthamoeba* was incubated in the human cord serum, heat-inactivated serum, or C1q-deficient serum. Ultrastructural examination of the serum-treated *Acanthamoeba* revealed the presence of a finely granular substance that surrounded the *Acanthamoeba*. Treatment of *Acanthamoeba* with enzymes or metabolic inhibitors prior to incubation in the normal human serum was performed to investigate the mechanism of the complement resistance. Cycloheximide or cytochalasin D pretreatment increased the susceptibility of *Acanthamoeba* to the complement lysis. Inhibition of serine protease activity by phenylmethylsulfonyl fluoride (PMSF) increased complement susceptibility of *Acanthamoeba*. Enzymatic removal of surface components from *Acanthamoeba*, with trypsin, neuraminidase, or phosphatidylinositol-specific phospholipase C, did not affect serum resistance. In contrast, phosphatidylinositol-specific phospholipase C treatment of *Acanthamoeba* significantly increased lysis by complement. The ability of *Acanthamoeba* species to activate the alternative complement pathway yet resist complement-mediated cellular lysis can be attributed to both the release of a transport-dependent extracellular matrix as well as the presence of complement inhibitory surface proteins (Toney and Marciano-Cabral, 1998).

3.7.7. Evasion of macrophages

The virulent *Acanthamoeba* destroys macrophage cultures *in vitro*. Although the activated primary peritoneal macrophages were susceptible to destruction by *Acanthamoeba*, they were more resistant to *Acanthamoeba*-mediated destruction than macrophage cell lines activated *in vitro*. However, the activated primary peritoneal macrophages also were capable of destruction of *Acanthamoeba* (Marciano-Cabral and Toney, 1998). The fact that these cells comprise the major cellular component of granulomatous suggests that they play an important role compared with the neutrophils. Furthermore, treatment of mice with macrophage-activating factors (immuno-potentiating agents, muramyl dipeptide and trehalose dimycolate) prior to infection protected up to 40 and 30% mice respectively against a lethal infection of *Acanthamoeba*.

3.7.8. Acanthamoeba-specific antibodies and activated macrophages, in combination show trophicidal effects

From above, it was clear that activated macrophages alone are susceptible to *Acanthamoeba*-mediated lysis. To determine whether *Acanthamoeba*-specific antibodies affect the function of activated macrophages, assays were performed. It is noteworthy that *Acanthamoeba*-specific antibodies increased the trophicidal activity of activated macrophages. Macrophage-mediated destruction was by contact-dependent cytolysis and by ingestion of *Acanthamoeba* (Marciano-Cabral and Toney, 1998).

3.7.9. Evasion of neutrophils

It is shown that the neutrophils alone lacked trophicidal activity. Similarly, in the presence of serum from the naïve animals, neutrophils showed no trophicidal activity. Even in the presence of serum, from immunized animals which contained anti-*Acanthamoeba* antibodies, neutrophils showed poor trophicidal activity. Pre-exposure of neutrophils to supernatants from peripheral blood mononuclear leukocytes (that secrete mediators known as lymphokines) had no significant effect on their trophicidal properties even in the presence of serum containing anti-*Acanthamoeba* antibodies. However, neutrophils pre-exposed to supernatants from peripheral blood mononuclear leukocytes, which had been stimulated with phytohemagglutinin showed significant trophicidal effects in the presence of serum containing anti-*Acanthamoeba* antibodies (Ferrante and Abell, 1986; Stewart *et al.*, 1994).

3.7.10. Natural killer cell activity in vivo

Using *in vivo* studies using mice, it was shown that mice infected with *Acanthamoeba* show elevated natural killer cell activity. The cytotoxic activity of natural killer cells of the *Acanthamoeba*-infected mice was significantly higher than that of non-infected mice from the 12 hour to the 48 hour after infection, showing the highest activity on the first day. On day 10 post-infection, the cytotoxic activity of natural killer cells was significantly suppressed as compared with that of the control group (Hyun *et al.*, 1992).

4. Effect of immune suppressive component in marijuana, cannabinoid delta-9-tetrahydrocannabinol on *Acanthamoeba* granulomatous encephalitis

Using a murine model of *Acanthamoeba* granulomatous encephalitis, it is shown that the major psychoactive and immune suppressive component in marijuana, cannabinoid delta-9-tetrahydrocannabinol exacerbates infection by these amoebae. Mice administered delta-9-tetrahydrocannabinol and infected with *Acanthamoeba* exhibited dose-related higher mortalities than the control group. The greater severity of disease for delta-9-tetrahydrocannabinol-treated mice was accompanied by decreased accumulation of macrophage-like cells at focal sites of infection in the brain. The cannabinoid act through cannabinoid receptor, CB2 receptor (CB2R) primarily expressed on microglia, a resident population of cells in the central nervous system. Microglia also expresses the cannabinoid receptor, CB1R. The CB1R is constitutively expressed at low levels while the CB2R is expressed at higher levels and is modulated in relation to cell activation state. The expression of the high levels of the CB2R corresponds with microglia being activated. In particular, the activated microglia exhibit response such as chemotaxis and antigen processing. Notably, the cannabinoid delta-9-tetrahydrocannabinol inhibits the chemotactic response of microglia to *Acanthamoeba*, through activation of the CB2R. These findings suggest that delta-9-tetrahydrocannabinol exacerbates infection by *Acanthamoeba* through CB2R. These findings also suggest that the CB2R plays a critical immune functional role in the *Acanthamoeba* granulomatous encephalitis. Furthermore, delta-9-tetrahydrocannabinol administration resulted in decreased levels of mRNA for the pro-inflammatory cytokines interleukin-1 alpha, interleukin-1 beta, and tumor necrosis factor alpha for neonatal rat microglia co-cultured with *Acanthamoeba*. These results indicate a potential for marijuana to alter the capacity of brain macrophage-like cells to mount a full complement of immune responsiveness to brain infection by opportunistic amoebae (Cabral and Marciano-Cabral, 2004; Cabral *et al.*, 2008).

From: *Acanthamoeba*: Biology and Pathogenesis. By: Naveed Khan.

Section F

Strategies Against *Acanthamoeba* Infections

1. Introduction

Antimicrobial chemotherapy remains the most widely used method of treatment against microbial infections. Antimicrobial agents are used to treat viral, bacterial, fungal and protozoan infections. However, protozoa are eukaryotes and they have metabolic processes close to the human host, compared with those of prokaryotic pathogens, thus protozoan infections are less easily treated than bacterial infections. Consequently, anti-protozoal drugs exhibit unwanted side effects. In addition, the acquisition of drug resistance is a constant threat due to current limited availability of drugs. In this chapter, I have described various antimicrobial agents that have been tested against *Acanthamoeba* spp. and their mode of action has been described further.

2. Chemotherapeutic agents and *Acanthamoeba*

A number of chemotherapeutic agents and plant extracts have been tested *in vitro* for anti-Acanthamoebic activity. However, there has been a great deal of conflicting results. This is probably due to the followings:

i) varied levels of growth rates, and/or cytopathic potential (which may result in varied levels of virulence) in *Acanthamoeba* isolates.
ii) the use of different isolates / species / genotypes.
iii) a lack of standardized *in vitro* method for the assay of trophicidal and cysticidal activity.

Perhaps it is the combination of the above (together with pharmacokinetics) that account for the lack of correlation between the *in vitro* activity and the *in vivo* efficacy of a given antimicrobial. Thus care should be taken in the interpretation of findings, before an antimicrobial compound is recommended for clinical use (Osato *et al.*, 1991). Further complexity is attributed due to the fact that the degree of virulence in *Acanthamoeba* isolates is not constant but it changes with time, culture conditions, etc. For example, long term axenic culturing attenuates *Acanthamoeba* virulence. No single drug has been shown to be effective against both the trophozoite stage and the cyst stage of *Acanthamoeba* of all known genotypes. Furthermore, *Acanthamoeba* infections are not readily recognized by clinicians or pathologists because many patients present with other underlying disease processes. Due to these reasons, in human infections, combination therapies have proven more successful than a single drug therapy. Apart from the limited usefulness of combination therapy, early treatment is highly desirable for the successful prognosis of *Acanthamoeba* infections.

The various antimicrobial agents tested against *Acanthamoeba* are described here. Because there are so many, it is convenient to group them according to their mode of action. It is noteworthy that any novel antimicrobial must be tested for the followings effects to determine its true potential as an anti-Acanthamoebic agent:

i) Trophicidal effects, i.e., killing of the amoebae trophozoites. The potency of such compounds is expressed as the "minimum trophicidal concentration" (i.e., lowest concentration of a test compound which when exposed for a specified time period to a suspension of trophozoites in culture leaves no culturable (i.e., viable) trophozoites).
ii) Cysticidal activities, i.e., killing of amoeba cysts. The potency of such compounds is expressed as the "minimum cysticidal concentration" (lowest concentration of a test compound which when exposed for a specified time period to a suspension of cysts leaves no culturable (i.e., viable) cysts. Concentration of organisms must be carefully controlled in these experiments.
iii) Trophistatic effects, i.e., growth inhibitory effects on trophozoites. The potency of such compounds is expressed as the "minimum trophistatic concentration" (i.e., lowest concentration of a test compound in the growth medium that allows no growth of a suspension of trophozoites. The effectiveness of a compound is described as inhibitory concentration (IC).
iv) "Cystistatic effects", i.e., inhibition of excystation. The potency of such compounds is expressed as the "minimum cystistatic concentration" (i.e., lowest concentration in culture medium of a test compound where a suspension of cysts shows no emergence of trophozoites (and/or growth, in the growth medium).

The above anti-Acanthamoebic effects should be examined to determine the potential use of an antimicrobial agent in the treatment of *Acanthamoeba* infections. Varieties of antimicrobials have been tested against *Acanthamoeba*. These included membrane-acting agents, the RNA and the DNA synthesis inhibitors or metabolic blockers as described below.

3. Membrane-acting agents

3.1. Polyhexamethylene biguanide

Polyhexamethylene biguanide (also known as polyaminopropyl biguanide or Baquacil by Zaneca Pharmaceuticals) is used as an environmental disinfectant, often used as a disinfectant in the swimming pools (at 45 parts per million) and as a preservative in the cosmetics and pharmacological formulations. *In vitro* assays revealed that polyhexamethylene biguanide (0.02%) showed both trophicidal and cysticidal effects. Importantly, toxicity to the ocular surface is shown to be limited with polyhexamethylene biguanide, compared with other drugs of choice. These findings suggested the potential use of polyhexamethylene biguanide in the treatment of *Acanthamoeba* keratitis. It is not surprising that polyhexamethylene biguanide is a key antimicrobial compound in the treatment of *Acanthamoeba* keratitis (Larkin *et al.*, 1992; Burger *et al.*, 1994). Indeed, the introduction of polyhexamethylene biguanide as a 0.02% topical solution, either alone or in combination therapy, has improved the prognosis of *Acanthamoeba* keratitis and studies have confirmed its *in vitro* and *in vivo* superiority over other treatments. Although the precise mode of action for polyhexamethylene biguanide is not known, however *Acanthamoeba* treated with this drug show structural and membrane damage and the cytoplasmic contents are severely depleted with the appearance of densely stained precipitates on the cell surface. In addition, polyhexamethylene biguanide induces a slight withdrawal of the cytoplasm from the *Acanthamoeba* cyst wall and, induces swelling of the cysts. These findings suggest that a major target site for polyhexamethylene biguanide is the plasma membrane. However, additional intracellular damage undoubtedly contributes to the lethal effects of this potent anti-Acanthamoebic compound.

3.2. Chlorhexidine

Like polyhexamethylene biguanide, chlorhexidine is a biguanide, and success has also been recorded with the topical chlorhexidine digluconate (0.02%), which has also been shown to have trophicidal and cysticidal activities. Chlorhexidine is a "membrane-acting" cationic biocide. At alkaline pH, the surface proteins of *Acanthamoeba* are negatively charged, interacting rapidly with this cationic biocide, inducing structural and permeability changes in the cell membrane leading to the leakage of ions, water and other cytoplasmic components resulting in the cellular damage (Perrine *et al.*, 1995). The concentration of chlorhexidine greater than 100 μg per mL produces shrinkage from the *Acanthamoeba* cyst wall. *In vitro* studies have shown that chlorhexidine is more toxic to human cells, than polyhexamethylene biguanide. Despite the value of these compounds in the treatment of *Acanthamoeba* keratitis, failures still occur resulting in infection recurrence, necessitating prolonged and intensive medical therapy, surgical intervention, and the possible permanent loss of visual acuity or enucleation of the eye.

3.3. Acriflavine hydrochloride and proflavine

Acriflavine hydrochloride (also used as protein kinase C inhibitor) binds to the plasma, and the nuclear membranes, and the cellular membranes because of its interaction with polyanionic substances. Thus, acriflavine hydrochloride and its related compound such as proflavine perturb the lipid bilayer and consequently perturb the function of the membrane-associated proteins, including protein kinase C, through modulation of the cell membrane integrity. Although the reported findings vary greatly between different isolates of *Acanthamoeba*, the minimum trophicidal concentration of acriflavine hydrochloride against *Acanthamoeba* was observed at 100 µg per mL against the virulent isolate, while cysticidal concentrations were significantly higher (Nagington and Richards, 1976).

The minimum trophicidal concentration of proflavine, a derivative of acriflavine, against *Acanthamoeba* trophozoites was observed at 100 µg per mL, while cysticidal concentration was more than 1,000 µg per mL, which is well over the recommended levels (Nagington and Richards, 1976).

3.4. Phospholipid analogues

Phospholipid analogues are shown to successfully inhibit the growth of *Acanthamoeba in vitro*. The relative activities of the phospholipid derivatives hexadecyl-PC, octadecyl-PC, elaidyl-PC, erucyl-PC and edelfosine, against *Acanthamoeba* and corneal epithelial cell cultures were tested. Both hexadecyl-PC and octadecyl-PC [IC50 (concentration that inhibits growth by 50%) values at 35 µM)] demonstrated considerable activity against *Acanthamoeba*, as did elaidyl-PC (IC50 values at 65 µM). In contrast, neither erucyl-PC nor edelfosine was inhibitory against *Acanthamoeba*. Notably, the growth of corneal epithelial cells was inhibited by octadecyl-PC, erucyl-PC and edelfosine. Notably, liposomal formulations of hexadecyl-PC and elaidyl-PC were less effective than free drug against *Acanthamoeba* (McBride *et al*., 2007). In addition, *Acanthamoeba* is shown to be sensitive to mefloquine (a phospholipid-interacting antimalarial drug) (Ferante *et al*., 1984; Schuster and Mandel, 1984).

3.5. Polymyxin B

Polymyxin B acts like cationic detergents to disrupt cell membranes. It is shown to alter the cytoplasmic membrane permeability by binding to a negatively charged site, which has an electrostatic attraction for the positively charged amino groups in the cyclic peptide portion and disrupts membrane integrity resulting in leakage of the cytoplasmic contents and inhibition of the cellular respiration, particularly against bacteria such as *Pseudomonas* spp.

Using *in vitro* assays, it was observed that polymyxin B exhibited trophicidal effects at a concentration of 50 µg per mL for clinical isolates of *Acanthamoeba*. While cysticidal concentrations of polymyxin B were more than a 1,000 µg per mL, which are well over the recommended levels (Nagington and Richards, 1976; Duma and Finley, 1976). Furthermore, unwanted toxicity of polymyxin B against human cells has limited its usefulness. Notably, the effects of polymyxin B vary significantly between different isolates / species / genotypes of *Acanthamoeba* spp.

3.6. Polymyxin E (colistin)

Polymyxin E (also known as colistin) is a polycationic and has both hydrophilic and lipophilic moieties. As for polymyxin B, this compound interacts with the microbial cytoplasmic membrane, changing its permeability and exhibits lytic effects. Using *in vitro* assays, it was observed that the minimum trophicidal concentration of polymyxin E at 48 hours is 62.5 µg per mL, while cysticidal concentration at 48 hours is more than 500 μg per mL, which is well over the recommended levels, however these findings vary significantly between different isolates / species / genotypes of *Acanthamoeba* spp. (Kilvington *et al*., 1990). Again, polymyxin E exhibits unwanted toxicity against human cells, which has limited its usefulness.

3.7. Caspofungin

At present, caspofungin is used as an anti-fungal drug, against infections with *Aspergillus*, *Histoplasma* and *Candida*. It is a member of a class termed the echinocandins and targets cell wall. It works by inhibiting the enzyme β-D-glucan synthase and thereby interferes with the β-glucan synthesis. The β-glucan is a basic subunit that is incorporated into the cell wall of fungi. Notably, the cyst stage of various protozoa also contains β-glucan as a basic constituent. For example, the inner wall of *Acanthamoeba* cysts contains cellulose, where the basic subunit is β-glucan. Thus interference with the β-glucan synthesis disturbs the integrity of the cell wall. Thus caspofungin may be a useful compound in the inhibition of encystation and may have potential in a combination therapy.

In vitro effectiveness of caspofungin against *Acanthamoeba* has been tested. It was shown that the trophozoites were susceptible *in vitro* to caspofungin at a concentration of 206 μM. Furthermore, this drug was shown to be cysticidal at a concentration of more than 400 μM against *Acanthamoeba*, while its effects on *Acanthamoeba* encystation have not been determined. However, at 50 μM, caspofungin showed toxicity against human cells (Bouyer *et al*., 2007).

3.8. Ergosterol inhibition

Acanthamoeba is capable of synthesizing its sterols *de novo* from acetate. One of the major sterols in the membrane of *Acanthamoeba* is ergosterol. In contrast, human cells possess cholesterol. These differences present a potential target for the rational development of therapeutic interventions. To this end, various azole compounds that target ergosterol biosynthesis have been tested including, ketoconazole, miconazole, and fluconazole. In general, the anti-Acanthamoebic effects of azole compounds against clinical isolates are trophistatic rather than trophicidal or cysticidal i.e., growth needs to continue for the effect to be shown. Given that ergosterol is a major constituent of *Acanthamoeba* cell membrane and may play an important role in the integrity of the cell, it is likely that ergosterol biosynthesis is important in the encystation of *Acanthamoeba*, supporting the use of azoles in therapy (that targeting ergosterol biosynthesis pathway may provide additive effects in the improved treatment of *Acanthamoeba* infections).

3.8.1. Ketoconazole

Ketoconazole is known to interfere with the biosynthesis of ergosterol, which is an important structural sterol component of the membrane of *Acanthamoeba*, thus leading to defective membranes, increased permeability, and leakage of ions from the cell. It targets the ergosterol pathway and has limited effects on the mammalian pathway because they synthesize cholesterol. As with other azole anti-fungal agents, ketoconazole works principally by inhibition of an enzyme, cytochrome P450 14-alpha-demethylase. This enzyme is in the sterol biosynthesis pathway that leads from lanosterol to ergosterol. *In vitro* studies revealed that ketoconazole exhibits limited trophicidal and cysticidal effects but showed trophistatic effects.

3.8.2. Miconazole

As for ketoconazole, miconazole is another sterol-targeting compound that inhibits 14 alpha-demethylase, an enzyme required for the biosynthesis of ergosterol, which is the main sterol constituent of the fungal cell membranes, affecting the membrane integrity. In addition, miconazole is shown to affect synthesis of triglycerides and phospholipids, as well as altered oxidative and peroxidative enzyme activities, which may contribute to the deterioration of sub-cellular organelles and to cell necrosis. *In vitro* studies have shown that miconazole is effective against another free-living amoeba, *Naegleria* at 80 μg per mL. In contrast, the minimal trophicidal concentration of miconazole nitrate when incubated for 48 hours is shown to be 125 μg per mL for clinical isolates of *Naegleria* and cysticidal concentrations were observed to be more than 500 μg per mL. However with *Acanthamoeba*, miconazole showed limited (anti-Acanthamoebic) effects (Duma and Finley, 1976; Nagington and Richards, 1976).

3.8.3. Clotrimazole

Clotrimazole essentially inhibits the synthesis of ergosterol by targeting fungal cytochrome P450, which decreases fungal cell wall integrity. In addition, it is shown to interfere with the cellular calcium homeostasis and inhibits the calcium ATPase. *In vitro* studies have shown that clotrimazole is effective against *Naegleria* at 50 μg per mL, however it had limited effects against *Acanthamoeba* (Nagington and Richards, 1976; Duma and Finley, 1976).

3.8.4. Fluconazole

The minimum trophistatic concentrations of fluconazole against *Acanthamoeba* observed at 48 hours was 62.5 μg per mL, while cysticidal concentration at 48 hours was observed to be more than 1,000 μg per mL, which is well over the recommended levels, suggesting its limited usefulness in the treatment of *Acanthamoeba* infections (Kilvington *et al*., 1990).

3.8.5. Amphotericin B

Unlike the azole compounds that target biosynthesis of ergosterol, amphotericin B binds to ergosterol in the membrane. This binding disrupts osmotic integrity of the cell membrane, resulting in leakage of the intracellular potassium, magnesium, sugars, and metabolites and then cellular death. Amphotericin B is the most effective drug against *Naegleria* (effective *in vitro* at concentration of 2 μg per mL). In contrast, amphotericin B showed trophistatic effects, when incubated for 48 h, at more than 8 μg per mL for clinical isolates of *Acanthamoeba*, but showed minimal trophicidal and cysticidal effects (Duma and Finley, 1976; Nagington and Richards, 1976).

3.9. Benzimidazole and Benzotriazole

Benzimidazole is a heterocyclic aromatic organic compound that consists of the fusion of benzene and imidazole, while benzotriazole is an aromatic azole containing two fused rings. These compounds are known to inhibit the growth of certain microorganisms. Although their precise mode of action is unknown, they are implicated to block protein kinase CK2, SARS Coronavirus Main Proteinase, increase oxidative stress associated with mitochondrial damage. Chloro-, bromo- and methyl- analogues of 1H-benzimidazole and 1H-benzotriazole and their *N*-alkyl derivatives were tested *in vitro* against *Acanthamoeba*. It was revealed that 5,6-dimethyl-1H-benzotriazole and 5,6-dibromo-1H-benzotriazole showed anti-Acanthamoebic effects (Kopańska *et al.*, 2004).

4. Inhibitors of DNA synthesis and polyamine metabolism

The polyamines are organic compounds having two or more primary amino groups, such as putrescine, spermidine, and spermine that are growth factors in both eukaryotic and prokaryotic cells. Polyamines are synthesized in cells via a highly regulated pathways, their actual function is not entirely clear but they are thought to bind to DNA. If cellular polyamine synthesis is inhibited, cell growth is stopped or severely retarded. The provision of exogenous polyamines restores the growth of these cells. Most eukaryotic cells have a polyamine transporter system on their cell membrane that facilitates the internalization of exogenous polyamines. This system is highly active in rapidly proliferating cells and is the target of some chemotherapeutics currently under development. Polyamines are also important modulators of a variety of ion channels, including NMDA receptors and AMPA receptors. They block inward-rectifier potassium channels so that the currents of the channels are inwardly rectified, thereby the cellular energy, i.e. K^+ ion gradient across the cell membrane, is conserved.

4.1. Diamidine derivatives

The exact mechanism of the anti-protozoal action of aromatic diamidines is unknown, though they are thought to act on polyamine synthesis, as well as may involve reactions with ubiquitin, and have been shown to inhibit DNA synthesis. When tested *in vitro*, it was observed that propamidine isethionate exhibited minimum trophicidal concentration within 48 hours of 10 μg per mL, and the minimum cysticidal concentration at 48 hours was 125 μg per mL against clinical isolates of *Acanthamoeba*. However, propamidine isethionate showed high levels of host cell toxicity. Interestingly, *in vitro*, the combination of cryotherapy and propamidine isethionate was more cysticidal than either treatment alone and eliminated detectable viable cysts (Matoba *et al.*, 1989; Larkin *et al.*, 1992; Kilvington *et al.*, 1990). Its use in the clinical settings, however, is discouraged because of the resistance of cysts, and possible neurotoxicity (Murdoch *et al.*, 1998 and Seal, 2003).

The minimum trophicidal concentration of pentamidine isethionate at 48 hours was 1.95 – 10 μg per mL and the minimum cysticidal concentration at

48 hours was 31.25 – 200 μg per mL for weakly cytopathogenic and clinical *Acanthamoeba* isolates, respectively (Duma and Finley, 1976; Kilvington *et al.*, 1990). In contrast, the minimal trophicidal concentration of hydroxystilbamidine isethionate against *Acanthamoeba* at 48 hours was 1.95 μg per mL, while cysticidal concentration at 48 hours is more than 250 μg per mL, which is well over the recommended levels (Nagington and Richards, 1976; Kilvington *et al.*, 1990).

Later studies tested various amidines against *Acanthamoeba*. The trophicidal and cysticidal efficiencies were found to be increased from propamidine to nonamidine; i.e., when the alkyl chain connecting the two benzene rings in their molecular structures was elongated. It was found that hexamidine, heptamidine, octamidine, and nonamidine showed significantly more effect on *Acanthamoeba* than propamidine, butamidine, or pentamidine. Based on these findings it was suggested that the latter three should be replaced with hexamidine as a prophylactic and therapeutic agent. In support the successful management of *Acanthamoeba* keratitis has been described with hexamidine di-isethionate (Desomedine) 0.1% eye drops (Perrine *et al.*, 1995; Brasseur *et al.*, 1994).

Other polyamine metabolism inhibitors that have been tested for anti-Acanthamoebic effects are as follows:

4.2. Methylglyoxal bis (guanyl hydrazone) (MGBG)

At 2.5 mM, MGBG inhibited multiplication of *Acanthamoeba* in peptone medium, accompanied by the disappearance of spermidine and a marked reduction in the level of diaminopropane. Moreover, MGBG completely inhibited growth in a chemically defined medium at 1 mM concentration. However, there was no decrease in the polyamine levels in the early stages of growth inhibition by MGBG. The addition of exogenous polyamines did not reverse the growth inhibitory action of MGBG. Inhibition of S-adenosylmethionine decarboxylase and decrease in polyamine synthesis do not seem to be the primary targets for MGBG.

4.3. Berenil

Berenil is a compound related to methylglyoxal bis(guanyl hydrazone) (MGBG) and the other diamidines like propamidine etc (see above). *In vitro* assays revealed that berenil inhibited *Acanthamoeba* growth in a chemically defined medium at 1 – 2 μM concentration. As for MGBG, there was no decrease in the polyamine levels in the early stages of growth inhibition and the addition of exogenous polyamines did not reverse the growth inhibitory action of berenil suggesting that inhibition of S-adenosylmethionine decarboxylase and decrease in polyamine synthesis may not be the primary targets for berenil.

4.4. Myristamidopropyl dimethylamine (MAPD)

The amidoamine myristamidopropyl dimethylamine exhibited trophicidal as well as cysticidal activities. The minimum cysticidal concentrations were 10 to 30 μg per mL, when incubated for 6 hours. While 50 and 100 μg per mL gave at least a 3-log cyst kill within 1 and 2 hours, respectively. The potential use of the amidoamine myristamidopropyl dimethylamine was demonstrated further with findings that it also possesses excellent anti-fungal and anti-bacterial activities. The amidoamine myristamidopropyl dimethylamine may represent a broad-spectrum therapeutic antimicrobial for keratitis and surgical prophylaxis and deserves further evaluation in these roles.

4.5. Folate biosynthesis inhibitors

Synthesis of folate (not carried out by mammalian cells) is necessary for nucleic acid synthesis in bacterial and eukaryotic microbial cells. In its absence, cells will be unable to divide. Thus folate synthesis inhibitors, such as sulphonamides would exhibit growth inhibitory effects. Sulphonamide mimics p-aminobenzoic acid inhibiting dihydropteroate synthesis vital for folate synthesis, thus affecting the growth of *Acanthamoeba*. Among the sulphonamides tested, *in vitro*, sulfamethoxazole and sulfadiazine were most effective, followed by sulphanilamide and sulphaguanidine. Inhibition by each sulphonamide was reversed by p-aminobenzoic acid as well as folic acid. 7-Methylguanosine, a pteridine synthesis-inhibitor (pteridine is a chemical compound composed of fused pyrimidine and pyrazine rings which is a precursor of dihydropterate), did not inhibit multiplication of *Acanthamoeba*. Among the dihydrofolate reductase inhibitors acting in the later part of the folate cycle,, pyrimethamine blocked amoebic growth, while trimethoprim and cycloguanil pamoate failed to cause significant inhibition of growth. Metoprine inhibited *Acanthamoeba* growth. Methotrexate (anti dhfr) and fluorouracil, an inhibitor of thymidylate synthetase (which uses the final product of the microbial folate cycle) inhibited growth strongly, with IC50 values (the concentration of the drug which causes 50% inhibition of the growth at 72 hours of 1.97 and 2.45 μg per mL, respectively. Inhibition by methotrexate, metoprine or 5- fluorouracil could not be reversed by folic acid, folinic acid, thymidine, or folinic acid plus thymidine. The results indicate that the anti cancer drugs methotrexate and 5- fluorouracil have at least as strong an interaction with *Acanthamoeba* dhfr and thymidylate synthase as they do against the equivalent human enzymes (unusual features in *Acanthamoeba* folate metabolism and may provide potential in the development of therapeutic interventions (Mehlotra and Shukla, 1993). As described above, one of the derivatives of sulphonamide is sulfadiazine that inhibits the production of folic acid. *In vitro* studies showed that *Acanthamoeba* is sensitive to sulfadiazine, which exhibited trophistatic effects. Moreover, *in vivo* studies showed that sulfadiazine is effective in protecting mice from *Acanthamoeba* granulomatous encephalitis as long as amoebae have not been established in the central nervous system (Rowan-Kelly *et al.*, 1982; Ferante *et al.*, 1984; Schuster and Mandel, 1984).

5. RNA synthesis inhibition

5.1. Rifampicin

Rifampicin inhibits DNA-dependent RNA polymerase by binding to its beta-subunit, thus preventing transcription to the RNA and subsequent translation to proteins. Its lipophilic nature makes it a good candidate to treat the central nervous infections, which requires distribution to the central nervous system and penetration through the blood-brain barrier. The protective or therapeutic effects of rifampicin against *Acanthamoeba* infections have been contradictory. In one study, *in vivo* rifampicin, given prophylactically exhibited partial protective effects in mice against *Acanthamoeba* granulomatous encephalitis, as assessed by the absence of symptoms of central nervous system malfunction and negative *Acanthamoeba* cultures from the brain, 10 days after inoculation. Treatment with rifampicin could be curative at high doses. After doses of 10, 25, and 50 mg per kg, two of six animals were free of symptoms and infection. This contrasts to some extent with other studies, which showed that rifampicin had no protective effect against *Acanthamoeba* granulomatous encephalitis (Ondarza *et al.*, 2006; Rowan-Kelly *et al.*, 1982; Das *et al.*, 1991).

5.2. 5-fluorocytosine

5-fluorocytosine acts in two possible ways: (i) once the drug is taken up, it is converted into the cytostatic flourouracil that undergoes further steps of activation and finally interacts as 5-fluorouridinetriphosphate with RNA biosynthesis and disturbs therefore the building of certain essential proteins, and (ii) the other mechanism is the conversion into 5-flourodeoxyuridine-monophosphate which inhibits the cellular DNA synthesis and defective RNA species.

Although 5-fluorocytosine (at concentrations of more than 10 μg per mL) exhibited trophicidal effects on weakly cytopathogenic *Acanthamoeba*, while lower concentrations had minimal effects, but significantly abrogated the encystation ability of these amoebae. In contrast, the clinical strains were capable of growing in the presence of 40 μg per mL. Moreover, 5-fluorocytosine appeared to be directed against both the DNA and the RNA in the weakly cytopathogenic strains, but the RNA appeared to be the main target in the clinical strain of *Acanthamoeba* as described above. Alternatively, 5-fluorocytosine may be metabolized to 5-fluoro-2'-deoxyuridine, which can starve the cell for thymidine by non-competitive inhibition of the enzyme thymidylate synthetase, thus inhibiting the DNA synthesis. Later studies reported that clinical *Acanthamoeba* was not susceptible to 5-fluorocytosine and this discrepancy was attributed to the use of different genotypes / species / isolates / strains of *Acanthamoeba*.

More importantly, for *Acanthamoeba* granulomatous encephalitis, *in vivo* studies showed that 5-fluorocytosine alone was of limited value in protecting mice infected with the virulent strains of *Acanthamoeba* (Duma and Finley, 1976).

6. Protein synthesis inhibition

6.1. Macrolide compounds

6.1.1. Rokitamycin

The mechanism of action of the macrolides such as rokitamycin is inhibition of protein biosynthesis by binding reversibly to the ribosome, thereby inhibiting translocation of peptidyl tRNA. This action is mainly growth inhibition, but can also be lytic at high concentrations. Generally, macrolides are used as anti-bacterials; however some of the derivatives have been tested for anti-Acanthamoebic effects. The growth curve evaluations revealed that rokitamycin exhibits trophistatic effects and at higher concentrations the effects also were cysticidal. The combination of rokitamycin with chlorpromazine or amphotericin B exhibited enhanced trophicidal and cysticidal activities. Furthermore, low doses of rokitamycin blocked *Acanthamoeba*-mediated host cell cytopathic effects suggesting their potential clinical use as anti-*Acanthamoeba* compounds (Mattana *et al.*, 2003)

6.1.2. Spiramycin

This is another macrolide antibiotic, which has been tested for its anti-Acanthamoebic effects. Using *in vitro* assays, it was observed that the minimum trophicidal concentration of spiramycin at 48 hours is 250 µg per mL. While the cysticidal concentration of spiramycin at 48 hours was more than 500 µg/mL, this is well over the recommended blood levels (Nagington and Richards, 1976; Duma and Finley, 1976).

6.2. Aminoglycosides

Aminoglycoside compounds contain an amino sugar, and amino-or guanido-substituted inositol ring which are attached by a glycosidic linkage to a hexose nucleus resulting in a polycationic and highly polar compound. They inhibit formation of the ribosomal initiation complex and also cause misreading of the messenger RNA, resulting in the inhibition of the protein. Generally, aminoglycosides are used as anti-bacterials; however some of the derivatives have been tested for anti-Acanthamoebic effects as follows:

6.2.1. Paromomycin

The minimal inhibitory concentrations of paromomycin sulfate against *Acanthamoeba* trophozoites at 48 hours is 15.6 µg per mL, while cysticidal concentrations are more than 62.5 µg per mL. Interestingly, the combination of cryotherapy and paromomycin was more cysticidal than either modality alone and eliminated detectable viable organisms. Between neomycin sulfate and paromomycin sulfate antibiotic solutions tested, paromomycin sulfate was the most effective (Matoba *et al.*, 1989; Nagington and Richards, 1976; Kilvington *et al.*, 1990).

6.2.2. Neomycin

Using *in vitro* assays, it was shown that the minimum inhibitory concentration of neomycin sulfate at 48 hours was 250 µg per mL; however the neomycin sulfate showed cysticidal effects at more than 500 μg per mL, which is well above the recommended levels. Of note, neomycin showed significant levels of host cell toxicity. The combination of cryotherapy and neomycin was more cysticidal (Matoba *et al.*, 1989; Larkin *et al.*, 1992; Nagington and Richards, 1976; Kilvington *et al.*, 1990).

6.2.3. Neosporin

The minimum trophistatic concentration of neosporin against *Acanthamoeba* trophozoites was shown at 1 μg per mL, while cysticidal concentrations were 1,000 μg per mL, which is well over the recommended levels (Nagington and Richards, 1976).

6.2.4. N-chlorotaurine

N-chlorotaurine acts by targeting proteins. Anti-*Acanthamoeba* activity of N-chlorotaurine, an endogenous mild antiseptic showed that it has anti-Acanthamoebic qualities, both in phosphate-buffered saline and in Acanthamoebic culture medium. At 20 mM N-chlorotaurine, the levels of trophozoites showed at least a 2-log reduction within 24 hours. The addition of ammonium chloride to N-chlorotaurine led to a faster decrease in the numbers of living cells and affected excystation. Moreover, N-chlorotaurine showed anti-Acanthamoebic activity at concentrations well tolerated by the human tissues and might be useful as a topical drug for the treatment of *Acanthamoeba* infections (Fürnkranz *et al.*, 2008).

7. Tricyclic neuroleptic agents (calmodulin inhibition)

7.1. Trifluoperazine dihydrochloride and chlorpromazine dihydrochloride (phenothiazine compound)

Both trifluoperazine and chlorpromazine are antipsychotic drugs that inhibit calmodulin. *In vitro* studies showed that the two tricyclic neuroleptic agents, chlorpromazine and trifluoperazine showed trophicidal and cysticidal effects. Both agents had the capacity to inhibit the putative disulfide reductase from the trophozoites at a concentration of 32 μg per mL during a 24-hour incubation. The association of chlorpromazine with rokitamycin or amphotericin B enhanced their activities, while low doses also were protective of parasite-mediated host cell cytotoxicity. Both trifluoperazine and chlorpromazine had the best inhibitory effects, followed by mepacrine, ketoconazole, pentamidine, miconazole, and amphotericin B (Ondarza *et al.*, 2006).

Among other compounds, *Acanthamoeba* are shown to be sensitive to primaquine (normally used to treat malaria, and affects the mitochondria of the exo-erythrocytic stages, but the mechanism is not clearly understood).

8. Artesunate

Artesunate is part of the artemisinin group of drugs that are used to treat malaria. Artemisinin, isolated from the traditional Chinese herb *Artemisia annua*, is increasingly being used in the treatment against malaria, however animals under prolonged exposure exhibit considerable neurotoxicity, including gait disturbance, loss of spinal and pain response reflexes, and prominent loss of brain stem and eye reflexes. Its mode of action is induction of oxidative stress, as observed as an increase of reactive oxygen species and of lipid peroxidation. Artemisinin showed trophistatic effects rather than trophicidal effects. At 100 – 500 μg per mL, that is much higher than active concentrations against malaria parasites, artemisinin caused more than 93% inhibition of growth. Moreover, excystation did not occur suggesting that the effects are trophistatic rather than trophicidal (Nacapunchai *et al.*, 2001). Study of the precise mechanisms of the artemisinin effects on *Acanthamoeba* may shed light on its possible therapeutic potential against *Acanthamoeba* infections.

Among other compounds, metronidazole tinidazole, trimethoprim (antifolate see above), G-418, clindamycin hydrochloride as well as its derivative U34728E), isoniazid, and rolitetracycline were shown to be ineffective or had limited effects against pathogenic *Acanthamoeba* (Duma and Finley, 1976; Nacapunchai *et al.*, 2001).

9. Antimicrobial compounds from natural products (in addition to the aforementioned antibiotics)

9.1. Animal-based natural products

9.1.1. Magainins

Novel naturally occurring compounds, magainins, are host defence peptides that were originally isolated from the African clawed frog and are known to possess a unique membrane-targeted mechanism of action which likely includes altering ion conductance across membrane barriers. Magainins are active against a broad range of Gram-positive and Gram-negative bacteria and possess anti-viral, anti-fungal and anti-parasitic activity. Two magainins tested (MSI-103 and MSI-94) showed trophistatic and trophicidal effects at 20 – 40 μg per mL. Both trophistatic and trophicidal activities were enhanced by combining the magainins with silver nitrate (200 μg per mL). These combinations have lytic activities against both trophozoites and cysts stages of *Acanthamoeba* and have promise as antimicrobial agents in the treatment (Schuster and Jacob, 1992).

9.2. Plant-based natural products (in addition to artemisinin, see above)

Various plants extracts have been tested for anti-Acanthamoebic effects. It was shown that extracts obtained from *Ipomoea* sp., *Kaempferia galanga*, and *Cananga odorata*) were trophicidal for *Acanthamoeba* and a fourth extract prepared from *Gastrochilus panduratum* showed limited trophicidal but strong trophistatic effects. Interestingly, these plant extracts were not lytic for normal macrophage cultures (Chu *et al.*, 1998).

The extracts of *Buddleia cordata* contain linarin and vanillic acid. Although linarin had potent trophistatic effects against various isolates of *Acanthamoeba*, but vanillic acid had no activity. However, acetyl vanillic acid had similar effects on *Acanthamoeba* to linarin. Threshold values of these two active compounds ranged from 31.25 μg per mL to 4 mg per mL and from 31.25 μg per mL to 8 mg per mL for linarin and acetyl vanillic acid, respectively. The varied threshold values were most like due to different strains / species / genotypes of *Acanthamoeba* (Rodríguez-Zaragoza *et al.*, 1999).

In vitro trophicidal activity of methanolic extract of *Thymus sipyleus subsp. sipyleus var. sipyleus* from Turkish flora against *Acanthamoeba* were observed at concentrations ranging from 1.0 to 32.0 mg per mL. However, it showed no cytotoxicity for the corneal cells at the concentration of 32 mg per mL. Thus polar subfraction of the methanolic extract of *Thymus sipyleus subsp. sipyleus var. sipyleus* could be concluded as a new natural agent for possible use in the treatment of *Acanthamoeba* infections (Polat *et al.*, 2007).

The *in vitro* trophistatic effect of the nonpolar subfraction of the methanol extract of garlic (*Allium sativum*) on *Acanthamoeba* trophozoites and cysts were determined. It was shown that garlic showed trophicidal and cysticidal effects in a dose- and time-dependent manner. Notably, it showed no cytotoxicity for the cornea cells at the concentration of 3.90 mg per mL. It was suggested that garlic alone, and in combination with other anti-Acanthamoebic agents, may be used in clinical practices after further investigations (Polat *et al.*, 2007).

In addition, the *in vitro* trophicidal and cysticidal effects of an ethanolic extract of propolis against *Acanthamoeba* were determined. At 2 – 6 mg per mL, the effects were trophistatic; at concentrations of more than 8 mg per mL, the effects were trophicidal; while at concentrations of more than 15 mg per mL, the effects were cysticidal within 72 hours. These findings suggested that the ethanolic extracts of propolis have trophicidal, as well as cysticidal, properties against *Acanthamoeba* (Topalkara *et al.*, 2007).

Agelasines and agelasimines are antimicrobial and cytotoxic purine derivatives isolated from marine sponges (*Agelas* sp.). Structurally simplified analogs of these natural products starting from beta-cyclocitral were synthesized. The agelasine and agelasimine analogs showed trophistatic effects on *Acanthamoeba* (Proszenyák *et al.*, 2007).

10. Drug resistance in *Acanthamoeba*

As indicated above, the chemotherapeutic approaches remain the most common method of treatment for *Acanthamoeba* infections. These include chemically synthesized antimicrobial compounds or natural products most commonly derived from other organisms that inhibit the growth of, or kill *Acanthamoeba*. These provide the most direct and cheapest way of controlling these infections. Many of the currently available anti-*Acanthamoeba* compounds were identified by screening large number of compounds. However, a major problem with drug therapy is the ability of a sub-population of parasites to develop drug resistance. Multiple drug-resistant *Acanthamoeba* isolates are therefore becoming a problem and the fear is that it will not be long before our limited arsenal of the anti-*Acanthamoeba* compounds become obsolete. The potential mechanisms of drug resistance may include the following:

i) conversion of the drug to an inactive form by an enzyme or failure to activate the drug.
ii) modification of a drug sensitive site.
iii) permeability changes leading to decreased influx or increased efflux.
iv) alternative pathways to bypass inhibited reactions or decreased requirement for product of inhibited reaction.
v) increased production of drug sensitive enzymes.

Many of these mechanisms have been observed in parasites such as alterations in the cell permeability, modifications of drug sensitive sites and increased quantities of the target enzymes. These modifications tend to arise in a population of parasites by various mechanisms such as;

i) physiological adaptations
ii) differential selection of resistant individuals from a mixed population of susceptible and resistant strains
iii) spontaneous mutations, followed by selection, and
iv) changes in gene expression

The appearance of resistance within a population has been observed to occur within 5 – 50 generations. One factor, which contributes to rapid acquisition of drug resistance, is the use of suboptimal concentrations of drug, resulting in more survivors. Given the limited availability of only a handful of available compounds, there is an urgent need to identify novel targets in pathogenic *Acanthamoeba*. A complete understanding of *Acanthamoeba* metabolism and life cycle should help identify novel targets for the rational development of anti-*Acanthamoeba* drugs. However, with the growing evidence of drug resistance in *Acanthamoeba*, other control measures should be explored in conjunction with chemotherapeutic approaches.

The most important reason for unsuccessful therapy seems to be the existence of a double-wall cyst stage that is highly resistant to the available treatments, resulting in recurrence of the infection. The double-walled structure of *Acanthamoeba* cysts comprise of an inner endocyst and an outer ectocyst, providing a physical barrier against chemotherapeutic agents. Additionally antimicrobial compounds may not be effective at the recommended concentrations against the cyst forms of *Acanthamoeba* due to their reproductive inactivity and/or very limited metabolic activity. Although, the ability of *Acanthamoeba* to develop resistance against chemotherapeutic agents is a significant threat in the successful treatment of *Acanthamoeba* infections, however the mechanisms of such drug resistance in *Acanthamoeba* remain incompletely understood. This is evident with observations that recurrence of *Acanthamoeba* infections is common and in some cases, *Acanthamoeba* may be non-responsive to the known anti-Acanthamoebic compounds at the recommended levels. In support there have been several reports of accurate diagnosis of *Acanthamoeba* keratitis cases, followed by treatment with anti-amoebic drugs, antibiotics, and corticosteroids, however despite temporary relief, infection is shown to progress, which was observed to be associated with perforation and iris prolapse, resulting in the loss of vision, or the eye had to be removed. The development of drug (propamidine isethionate) resistance during the course of therapy has been observed, which led to the recurrence of the infection. This may be due to the fact that although propamidine isethionate inhibits DNA synthesis, but at the recommended concentrations it may not be active against the cyst forms of *Acanthamoeba*. Diamidines like propamidine also stimulate encystment.

Furthermore, axenic cultures in the presence of drugs are shown to develop resistance against antimicrobial compounds. To this end, studies showed that *Acanthamoeba* develops resistance by stepwise selection against cycloheximide. This was achieved by culturing *Acanthamoeba* using 40 μg per mL of the mutagen N-methyl-N'-nitro-N-nitrosoguanidine for 30 min, followed by the addition of 50 μg per mL cycloheximide. *Acanthamoeba* was cultured in the growth medium for up to 6 days, and then cycloheximide was added at a final concentration of 100 μg per mL. After 24 hours, about 50 – 70% *Acanthamoeba* underwent morphological changes and detached from the flasks. By 48 hours, nearly all cells became round and were detached from the flasks. After several days, only resistant cells were present. The resistant cells were cloned by culturing them in the presence of 7.5 μg per mL of cycloheximide. The clones were cultured for several generations without cycloheximide before further characterizations. It was shown that modification in the resistant mutant is cytoplasmic, and modified element is most likely ribosomal component. In fungi, cycloheximide resistance was shown to be associated with mutation of gene for a ribosomal protein is mutated (Chisholm and Vaughan, 1979). Other studies showed that *Acanthamoeba* develops resistance by stepwise selection against erythromycin, chloramphenicol and oligomycin (Seilhamer and Byers, 1978).

Several lines of evidence suggest that the clinical isolates of *Acanthamoeba* are generally more resistant to antimicrobial compounds, compared with the weakly or non-cytopathogenic pathogens. Thus it is important to test the potency of novel antimicrobial compounds using the clinical isolates of *Acanthamoeba*. For example, the resistance of a clinical isolate of *Acanthamoeba* was shown to be greater than 1,000 μg per mL levels of paromomycin, polymyxin B-bacitracin-neomycin, acriflavine, 5-fluorocytosine, amphotericin B, gentamicin and trimethoprim-sulfamethoxazole (Ma *et al.*, 1981). However this isolate was shown to be susceptible to pimaricin (0.5%), thus patient was treated with pimaricin and responded well.

The gain or loss of resistance to drugs was also demonstrated by culturing *Acanthamoeba* in the presence of 40 μg per mL of 5-fluorocytosine. It was shown that the clinical isolates of *Acanthamoeba* were capable of growing in the presence of 40 μg per mL of 5-fluorocytosine after an initial period of susceptibility. Drug-treated *Acanthamoeba* cultures exhibited normal growth when transferred to fresh growth medium but did not retain resistance to the drug beyond 15 subcultures in a drug-free growth medium indicating that resistance is drug-induced rather than constitutive resistance. One possible explanation of the drug resistance was thought to be due to the ability of *Acanthamoeba* to alter membrane permeability and thus lower uptake.

Other studies showed that *Acanthamoeba* developed resistance against oligomycin. The drug resistance was shown to be associated with mitochondrial ATPase. Other possible mechanisms of resistance may be smooth endoplasmic reticulum that is involved in the synthesis of lipids and thus it may play a role in the inactivation or detoxification of drugs that might otherwise be toxic or harmful to the cell. Furthermore, *Acanthamoeba* contains soluble glutathione transferase enzyme that may play an important role in the detoxification.

Additionally, the dangers of increased selection pressure induced by continuous drug exposure should not be ignored. A precise understanding of these mechanisms is crucial for the rationale development of much needed drugs against this serious infection. It is noteworthy that several compounds at the concentration of 100 μg per mL including, acetarsol, diloxanide furoate, oxytetracycline, benzyl penicillin, streptomycin sulphate, neomycin, kanamycin, paromomycin, chloroquine, nystatin, metronidazole, polymyxin B, sulphonamides, trimethoprim lactate, pyrimethamine sulphate, trimethoprim + sulphamethoxazole, trimethoprim + polymyxin B, pyrimethamine + sulphormethoxine, trimethoprim + pyrimethamine, amphotericin B exhibited no or limited effects against the clinical isolates of *Acanthamoeba*. In contrast, hydroxystilbamidine isethionate (a diamidine like propamidine) and 5-fluorocytosine exhibited trophicidal effects (Hamburg and De Jonckheere, 1980; Ma *et al.*, 1981; Johnson and Thomas. 2002; Neff and Neff, 1969; Turner *et al.*, 2000; Casemore, 1970; Stevens and O'Dell, 1974; Dierickx *et al.*, 1990; Seilhamer and Byers, 1982).

11. Disinfectants and *Acanthamoeba*

As for chemotherapeutic drugs, the existence of a double-wall cyst stage appears to be a key feature of the inefficiency of disinfectants against *Acanthamoeba*. Cysts are highly resistant to various disinfectants and other physiological conditions and generally high concentrations of these agents are required to kill the viable cysts, as described below.

11.1. Disinfection agents

The detergent, BEN22 (50:50 mixture of L-alpha-L-rhamnopyranosyl-beta-hydroxydecanoyl-beta-hydroxydecanoate and 2-O-alpha-L-rhamnopyranosyl-alpha-L-rhamnopyranosyl-beta-hydroxydecanoyl-beta-hydroxydecanoate) detach *Acanthamoeba* from the contact lenses. It was found that a concentration of 0.05% detached the trophozoites to a statistically significant greater extent than saline, but commercial ReNu Multi-Purpose Solution (Bausch and Lomb, Italy) and BioSoak (Finnsusp Ltd., Finland) did so as well. The ReNu Multi-Purpose Solution was more effective than 0.005% BEN22 in detaching the trophozoites of *Acanthamoeba* from the contact lenses. After the 2 hour immersion period, a maximum of 97% of the initial trophozoites were detached. The variation between individual lenses was significantly greater than that within the different areas of one lens (Raali *et al.*, 2001).

Chlorine is the most widely used disinfectant worldwide. Chlorine was introduced as a water disinfectant in the early 1900s and since then it has been effectively used, especially for potable water. It is highly effective as a germicide due to its ease of application, measurement, control and economy, and its reasonable persistence in waters makes it a suitable disinfectant. The pathogenic *Acanthamoeba* spp. are highly resistant to chlorine. For example, it is shown that the pathogenic *Acanthamoeba* cysts are viable after 3 hours of contact with a residual concentration (usual test used) of 40 μg of free chlorine per mL. In comparison, cysts of the pathogenic *Naegleria fowleri* were destroyed at a residual concentration of 0.8 μg of free chlorine per mL. These findings indicate that the resistance of cysts of pathogenic *Acanthamoeba* to chlorine is far higher than levels being used in for chlorination of water supplies, which may be as high as 1 μg per mL. Other studies demonstrated that *Acanthamoeba* trophozoites survived exposures to free chlorine residuals at concentrations of up to 10 mg per litre for 24 hours. However *Acanthamoeba* encysted during the incubation (King *et al.*, 1988).

In addition, cysts of clinical isolates of *Acanthamoeba* were not destroyed by other halogens such as bromine (at concentrations of 0.4 to 1 μg per mL), iodine (at concentrations of 2 to 5 μg per mL), and iodophore (at concentrations of 2 to 5 μg per mL) after 24 hours of contact time, as well as polyquaternium-1 disinfectant (De Jonckeere and DeVoorde, 1976). Antiseptic solutions containing ammonium derivatives or the biguanides chlorhexidine, sorbate and polyaminopropyl biguanide, or polyquaternium-1, killed trophozoites but had limited cysticidal effects. While 0.005% quarternary benzalkonium chloride did not show 100% cysticidal effects against *Acanthamoeba* (Georges and Perrine, 1989; Silvany *et al.*, 1990). Other disinfectants, chlorine dioxide, deciquam 222 (de-decyldimethyl-ammonium bromide), and ozone (produced by electrical discharge of oxygen by passing dried oxygen through an Mk II ozonizer and collecting the dissolved ozone in sterile phosphate buffered saline) also showed anti-Acanthamoebic properties. The effects of modern biocides in cooling systems found that among cooling tower biocides including, a thiocarbamate compound, tributyltin neodecanoate mixed with quarternary ammonium compounds, and an isothiazolin derivative, the isothiazolin derivative was least effective among various compounds tested (Sujata and Berk, 1993).

In the majority of studies, it was shown that hydrogen peroxide solutions (at 3% concentration) with exposure for 2 – 3 hours killed *Acanthamoeba* cysts. However, several reports showed that 3% hydrogen peroxide is ineffective against *Acanthamoeba* cysts. This discrepancy may be due to different genotypes / species / isolates of *Acanthamoeba*. Thus, the use of hydrogen peroxide needs addressing as issues such as time of the exposure, post-neutralization residuals, and lens stability remains unclear. The current data indicate that the currently recommended short exposure time (10 to 15 min) is inadequate for protection against *Acanthamoeba* and that at least 2 hour is necessary. The addition of weekly "enzyme" i.e., for lens cleaning is also essential in decreasing the incidence of the contact lens-related conjunctivitis when lenses are not replaced frequently (Aksozek *et al.*, 2002; Georges and Perrine 1989; Holden, 1990; Silvany *et al.*, 1990).

Later studies compared the anti-amoebic activity of a new multi-purpose disinfection solution (i.e., OPTI-FREE EXPRESS with ALDOX) with a 3% hydrogen peroxide disinfecting system. The antimicrobial ingredients in the new solution are the quaternary, polyquaternium-1 and myristamidopropyl dimethylamine. It was observed that the OPTI-FREE EXPRESS with ALDOX (EXPRESS MPDS) Multi-Purpose Disinfecting Solution provided anti-*Acanthamoeba* activity for disinfection of contact lenses. OPTI-FREE EXPRESS with ALDOX showed activity in the range similar to 3% hydrogen peroxide system. However, later studies demonstrated that Opti-Free express had limited cysticidal effects and no killing was observed with mature cysts (Rosenthal *et al.*, 1999; Rosenthal *et al.*, 2000; Kilvington and Anger, 2001). Among other disinfectants, it was found that cysts can survive in ophthalmic saline solution containing 0.25% sorbic acid, 0.13% potassium sorbate, 0.1% EDTA, 0.00005% polyaminopropyl biguanide, 0.001% polyquaternium-1 and 0.004% thimerosal (some of these purchased from different companies including Alcon, Allergen, Bausch and Lomb, Barnes and Hinde; Coopervision) from a minimum of few days to up to 90 days of exposure. However, thimerosal 0.004% when combined in solution with EDTA was effective. The differences were perhaps due to varying degree of cytopathogenicity between different isolates. In addition, cysts can survive in ophthalmic disinfectant solutions containing 0.005% chlorhexidine (purchased from American Optical, Allergen, Alcon, Bausch and Lomb, Barnes and Hinde) from a minimum of 6 hours to 14 days of exposure and benzalkonium chloride (0.001% and 0.004%) (Brandt *et al.*, 1989; Connor *et al.*, 1989; Silvany *et al.*, 1991).

Later studies compared the anti-Acanthamoebic properties of MiraFlow, an extra strength cleaner containing 20% isopropyl alcohol and other disinfecting solutions and heat. MiraFlow was better than disinfecting solutions containing hydrogen peroxide (3%), chlorhexidine, or benzalkonium chloride. Thermal disinfection of *Acanthamoeba* cysts occurred rapidly at temperatures above 70°C. These findings suggest that either thermal disinfection or cleaning lenses with MiraFlow followed by disinfection solution such as 3% hydrogen peroxide, should inactivate *Acanthamoeba* cysts. Furthermore, MiraFlow was the only cleaner that killed trophozoites and cysts on all lenses during the cleaning step (Ghajar *et al.*, 1989; Penley *et al.*, 1989).

The use of MiraFlow as a disinfectant against *Acanthamoeba* was supported further when thirteen commercially available contact lens solutions were tested for their ability to kill the cysts of *Acanthamoeba*. Again, Miraflow, which contains 20% isopropyl alcohol, was the most effective at killing the cyst (>94%), followed by solutions containing thimerosal (89%). The rigid gas permeable lens solutions in general were more effective than soft lens solutions. None of the solutions tested were completely cidal, but these data do suggest a combination of a good daily cleaner and disinfecting solution may be effective in reducing *Acanthamoeba* exposure. These findings should provide guidelines for the practitioner in selecting the best disinfection system for the contact lens patient (Connor *et al.*, 1991).

In a comparative study, six multipurpose contact lens solutions [All-in-One, All-in-One (Light), ReNu MultiPlus, Optifree Express, Complete, and Solo-care soft] were tested for their efficacies against *Acanthamoeba* trophozoites and cysts. Against trophozoites, All-in-One, ReNu Multiplus, and Optifree Express achieved total kill after the manufacturer's minimum recommended disinfection time, with the remaining solutions failing to reach a 100% kill. After 24 hours of exposure, all solutions proved anti-Acanthamoebic, achieving total kill, with the exception of Complete. Against cysts, All-in-One produced the best anti-Acanthamoebic properties within the minimum recommended disinfection time. After 24 hours of exposure, All-in-One achieved total kill of cysts, while other solutions showed less than 100% kill. Overall, these findings clearly showed that the multipurpose solutions tested, ReNu MultiPlus and Optifree Express, demonstrated effective trophicidal activities within the recommended disinfection times; however, only All-in-One proved effective against both trophozoites and cysts over the same time period (Beattie *et al.*, 2003).

To overcome the unreliability of hydrogen peroxide as a disinfectant, the efficacy of one-step hydrogen peroxide (3%) disinfecting solutions including the Silver Sept (platinum and silver disk for intensifying disinfection) and Blue Vision (newly composed catalytic tablet) was compared to the efficacy of 2 two-step system such as Titmus hydrogen peroxide (0.6% and 3% hydrogen peroxide) against cysts of *Acanthamoeba*. After a soaking time of 8 hours (overnight soaking of contact lenses) the 2 two-step systems completely destroyed the cysts of *Acanthamoeba*, while the one-step systems showed weaker effects. One-step hydrogen peroxide systems do not have sufficient effects on *Acanthamoeba* cysts and therefore may not protect the contact lens user from a possible infection of the eye (Hiti *et al.*, 2005).

Overall, various anti-infective agents possessing *in vitro* cysticidal activity have been described, but no agent has been shown to be uniformly effective against all isolates / genotypes of *Acanthamoeba*. Among other compounds, polyhexamethylene biguanide is used as a disinfectant in swimming pools (at 45 parts per million) and as a preservative in cosmetics and pharmacological formulations. *In vitro* assays revealed that polyhexamethylene biguanide (0.02%) showed both trophicidal and cysticidal effects.

Clotrimazole at high concentration (i.e., 0.1%, inappropriate for treatment) is effective in killing both trophozoites and cysts, and may be of value as a disinfectant (Aksozek *et al.*, 2002; Larkin *et al.*, 1992; Burger *et al.*, 1994). In addition, either 0.001% chlorhexidine for 4 hours, or 0.004% chlorhexidine for 1 hour showed effective cysticidal effects.

11.2. Freeze-thaw

The repeated freeze-thawing for up to 5 times had limited effects on the viability of *Acanthamoeba* cysts (Aksozek *et al.*, 2002).

11.3. Heat disinfection

The heat disinfection at more than 60°C (i.e., 70 to 80°C) for 10 min showed potent trophicidal and cysticidal effects.

11.4. Microwave irradiation

The use of microwave irradiation to disinfect contact lenses was shown with findings that trophozoites, as well as cysts of different *Acanthamoeba* strains / genotypes / species, were effectively killed, even by only 3 minutes of microwave irradiation, and there were no negative effects of irradiation on the contact lens cases themselves. These findings suggested that microwave treatment is a very effective, easy, and cheap method to keep contact lens cases free of *Acanthamoeba*, thus may have potential to considerably reduce the risk of *Acanthamoeba* keratitis (Hiti *et al.*, 2001).

11.5. Ultraviolet light irradiation

In a comparative study, the germicidal effects of a 253.7 nm ultraviolet light with an intensity of 1,100 microW per cm^2 were tested. The exposure time necessary to reduce a concentration of organisms from 10^6 per mL to less than 10 per mL was 30 seconds for *Staphylococcus aureus*, 60 seconds for *Pseudomonas aeruginosa*, and 84 seconds for *Candida albicans*. The time necessary to sterilize a suspension of 10^4 per mL *Acanthamoeba* trophozoites was approximately three minutes with this technique. Furthermore, a 15 min cleaning disinfection cycle with UV was highly effective on contact lenses contaminated with *Bacillus pumilus*, *Aspergillus niger*, *Pseudomonas aeruginosa*, and *Acanthamoeba* trophozoites subjected. Notably, the soft contact lenses exposed to the ultraviolet light under the aforementioned conditions for over eight hours showed no effect on their appearance, comfort, or refraction.

Other studies examined the effects of prolonged ultraviolet light of different wavelength on soft and rigid gas permeable contact lenses contaminated with *Acanthamoeba* trophozoites. It was revealed that the ultraviolet light lamp (253.7 nm, 250 mW per cm^2) was germicidal for *Acanthamoeba* within 20 min, but caused destruction of the soft lens polymers within 6 hours of exposure. The ultraviolet light also caused damage to the rigid gas permeable lenses in less than 100 hours. While the ultraviolet light (290-310 nm, 500 mW per cm^2) also killed *Acanthamoeba* trophozoites with a 180 min exposure but caused less severe changes in the soft lens polymers than did the 253.7 nm lamp, although exposure of 300 hours did substantially weaken the soft lens material. Rigid gas permeable materials were minimally affected by exposure to 300 h of 290-310 nm lamp. In addition, it is shown that *Acanthamoeba* cysts are resistant to up to 800 mJ per cm^2 of the ultraviolet radiation. The ultraviolet light is an effective germicidal agent but is injurious to soft lens polymers; its possible utility in the sterilization of rigid gas permeable lenses and lens cases deserves further study. Overall, these studies demonstrated the potential of ultraviolet light as a possible alternative to existing contact lens care systems (Aksozek *et al.*, 2002; Dolman and Dobrogowski, 1989; Admoni *et al.*, 1993; Gritz *et al.*, 1990).

11.6. Gamma irradiation

Acanthamoeba cysts are shown to be resistant to 250 K rads of gamma irradiation (Aksozek *et al.*, 2002).

Overall, these results suggest that *Acanthamoeba* cysts show enhanced survival because of their resistance to very harsh environmental conditions. An important feature of the use of various disinfectants, the requirement that any disinfectant remaining on lenses must be neutralized before use to avoid pronounced stinging and possible corneal damage.

12. Future prospects for treatment

12.1. Antibody-based therapy

The *in vitro* microbicidal activity of a murine monoclonal anti-idiotypic antibody against *Acanthamoeba* was tested. In addition, a synthetic killer mimotope, which mimics a yeast killer toxin characterized by a wide spectrum of antimicrobial activity through interaction with specific cell wall receptors, mainly constituted by beta-glucans was tested for amoebicidal activity being investigated after incubation of trophozoites under different experimental conditions with laminarinase, monoclonal anti-idiotypic antibody, killer mimotope and scrambled decapeptide as a control. To confirm the specific interaction of killer mimotope with beta-glucans, the experiments were also carried out in the presence of laminarin (beta $1\rightarrow 3$ glucan) or pustulan (beta $1\rightarrow 6$ glucan); both glucan molecules were co-incubated with killer mimotope or scrambled decapeptide. It was shown that monoclonal anti-idiotypic antibody and killer mimotope exhibited a time-dependent killing activity, in comparison with scrambled decapeptide or heat-inactivated monoclonal anti-idiotypic antibody; this activity was completely abolished by pre-incubation with laminarin, but not by pustulan. Notably, *in vitro* amoebicidal activity was observed in the presence of laminarinase, an enzyme that specifically hydrolyses beta-glucans. Furthermore, killer mimotope specifically inhibited the growth of *Acanthamoeba* on the infected contact lenses and the remaining adherent killer mimotope-treated trophozoites appeared strongly damaged. These results indicate that the expression of beta ($1\rightarrow 3$) glucan receptors in the cell membrane is probably modulated during cell growth of *Acanthamoeba* and is critical for the killing activity of yeast killer toxin-like molecules. These data confirm the broad antimicrobial spectra of monoclonal anti-idiotypic antibody and killer mimotope, emphasize the crucial role of beta ($1\rightarrow 3$) glucan in microbial physiology and suggests the potential use of monoclonal anti-idiotypic antibody and killer mimotope in the prevention and therapy of *Acanthamoeba* infections or in preventing *Acanthamoeba* contamination during storage of the contact lenses (Fiori *et al.*, 2006).

12.2. Using a carrier for the drug delivery

Whether dimethylsulfoxide can increase the permeability of cyst membranes so drugs can be delivered to the amoeba within more easily, thus enhancing killing was tested. When applied to a cyst-only population of *Acanthamoeba*, none of three standard drugs, propamidine isethionate 0.1%, neomycin 1%, or miconazole 1%, was cysticidal. In contrast, when combined with dimethylsulfoxide 30%, propamidine isethionate proved to be cysticidal. These studies suggested that the use of a carrier for the known anti-Acanthamoebic drugs may increase their penetration into the cyst form of the organism, which is normally refractory to drug treatment.. Because dimethylsulfoxide has been used topically in the past and shown to be quite safe, this may be a viable new approach to therapy for this difficult condition (Saunders *et al.*, 1992)

12.3. Use of a liposome for drug delivery

Depending on their phospholipid composition, liposomes are either endocytosed or fused with the plasma membrane of *Acanthamoeba*. For example, unilamellar egg lecithin vesicles are endocytosed and uptake is inhibited at 4°C. In contrast, unilamellar dipalmitoyl lecithin vesicles fuse with the plasma membranes of *Acanthamoeba* and increase the surface area of the cell, while their contents are found in the cytoplasm as well as the medium. However, unilamellar dipalmitoyl lecithin liposomes fuse with the plasma membranes of *Acanthamoeba* and increase the surface area of the cell, while their contents are released in the cytoplasm of the cell (Batzri and Korn, 1975).

12.4. Photodynamic chemotherapy against *Acanthamoeba*

To develop alternative approaches for medical treatment in the control of *Acanthamoeba* infections, photodynamic treatment may be advantageous over conventional methods due to its localized use, in particular for *Acanthamoeba* keratitis.

12.4.1. Photo-inhibition of *Acanthamoeba*

With the findings that visible light inhibits growth of *Paramecium* or *Tetrahymena*, photo-inhibition in *Acanthamoeba* was studied (Dolphin, 1970). It was shown that light from 350 to 680 nm at intensities of up to 1.62×10^5 ergs per sec per cm^2 inhibits *Acanthamoeba* growth (Dolphin, 1970). Photo-inhibition of growth was attributed to a direct effect of light on *Acanthamoeba* and not to photo-degradation of the medium. However, the growth medium that was previously irradiated showed some inhibitory effects on its growth supporting properties and these effects were reversed by the addition of proteose peptone and yeast extract. The reduced growth rate in the irradiated *Acanthamoeba* cultures was not due to the cyst formation or development of multinuclearity. It was concluded that the light affects rate-limiting processes in *Acanthamoeba*, most likely through direct absorption by intracellular pigments (Dolphin, 1970).

12.4.2. Use of photosensitizers against Acanthamoeba

Incubation of *Acanthamoeba* cells with a tetracationic phthalocyanine (RLP068) at concentrations ranging between 0.2 and 1.0 μM, caused a ready uptake of the photosensitizer with recoveries of the order of 0.5 – 2.5 nmoles per mg of the cell protein. The amount of cell-bound phthalocyanine did not appreciably change with incubation times ranging between 0.5 and 3 hours. Fluorescence microscopic investigations showed an obvious accumulation of the phthalocyanine at the level of the vacuolar membranes. A nearly complete photo-induced cell death occurred upon irradiating *Acanthamoeba* cells with 600 – 700 nm light with a total energy of 15 – 30 J per cm^2 using 1.0 μM of the phthalocyanine in the incubation medium. DAPI staining of the photosensitized cells indicates significant damage of the nucleus. On the other hand, photosensitization of the protozoan cells does not directly involve the mitochondria as shown by the lack of photo-induced decrease in the activity of typical mitochondrial enzymes, such as NADH dehydrogenase and citrate synthase (Kassab *et al.*, 2003).

Later studies tested photodynamic treatment with a tetracationic Zn(II)-phthalocyanine (RLP068). Incubation of cysts with RLP068 for 1 hour caused an accumulation of readily detectable concentrations of the phthalocyanine, even at doses as low as 0.5 μM per litre. The RLP068 exhibited no dark toxicity towards cysts up to 5 μM per litre concentration. A decrease of approximately 50% in the cyst survival in comparison with the control group was measured upon incubation of the cysts with 0.5 μmoles per litre. RLP068, followed by exposure to light (600 – 700 nm) for 20 min at a fluence rate of 60 J per cm^2. After incubation with 3 and 5 μmoles per litre RLP068 and irradiation, the cysts lost their excystation ability as early as day 5 after exposure and up to day 10, and were clearly damaged when observed under an interference contrast microscope. These data indicate the possible use of RLP068 in photodynamic treatment of diseases caused by pathogenic *Acanthamoeba* in the initial disinfection of wastewaters. These findings are significant in that rapid and extensive photo-damage may be induced in the highly resistant cystic stages by means of 600 – 700 nm light sources and future studies will determine its usefulness as a clinical approach (Ferro *et al.*, 2006).

From: *Acanthamoeba*: Biology and Pathogenesis. By: Naveed Khan.

Section G

Acanthamoeba: Trojan Horse of the Microbial World

1. Introduction

Acanthamoeba contains diverse microbial endosymbionts including viruses, yeast and bacteria, some of which are potential human pathogens. An endosymbiont is any organism that lives within the body or cells of another organism, i.e. forming an endosymbiosis (Greek: "endo" means "inner", "sym" means "together" and "biosis" means "living"). If the endosymbiont or the host can not live independent of each other, it is called "obligate endosymbiosis". However, not all endosymbiosis are obligate. Endosymbiosis can also be harmful to either of the organisms involved. For *Acanthamoeba*, the exact nature of these symbionts and the benefit they represent for the amoeba host still have to be clarified. Of interest, *Acanthamoeba* has been shown to host / contain various microbes, including viruses, bacteria, yeast and protozoa as described below (Fig. 1).

2. A host for viruses

2.1. Mimivirus

Acanthamoeba is known to host the largest known viruses, such as Mimivirus. This virus has a particle size of 400 nm, comparable to mycoplasma. Mimivirus is a double-stranded DNA virus growing inside *Acanthamoeba*. It has a genome size of 1,181,404 base pairs (1.2 million bp), and huge gene repertoire (911 protein coding genes), 10% of which exhibit a similarity to proteins of known functions, all contribute to blur the established boundaries between viruses and the smallest parasitic cellular organisms, a finding that may have huge implications in the evolutionary studies. In addition to the exceptional genome size, Mimivirus exhibits many features that distinguish it from other nucleocytoplasmic large DNA viruses. The most unexpected is the presence of numerous genes encoding central protein-translation components, including four amino-acyl transfer RNA synthetases, peptide release factor 1, translation elongation factor (EF) Tu, and translation initiation factor 1. The translation elongation factor (EF) Tu promotes codon-directed binding of aminoacyl-tRNA into the ribosomal A site from an

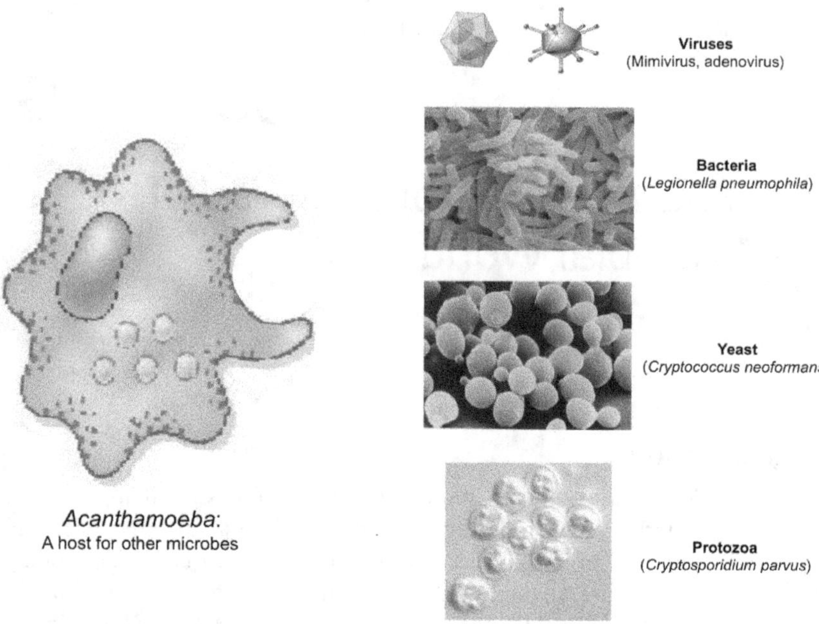

Figure 1. *Acanthamoeba* feeds on other microbes, and acts as a host for microbial organisms. Notably, *Acanthamoeba* acts as a biological host for some microbial pathogens, i.e., pathogen multiplication inside *Acanthamoeba*.

EF-Tu·GTP·aminoacyl-tRNA ternary complex with concomitant hydrolysis of GTP. Mimivirus genome also exhibits six tRNAs. Other notable features include the presence of both type I and type II topoisomerases, components of all DNA repair pathways, and many polysaccharide synthesis enzymes, and one intein-containing gene (an intein is a segment of a protein that is able to excise itself and rejoin the remaining portions, the exteins, with a peptide bond). Inteins have also been called "protein introns".

2.1.1. An evolutionary mystery

The discovery of Mimivirus constitutes an evolutionary mystery. The size and complexity of the Mimivirus genome, ranged in between viruses and the cellular organisms. Phylogenetic analyses suggested that the Mimivirus lineage could have emerged prior to the individualization of the cellular organisms. The initial analysis of the genome of the Mimivirus predicted a proteome of size and complexity more akin to small parasitic bacteria than to other nucleocytoplasmic large DNA viruses and identified numerous functions never before described in a virus. These studies led to the conclusion that the Mimivirus lineage could have emerged before the individualization of the cellular organisms from the three domains of life. An exhaustive *in silico* analyses of the non-coding moiety of all known viral genomes now uncovers the unprecedented perfect conservation of an AAAATTGA motif in close to 50% of the Mimivirus genes. This motif preferentially occurs in genes transcribed from the predicted leading strand and is associated with functions required early in the viral infectious cycle, such as transcription and protein translation. A comparison with the known promoter of the unicellular eukaryotes, amoeba in particular, strongly suggests that the AAAATTGA motif is the structural equivalent of the TATA box core promoter element. This element is specific to the Mimivirus lineage and may correspond to an ancestral promoter structure predating the radiation of the eukaryotic kingdoms. This unprecedented conservation of core promoter regions is another exceptional feature of the Mimivirus that again raises the question of its evolutionary origin. Furthermore, its unique morphological and genomic characteristics have led to propose Mimivirus as a member of a new distinct nucleocytoplasmic large DNA virus family, the Mimiviridae. This new data might help shed new light on the origin of the DNA viruses and their role in the early evolution of eukaryotes. In addition, these findings have raised questions on the existence of other possible endosymbionts within *Acanthamoeba* and other protists.

2.1.2. Do endosymbionts contribute to Acanthamoeba infections?

Mimivirus was originally suggested as a new pneumonia-associated human pathogen. This was shown by inoculating Mimivirus in the adult BALB/c mice (an albino strain of laboratory mouse that are distributed globally, and among the most widely used inbred strains used in animal experimentation) and C57BL/6 mice (referred to as "C57 black 6" is a common inbred strain of lab mouse and among the most widely used lab mouse strain). Mimivirus was inoculated via intracardiac route. Interestingly, C57BL/6 mice developed histopathological features of pneumonia by day 3, post-infection and BALB/c mice, by day 7. The histopathological features of pneumonia, characterized by the presence of thickened alveolar walls with cellular infiltrates comprising mononuclear leukocytes, macrophages and lymphocytes, diffuse alveolar damages with the formation of hyaline membrane and erythrocytes in the alveolar lumen, were most marked by day 7. These findings led to the conclusion that *Acanthamoeba* Mimivirus is pathogenic in mice under experimental conditions (La Scola *et al.*, 2003; Raoult *et al.*, 2004; Suhre *et al.*, 2005; Suzan-Monti *et al.*, 2006; Ghedin and Claverie, 2005; Khan *et al.*, 2007). These findings have also raised questions on the contribution of endosymbionts (Mimivirus) in the pathogenicity of *Acanthamoeba* infections, ecological distribution of the host as well as the endosymbionts, and their effects on the host biology. For example, is keratitis or encephalitis purely due to *Acanthamoeba* or do endosymbionts contribute to these serious infections? If so how? Future studies in the identification of virulence factors of the endosymbiont and of the host, and their precise role in disease should clarify these issues, i.e., once the complete genome of the host as well as the endosymbionts is known then we can associate a particular virulence factors with a particular genome.

2.2. Coxsackievirus

Later studies determine whether *Acanthamoeba* could play a role in the survival and the transmission of coxsackieviruses and focused on *in vitro* interactions between *Acanthamoeba* and the coxsackie B3 viruses. Residual virus titer evaluations and the immunofluorescence experiments revealed a remarkable coxsackie B3 viruses adsorption on *Acanthamoeba* surfaces and their accumulation inside *Acanthamoeba*. The survival of viruses was independent of the dynamics of *Acanthamoeba* replication and encystation. In addition, it was shown that virus-infected *Acanthamoeba* can release infectious viruses during interaction with the human macrophages. Based on these data, it can be concluded that *Acanthamoeba* appears to be a potential promoter of the survival of coxsackieviruses and their transmission to the human hosts (Mattana *et al.*, 2006).

2.3. Adenovirus

236 potentially pathogenic *Acanthamoeba* strains, isolated from water sources in the Canary Islands, were surveyed for the presence of human adenoviruses using a polymerase chain reaction-based typing assay. A total of 34 of these strains were found to be positive for adenovirus belonging to 4 different human adenoviruses serotypes (HAdV-1, 2, 8, and 37). It was found that human adenoviruses (HAdV)-2 was the most frequently encountered serotype amongst the *Acanthamoeba* strains, and their identification was confirmed by a nested PCR specific for this serotype. Furthermore, it was shown that *Acanthamoeba* isolates belonging to the T4 genotype was highly associated with the serotype HAdV-2, whereas *Acanthamoeba* isolates of the genotype T3 was most often associated with adenovirus serotypes related to the ocular diseases. Based on these data, it was suggested that *Acanthamoeba* should be considered as a potential reservoir and perhaps even a transmitter of adenoviruses to human and other secondary hosts (Lorenzo-Morales *et al.*, 2007).

2.4. Poliovirus, echovirus, enterovirus, or vesicular stomatitis virus

It was shown that trophozoites of *Acanthamoeba* are unable to adsorb poliovirus or vesicular stomatitis virus. Furthermore, after encystation in the medium containing these viruses, the cysts did not contain viruses. In other studies, *Acanthamoeba* were incubated with vaccination poliovirus type 1 and type 3 and with echovirus type 4 and echovirus type 30. However, no remarkable virus accumulation was observed on the surface or inside *Acanthamoeba* cells for up to 21 days. Enteroviruses were most probably present only on *Acanthamoeba* surfaces. In contrast, echoviruses bound to *Acanthamoeba* and persisted even after 52 to 75 days. It was concluded that the *Acanthamoeba* tested played only the role of a solid-like carrier in this survival of echoviruses (Danes and Cerva, 1981; Baron *et al.*, 1980).

3. A host for yeast

Cryptococcus neoformans is phagocytosed by and replicates in *Acanthamoeba*, which leads to the death of *Acanthamoeba*. An acapsular strain of *Cryptococcus neoformans* did not survive when incubated with *Acanthamoeba*, but melanization protected these cells against killing by *Acanthamoeba*. A

Figure 2. The 19-storey building of the Bellevue-Stratford Hotel, Philadelphia, USA.

phospholipase mutant had a decreased replication rate in *Acanthamoeba* compared with isogenic strains. These observations suggest that cryptococcal characteristics that contribute to mammalian virulence also promote fungal survival in *Acanthamoeba*. Intracellular replication was accompanied by the accumulation of polysaccharide containing vesicles similar to those described in *Cryptococcus neoformans*-infected macrophages. The results suggest that the virulence of *Cryptococcus neoformans* for mammalian cells is a consequence of adaptations that have evolved for protection against environmental predators such as *Acanthamoeba* and provide an explanation for the broad host range of this pathogenic fungus.

4. A host for protozoa

The capacity of *Acanthamoeba* to predate *Cryptosporidium* oocysts was demonstrated. A maximum of six oocysts per *Acanthamoeba* trophozoite were detected, and a slow elimination of the internalized oocysts to the surrounding culture medium was observed. It was suggested that the free-living amoeba, *Acanthamoeba* may act as a carrier of *Cryptosporidium* oocysts and, thus, may play an important role in the transmission of cryptosporidiosis (Gómez-Couso *et al.*, 2007).

5. *Acanthamoeba* and bacteria interactions

5.1. Story of *Acanthamoeba* and Legionnaires' disease

During July 21-24, 1976, the Bellevue-Stratford Hotel Philadelphia, USA) (Fig. 2) hosted the 58th State convention of the American Legion Department of Pennsylvania. When the participants and their families returned home, many of them felt sick and some died. Within days, a mystery disease had killed 34 and sickened 221 people in a matter of days, all of whom had spent time at the hotel. Thousands of veterans were in a state of shock with many families isolated and not welcome in their communities. People generally developed pneumonia-like symptoms with a cough, headache, and a high temperature. However, later stages were much more aggressive and patients

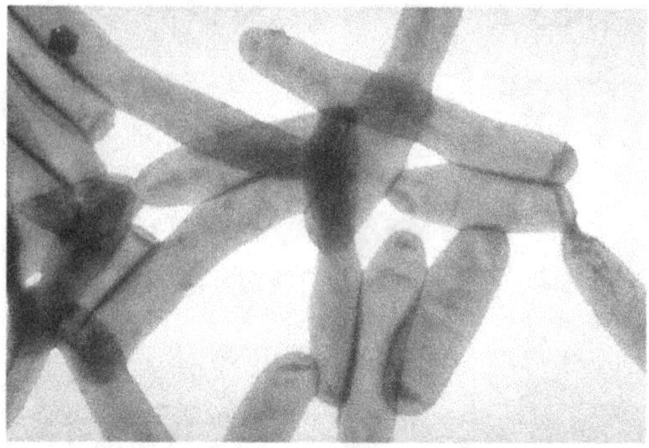

Figure 3. Scanning electron micrograph of *Legionella pneumophila*.

hospitalized for weeks to months. At the time, the disease responsible for so much suffering was a mystery killer to the medical community and it was given an identity linked to those it affected the most, hence called Legionnaires' disease. Not surprisingly, some of the survivors (legionnaires) are upset with the name given to this disease as it did cast a dark cloud over the American Legion and they did not want to be affiliated with it. Moreover, the name suggests that only the veterans would get the disease. At the time, no one knew of how and why so many people were getting sick or dying. All sorts of questions were raised with the quality of air within the Bellevue-Stratford hotel, the abundance of pigeon droppings outside the hotel, and the water supplies that were held in galvanized canisters throughout the convention.

On January 18, 1977, the Centers for Disease Control in Atlanta, USA (http://www.cdc.gov/) identified the causative agent as a previously unknown bacterium, subsequently named *Legionella* (Fig. 3). Later work identified the source of bacterium as potable or standing water and most often found in the water systems of large buildings as well as in the air-conditioning units. Although an accurate source of this outbreak was not clearly described to convince legionnaires, one of the survivors reported that he did **smell a bad odour from the air-conditioning unit, while he was in bed.** However, the question remains that how do this bacterium transmit from the water/aerosols into the body. In this regard, recent studies have shown that *Legionella* bacterium infects free-living amoebae, such as *Acanthamoeba*, which is abundant in the water supplies and are know to supports bacterial growth (Fig. 4). In addition, *Acanthamoeba* cysts are resistant to many disinfectants, which would allow bacteria to survive in the water supplies and they can be air-borne. Thus *Acanthamoeba* acts both as a biological host and a vector to transmit *Legionella* to the susceptible hosts. Several lines of evidence suggest that inhalation of airborne contaminated water droplets enter the lungs resulting in disease. This often occurs in poorly ventilated areas such as prisons where a condensating air conditioner can spread it throughout the entire room, infecting anyone not immune to the strand of bacteria. Potential sources of such contaminated water include cooling towers used in industrial cooling water systems as well as in large central air conditioning

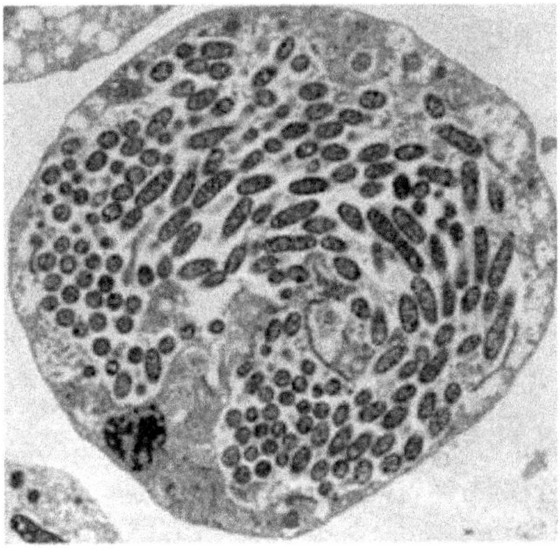

Figure 4. Transmission electron micrograph of an amoeba-filled with *Legionella pneumophila*.

systems, evaporative coolers, hot water systems, showers, whirlpool spas, architectural fountains, room-air humidifiers, ice making machines, misting equipment, and similar disseminators that draw upon a public water supply. The disease may also be spread in a hot tub if the filtering system is defective. Freshwater ponds, creeks, and ornamental fountains are also potential sources of *Legionella*. The disease is particularly associated with hotels, cruise ships and hospitals with old, poorly maintained pipe work and cooling systems.

Today, Legionnaires' disease has low mortality and can be treated with early diagnosis. The patients suffering from Legionnaires' disease usually have fever, chills, and a cough, which may be dry or may produce sputum. Some patients also have muscle aches, headache, tiredness, loss of appetite, loss of coordination (ataxia), and occasionally diarrhea and vomiting. Laboratory tests may show that patients' renal functions, liver functions and electrolytes are deranged, including hyponatremia (sodium concentration in the plasma falls dangerously below 135 mM per L and may result in water intoxication). Chest X-rays often show pneumonia with bi-basal consolidation. It is difficult to distinguish Legionnaires' disease from other types of pneumonia by symptoms or radiologic findings alone and laboratory tests are required for accurate diagnosis. The time between the patient's exposure to the bacterium and the onset of illness for Legionnaires' disease is 2 to 10 days.

5.2. Method for isolating *Legionella pneumophila* and amoebae from water samples

As indicated above, *Acanthamoeba* not only harboured *Legionella pneumophila* but bacteria also multiplied within *Acanthamoeba* suggesting that bacteria use amoeba both as a host and a reservoir.

To isolate *Acanthamoeba* and *Legionella* from the ground-water, drinking water supplies and whirlpools, volumes of 10 to 1,000 mL were concentrated by membrane filtration. *Legionella pneumophila* was detected on buffered charcoal yeast extract (BCYE) agar, and *Acanthamoeba* by inverting filters on non-nutrient agar plates seeded with *Escherichia coli* that were incubated at 37°C for up to 12 days. In 65% of the samples positive for *Legionella pneumophila*, *Acanthamoeba* was also detected. *Legionella pneumophila* and *Acanthamoeba* were detected together in 38% of warm drinking water samples. The highest isolation temperature for *Acanthamoeba* was 57°C, but fewer amoebae were detected above 50°C. In cold drinking water, amoebae were found in 88% of samples. The presence of *Legionella pneumophila* and *Acanthamoeba* in whirlpool waters (42%) presents a risk for man. Fresh environmental isolates of an *Acanthamoeba* species and *Legionella pneumophila* serogroup 4 were used for laboratory experiments. The *Acanthamoeba* supported intracellular multiplication of *Legionella pneumophila* in Chang's medium and autoclaved tap water, as shown by colony-forming unit counts, direct fluorescent antibody test and Giemsa staining. The results further confirmed that interaction between *Legionella pneumophila* and *Acanthamoeba* could occur in nature, and that the latter could act as hosts for *Legionella* and support their growth (Henke and Seidel 1986).

Overall, it is the ability of *Acanthamoeba* to host some of the most notorious bacterial pathogens that has gained particular attention from the scientific and the medical community. *Acanthamoeba* was first shown to be infected and lysed by bacteria in 1954, (Drozanski, 1956) and to harbor bacteria as endosymbionts in 1975 (Proca-Ciobanu *et al.*, 1975). Later studies revealed that *Acanthamoeba* acts as a reservoir for pathogenic facultative mycobacteria (Krishna-Prasad and Gupta, 1978). As described above, *Acanthamoeba* has been shown to harbour virulent *Legionella* spp. associated with Legionnaires' disease (Rowbotham, 1980). At the same time, it is well-established that *Acanthamoeba* consumes bacteria in the environment, so the interactions of *Acanthamoeba* and bacteria are highly complex and dependent on the virulence of *Acanthamoeba*, the virulence of bacteria, and the environmental conditions. The outcome of these convoluted interactions may be beneficial to *Acanthamoeba* or to bacteria or may result in the development of a symbiotic relationship (Fig. 5). Adding to this complexity, *Acanthamoeba* is known to interact with various Gram-positive and Gram-negative bacteria resulting in a range of outcomes. In a survey, 57 environmental and clinical isolates of *Acanthamoeba* were tested for the presence of bacteria. It was shown that 14 of 57 (24%) axenically grown *Acanthamoeba* contain intracellular bacteria. The rod-shaped bacteria were identified in 5 of 23 clinical *Acanthamoeba*, 4 of 25 environmental *Acanthamoeba* isolates, and 2 of 9 ATCC *Acanthamoeba* isolates. Previously these amoebae were unrecognized as having endosymbionts. Interestingly, the isolated bacteria could not be cultured using routine methodologies suggesting that these bacteria are endosymbionts which have an obligate need to multiply within their amoebic hosts. Overall, these studies suggested that *Acanthamoeba* might commonly possess bacterial endosymbionts (Fritsche *et al.*, 1993).

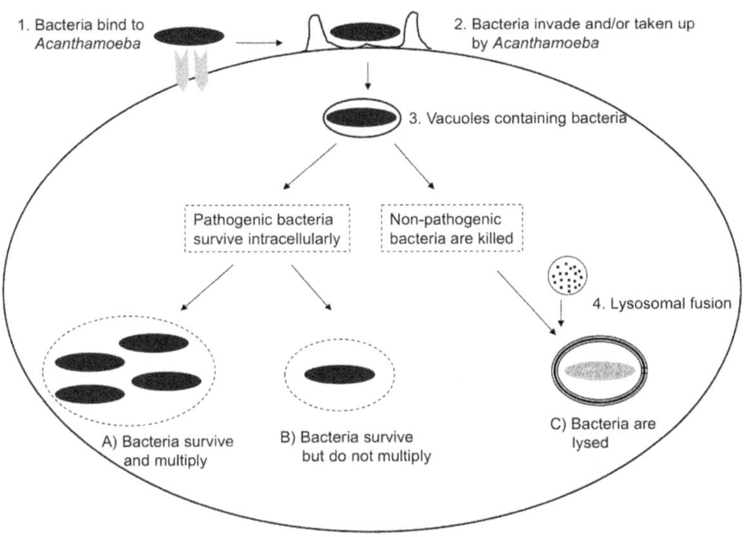

Figure 5. Bacterial interactions with the clinical isolate of *Acanthamoeba*. Bacteria invade into and/or taken up by *Acanthamoeba*. Once intracellular, non-pathogenic bacteria are most likely killed and used as food source, while pathogenic bacteria possess the ability to evade intracellular killing mechanisms, and either survive or multiply within *Acanthamoeba*. **(A)** *Acanthamoeba* act as host and/or reservoir for bacteria, **(B)** *Acanthamoeba* act as Trojan horse for bacteria, and **(C)** *Acanthamoeba* act as potent predator of bacteria.

5.3. Endosymbiosis and pathogenicity

For the last several decades, it has been of interest, whether bacterial or other microbial endosymbionts enhance the pathogenesis of *Acanthamoeba*. Although, the results have been inconclusive, there are few reports suggesting that the bacterial endosymbionts may enhance the pathogenicity of *Acanthamoeba*. For example, it is shown that the endosymbiont-infected *Acanthamoeba* produced a statistically significant enhancement in the cytopathic effects in comparison to the uninfected *Acanthamoeba*. In contrast, endosymbionts alone on monolayers produced no cytopathic effects. This report suggested that the obligate bacterial endosymbionts might enhance *Acanthamoeba* pathogenic potential *in vitro* by some as-yet unknown mechanism (Fritsche *et al.*, 1998).

For simplicity and for researchers new to this area, these interactions are discussed in the sections below.

5.4. *Acanthamoeba* as a bacterial predator

This group includes bacteria that are used as food source by *Acanthamoeba*. Bacteria are taken up by phagocytosis, followed by their lysis in phagolysosomes. Although, *Acanthamoeba* consumes both Gram-positive and Gram-negative bacteria, they preferentially graze on Gram-negative bacteria, which are used widely as a food source in the isolation of *Acanthamoeba*. However, the ability of *Acanthamoeba* to consume bacteria is dependent on the virulence properties of bacteria and the environmental conditions. For example, recent studies have shown that in the absence of nutrients, virulent strains of *Escherichia coli* K1 invade *Acanthamoeba* and remain viable intracellularly. And upon the availability of nutrients, *Escherichia coli* K1 escape *Acanthamoeba*, grow exponentially and lyse the host *Acanthamoeba*. Other regulatory factor is the density of bacteria. At low density, Gram-negative bacteria support the growth of *Acanthamoeba*. However at high densities (i.e., >10 to 1 ratio), bacteria inhibit the growth of *Acanthamoeba*.

By contrast, in the absence of nutrients the avirulent strains of *Escherichia coli* K-12 are phagocytosed (instead of bacterial invasion into the amoeba) by *Acanthamoeba* and are killed. This demonstrates clear differences in the ability of *Acanthamoeba* to interact with *Escherichia coli* that are dependent on the virulence properties of the *Escherichia coli* and the environmental conditions. Later studies compared the growth and starvation responses of *Acanthamoeba* in the presence of suspended and attached *Escherichia coli* K-12. The *Acanthamoeba* perceived all the suspended systems to be unfavorable for growth, despite being challenged with high levels of prey, and as a consequence, they exhibited a starvation response. This resulted in *Acanthamoeba* producing characteristic cysts. The amoebic cysts were re-activated into the feeding trophozoites in the presence of bacterial aggregates, which formed in the suspended systems after 68 hours of incubation. In contrast, both species of *Acanthamoeba* grew well in the presence of attached *Escherichia coli* at a concentration of 10^6 cells per cm^2 of agar and yielded specific growth rates of approximately 0.04 per hour. Starvation responses were induced at the end of the growth phase, and these were equivalent to those recorded in the suspended systems. It was concluded that, when suspended, *Acanthamoeba* in the "floating form" cannot feed effectively on the suspended prey, and hence the starvation response is initiated. Thus the majority of *Acanthamoeba* feeding is via trophozoites grazing of attached bacterial prey.

Future studies should examine the basis of these differences as well as how bacteria are taken up (invasion *versus* phagocytosis) and the associated molecular mechanisms (Bottone *et al.*, 1994; Alsam *et al.*, 2005c; Pick *et al.*, (2007).

5.5. *Acanthamoeba* as a host for bacterial pathogens (role as a reservoir or a Trojan horse or a vector)

The ability of *Acanthamoeba* to act as a host for bacterial pathogens has gained particular attention, because the majority of the bacteria involved are human pathogens. The term reservoir is used to describe bacterial pathogens that are present inside *Acanthamoeba* as well as multiply within it. In contrast, the term "Trojan horse" is used to describe bacterial presence inside *Acanthamoeba* but without any multiplication. The ability of bacteria to survive inside *Acanthamoeba* may be used to survive harsh environments during transmission from one host to another. The term "vector" is used to describe the role of *Acanthamoeba* as a "carrier" which may be mere attachment/adsorption of bacteria on the surface of amoeba. However, at present it is difficult to differentiate between the three categories, hence I have indicated all bacterial pathogens that are known to interact or multiply within *Acanthamoeba* in this action. More research will enable us to better understand the properties of bacteria that allow them to interact with *Acanthamoeba* in more than one ways.

5.5.1. Other bacterial pathogens and Acanthamoeba

The following bacteria are known to interact and/or multiply within *Acanthamoeba*.

- *Escherichia coli* O157: causative agent of bloody diarrhea and occasionally result in kidney failure, especially in young children and the elderly people.
- *Coxiella burnetii*: causative agent of Q fever (Q stands for "query fever" because originally the causative agent was unknown). The Q fever involves respiratory symptoms.
- *Salmonella typhimurium*: causative agent of gastroenteritis, referred to as salmonellosis involving diarrhea, fever, vomiting, and abdominal cramps with potentially fatal consequences, while *Salmonella enterica* serovar Typhi is known to produce typhoid fever.
- *Francisella tularensis*: causative agent of tularemia or rabbit fever that may result in ulceroglandular, or potentially lethal pneumonic tularemia.
- *Pseudomonas aeruginosa*: causative agent of keratitis. In addition, it infects the pulmonary tract, urinary tract, burns, wounds, and also causes other blood infections.
- *Vibrio cholerae*: causative agent of cholera [cholera is a form of infectious gastroenteritis (intestinal infection) that results in frequent watery stools, cramping abdominal pain, and eventual collapse from dehydration].
- *Helicobacter pylori*: causative agent of gastric ulcers, gastritis (infection or inflammation of the stomach), duodenitis (infection or inflammation of the duodenum), and cancers.
- *Simkania negevensis*: causative agent of pneumonia-like symptoms.
- *Listeria monocytogenes*: causative agent of listeriosis and may include septicemia, meningitis, encephalitis (or meningo-encephalitis), corneal ulcer, and pneumonia.
- *Burkholderia pickettii*: biovar 2; a facultative pathogen that has been associated with septicemia and respiratory tract infections.
- *Burkholderia* spp.: an opportunistic pathogen associated with respiratory tract infections, in particular in cystic fibrosis patients.
- *Chlamydophila pneumoniae*: causative agent of pneumonia and has been associated with atherosclerosis and with asthma.
- *Rickettsia*: causative agents of typhus, rickettsialpox, Boutonneuse fever, African tick bite fever, Rocky Mountain spotted fever, Australian tick typhus, Flinders Island spotted fever and Queensland tick typhus.
- *Porphyromonas gingivalis*: implicated in periodontal disease, as well as the upper gastrointestinal tract, respiratory tract, and in the colon.
- *Prevotella intermedia*: implicated in the periodontal disease.
- *Shigella sonnei*: causative agent of shigellosis, also known as bacterial dysentery.

- *Pasteurella multocida*: causative agent of fowl cholera in birds. It can also cause a zoonotic infection in humans, resulting in cellulitis, and may involve the respiratory tract, bacteremia, can cause osteomyelitis or endocarditis. The bacteria may also cross the blood-brain barrier and cause meningitis.
- *Campylobacter jejuni*: commonly associated with poultry and naturally colonizes the gastrointestinal tract of many bird species causing enteritis. It also can cause human gastroenteritis.
- *Francisella tularensis*: infects a number of mammals such as voles, rabbits, and muskrats. It also causes a zoonotic infection in humans causing tularemia.
- *Aeromonas* spp.: suspected to cause gastroenteritis ranging from mild enteritis to cholera-like diarrhoea. *Aeromonas* species have also been reported to cause infections such as septicaemia, endocarditis, osteomyelitis, myonecrosis (necrosis of muscle), haemolytic uraemic syndrome, meningitis, peritonitis, respiratory tract disease and ocular infections as well as causing infections in other animals.
- *Flavobacterium* spp.: causative agent of septicaemia and pneumonia.
- *Cytophaga* spp.: causative agents of abdominal, pelvic and blood infections.
- *Candidatus Odyssella thessalonicensis'* gen. nov., sp. nov.,: suspected to produce pneumonia-like infections.
- Methicillin-resistant *Staphylococcus aureus* (MRSA): causative agent of a variety of infections including mild boils to cellulitis, septicaemia, toxic shock syndrome, joint infections, endocarditis, meningitis, and respiratory tract diseases.
- *Mycobacterium avium*: patients suffer from anemia and neutropenia if bone marrow is involved. Pulmonary involvement is similar to tuberculosis, while diarrhea and abdominal pain are associated with gastrointestinal involvement. The interaction of *Mycobacterium avium* with *Acanthamoeba* are similar to that shown for mycobacteria within macrophages, *Mycobacterium avium* inhibits lysosomal fusion and replicates in vacuoles that are tightly juxtaposed to the bacterial surfaces within amoebae. Furthermore, *Acanthamoeba*-grown *Mycobacterium avium* was also more virulent in the beige mouse model of infection. These data suggest a role for protozoa present in water environments as hosts for pathogenic mycobacteria, particularly *Mycobacterium avium*.

In addition, novel bacterial endosymbionts in amoeba have been identified. Based on 16S rDNA sequences revealed that the endosymbionts are related to *Caedibacter caryophilus*, *Holospora elegans* and *Holospora obtuse*, which were proposed to classify as '*Candidatus Caedibacter Acanthamoeba*e', '*Candidatus Paracaedibacter Acanthamoeba*e' and '*Candidatus Paracaedibacter symbiosus*'.

This property of *Acanthamoeba* to act as a reservoir (for the aforementioned bacteria) and/or a Trojan horse is particularly important as these bacterial pathogens may not only survive intracellularly but multiply within them. This allows bacteria to transmit throughout the environment, evade the host defences and/or chemotherapeutic drugs, and reproduce in sufficient number to produce disease. Upon favourable conditions, the increasing bacterial densities lyse their host *Acanthamoeba* and infect new *Acanthamoeba* and/or produce disease. In the long-term, these *Acanthamoeba*-bacteria interactions can be considered as "parasitic" as they result usually in the death of the amoeba. It is important to recognize that the virulence determinants responsible for bacterial invasion of *Acanthamoeba*; their intracellular survival by inhibition of phagolysosome formation or growth at the acidic pH; and their escape from *Acanthamoeba* vary between different bacteria (Krishna-Prasad and Gupta, 1978; Rowbotham, 1980; Tyndall and Domingue 1982; Anand *et al.*, 1983; Ly and Muller, 1990; Vandenesch *et al.*, 1990; Kilvington and Price 1990; Thom *et al.*, 1992; Bareker *et al.*, 1993; Michel *et al.*, 1995; Michel and Hauröder 1997; Essig *et al.*, 1997; Steinert *et al.*, 1998; Barker *et al.*, 1999; Cirillo *et al.*, 1999, 1997, 1994; Marolda *et al.*, 1999; Fritsche *et al.*, 1999; Müller *et al.*, 1999; Lenders *et al.*, 2000; Horn *et al.*, 2001; La Scola and Raoult, 2001; Winiecka-Krusnell *et al.*, 2002; Kahane *et al.*, 2001; Greub and Raoult, 2002, 2004; Abd *et al.*, 2003, 2005; Hundt and Ruffolo 2005; Snelling *et al.*, 2005; Huws *et al.*, 2006; Wagner *et al.*, 2006; Jeong *et al.*, 2007; Rahman *et al.*, 2008).

5.6. Can bacterial pathogens survive the encystation of *Acanthamoeba*

Interestingly, it is shown that although *Listeria monocytogenes* cells are ingested by *Acanthamoeba* spp. survive within and multiply intracellularly, but they are destroyed in cysts. Similarly it is shown that *Vibrio cholerae* survived and multiplied in *Acanthamoeba*, however, *Vibrio cholerae* did not survive encystation process. This property seems to be limited to *Legionella* which can survive in both the trophozoite and the cyst stage suggesting that perhaps *Legionella* are more equipped to evolutionary interact with *Acanthamoeba* (Ly and Müller (1990; Thom *et al.*, 1992; Cirillo *et al.*, 1997; Steinert *et al.*, 1998).

A comparison of *Acanthamoeba* infected with *Legionella pneumophila* and *Acanthamoeba* infected with *Mycobacterium avium* by electron microscopy demonstrated that there were striking differences in the locations of the bacteria within *Acanthamoeba* cysts. While *Legionella pneumophila* resided within the cysts, *Mycobacterium avium* was found within the outer walls of the double-walled cysts of *Acanthamoeba*.

Overall, these findings suggest that *Acanthamoeba* facilitate bacterial transmission and/or provide protection against the human immune system. In support of this, it was shown that *Mycobacterium avium* showed increased colonization of mice, when inoculated in the presence of *Acanthamoeba*. The ability of *Acanthamoeba* to resist harsh conditions (such as extreme temperatures, pH, and osmolarity), especially during their cyst stage, suggest their usefulness as bacterial hosts. In particular, *Acanthamoeba* cysts are notoriously resistant to chlorine (a key and sometimes the only compound used in cleaning water systems). This poses clear challenges in eradicating bacterial pathogens from public water supplies, especially in developing countries. In addition, *Acanthamoeba*-bacteria interactions also affect bacterial virulence. For example, as described above, *Legionella pneumophila* grown within *Acanthamoeba* exhibited increased motility, virulence and drug resistance compared with axenically grown *Legionella*.

5.7. *Acanthamoeba* as evolutionary precursor of pathogenicity in bacteria

In addition, it is shown that the cysts produced from infected trophozoites (infected with *Legionella*) were found to protect the *Legionella* from at least 50 mg per mL free chlorine. The ability of *Legionella pneumophila* to survive within the cysts of *Acanthamoeba* is suggested as a possible mechanism by which the organism evades disinfection and spreads to colonize new environments.

It is thought that bacteria isolated from *Acanthamoeba* (intracellular bacteria) are more pathogenic to humans (as discussed above) and more resistance to harsh environmental conditions. For latter, the effect of free chlorine on various bacteria including, *Escherichia coli*, *Citrobacter freundii*, *Enterobacter agglomerans*, *Enterobacter cloacae*, *Klebsiella pneumoniae*, *Klebsiella oxytoca*, *Salmonella typhimurium*, *Yersinia enterocolitica*, *Shigella sonnei*, *Legionella gormanii* and *Campylobacter jejuni* were tested. It was observed that all bacteria were sensitive to 0.5 mg of free chlorine residual per litre in less than 1min. When co-cultured with *Acanthamoeba*, bacteria were 10 to 50-fold more resistant to free-chlorine suggesting this relationship as a possible mechanism for survival of bacteria in inhospitable environments.

To determine increased resistance of *Legionella pneumophila* to biocides, bacteria were exposed to polyhexamethylene biguanide, pre- and post-intracellular in *Acanthamoeba*. It was observed that polyhexamethylene biguanide killed 99.99% of bacteria grown in broth cultures within 6 hours, but only 90% of bacteria grown in amoebae suggesting that activity of biocide was greatly reduced against *Legionella pneumophila* grown in *Acanthamoeba* (Barker *et al*., 1992; King *et al*., 1988).

Importantly, it is shown that similar mechanisms are utilized by *Legionella pneumophila* to replicate within two evolutionarily distant hosts, mammalian and *Acanthamoeba*, suggesting that pathogenicity may have been an evolutionary acquisition. *Legionella pneumophila* grown in *Acanthamoeba* were found to be 100-fold more invasive for epithelial cells and 10-fold more invasive for macrophages than *Legionella pneumophila* grown on agar. Also, *Acanthamoeba*-grown virulent *Legionella pneumophila* displays increased replication in monocytes and is more virulent in A/J, C57BL/6 Beige, and C57BL/6 mice. These data demonstrate that the intra- *Acanthamoeba* growth environment affects the entry mechanisms and virulence of *Legionella pneumophila*.

However, it was shown that the morphology and structure of *Legionella* cells grown in co-culture of *Acanthamoeba* was modified after 20 days of co-culture: viable bacteria showed large fatty cytoplasmic inclusions, gas liquid chromatography analysis demonstrated a decrease in the 16:0 fatty acid ratio. Other studies revealed that *Legionella pneumophila* grown intracellular in *Acanthamoeba* exhibit different surface properties compared with the *Legionella pneumophila* grown *in vitro*. Notably, intracellular bacteria contained a 15 kDa outer membrane protein and a monosaturated straight chain fatty acid. These compounds were also found in quantities in the host *Acanthamoeba* suggesting that 15 kDa is of *Acanthamoeba* origin.

Analyses of protein expression on the two bacteria revealed differences in protein profiles. It was suggested that replication of *Legionella pneumophila* in *Acanthamoeba* present in domestic water supplies may be necessary to produce bacteria that are competent to enter mammalian cells and produce human disease.

5.8. Benefit for the host

The interactions between *Acanthamoeba* and bacteria are highly complex and depend on the virulence properties of the bacteria and the amoeba. At least for the pathogenic bacteria, research so far has shown that *Acanthamoeba* acts as a mere host and bacteria manipulate the host to their benefit. At present, there is limited evidence of beneficial effect of hosting bacteria for *Acanthamoeba*. In one study *Acanthamoeba* keratitis patient, whose contact lens care system contained many *Acanthamoeba* mixed with *Xanthomonas maltophilia*, and many of the bacteria were adherent to amoebae. The growth of *Acanthamoeba* in the contact lens care system, in the presence of bacteria such as *Xanthomonas maltophilia*, *Flavobacterium breve*, and *Pseudomonas paucimobilis* showed enhanced *Acanthamoeba* growth (1.5X) in the presence of bacteria (Bottone *et al.*, 1992). With the remarkable implications of parasite-parasite interactions, which may contribute to the evolution of one (either bacteria or *Acanthamoeba*) or both parasites to become successful human and animal pathogens, this area of research should get increasing attention from the scientific and the medical community in future.

From: *Acanthamoeba*: Biology and Pathogenesis. By: Naveed Khan.

Section H

Conclusions

1. A model organism with the pathogenic potential

Studies involving *Acanthamoeba* have grown exponentially. This organism has gained increasing attention from the broad scientific and the medical community, studying cellular microbiology, environmental biology, physiology, cellular interactions, molecular biology, biochemistry and the evolutionary studies. This is due to their versatile roles in the ecosystem and their ability to capture prey by phagocytosis (similar to macrophages), act as vectors, reservoirs and Trojan horses for microbial pathogens, and to produce serious human infections, including a blinding keratitis and a fatal encephalitis. In addition, this unicellular organism has been used extensively to understand the molecular biology of the motility.

Acanthamoeba does not differ greatly at the ultrastructural-level from a mammalian cell, thus presents an excellent model for the cell biology studies. *Acanthamoeba* trophozoites can be permeabilized using digitonin. The digitonin permeabilization alters neither the morphology nor the motility of the cell, but favours the penetration of phalloidin and viroisin. Furthermore, the ability of *Acanthamoeba* to switch phenotypes makes it an attractive model for the study of cellular differentiation processes. The increasing numbers of HIV/AIDS patients, contact lens wearers, and warmer climates, will add to the increasing importance of *Acanthamoeba* infections. Notably, the burden of *Acanthamoeba* infections is not truly appreciated. This is due to the lack of awareness, difficulty in diagnosis, and most distressingly a lack of effective treatment. A vast number of *Acanthamoeba* cases have most likely been undetected, and the actual number must be significantly higher (Khan, 2006). For example, there is no single report of *Acanthamoeba* granulomatous encephalitis in Africa, despite millions of HIV-infected individuals, and who are susceptible hosts to the opportunistic pathogens, and live in a warm climate, where there is probable frequent environmental exposure and subordinate sanitation in many places. Our understanding of *Acanthamoeba* pathogenesis, both at the molecular-level and at the cellular-level, as well as its ability to transmit, adapt to diverse conditions, overcome the host barriers and emerge as an infective trophozoite will provide targets for therapeutic interventions.

2. New approaches in the study of *Acanthamoeba*

2.1. Genome and transfection assay

Previous studies in the field of *Acanthamoeba* have been challenging due to the unavailability of the genome information of this pathogen, and due to difficulties in the development of assays to genetically manipulate this organism. To this end, *Acanthamoeba* genome project is partially completed that will pave the way for our much needed understanding of this fascinating organism, that also has a unique phylogenetic position. In addition, a simple method for the stable transfection of *Acanthamoeba* using plasmids, which confer resistance to the neomycin G418, has been developed. The expression of neomycin phosphotransferase is driven by the *Acanthamoeba* TATA-binding protein (TBP, that is a transcription factor that binds specifically to a DNA sequence called the TATA box) gene promoter, and can be monitored by cell growth in the presence of neomycin G418 or by Western blot analysis. The transfected *Acanthamoeba* can be passaged in the same manner as the control cells and can be induced to differentiate into cysts, in which form they maintain resistance to neomycin G418 for at least several weeks, although expression of neomycin phosphotransferase is repressed during encystation. The expression of EGFP or an HA-tagged EGFP-TBP fusion can be driven from the same plasmid, using an additional copy of the *Acanthamoeba* TBP gene promoter or a deletion mutant. The TBP-EGFP fusion is localized to the nucleus, except in a small proportion of presumptive pre-mitotic cells. The EGFP expression can also be driven by the cyst-specific protein 21 (CSP21) gene promoter, which is completely repressed in growing cells but strongly induced in differentiating cells. The transfected amoebae maintain their phenotype for several weeks, even in the absence of neomycin G418, suggesting that transfected genes are stably integrated within the *Acanthamoeba* genome. These developments demonstrate the utility of the neomycin resistance-based plasmids for stable transfection of *Acanthamoeba*, and should assist investigations in the biology and pathogenesis of *Acanthamoeba* (Peng *et al*., 2005).

Overall, the availability of the *Acanthamoeba* genome, together with the recently developed transfection assays (Peng *et al.,* 2005; Yin and Henney, 1997), and the RNA interference methods (Lorenzo-Morales *et al.,* 2005), will undoubtedly increase the pace of our understanding of this complex but fascinating organism.

2.2. Drug discovery and delivery

The involvement of the central nervous system in *Acanthamoeba* infection almost always results in death. This is due to the unavailability of effective and/or recommended drugs, and the inability of drugs to traverse the blood-brain barrier to gain entry into the brain to kill parasites. Current treatment usually involves a combination drug approach, given intravenously, by including various drugs such as ketoconazole, fluconazole, sulfadiazine, pentamidine isethionate, amphotericin B, azithromycin, itraconazole, or rifampicin that rarely leads to improved prognosis and heavily depends on early diagnosis, followed by aggressive treatment. Thus, there is an urgent need to initiate both drug discovery and drug delivery programmes.

Given the urgency of finding new anti-*Acanthamoeba* compounds, drug discovery should be attempted by identifying active compounds from the natural products (traditional approach) and by understanding how disease and infection are controlled at the molecular and physiological-level, thus identifying specific target, which should aid in the rational development of drugs.

Until novel anti-*Acanthamoeba* compounds are available with increased potency, perhaps improved drug delivery for the known anti-*Acanthamoeba* compounds may have potential in the enhanced treatment of *Acanthamoeba* infections. Currently, the delivery system for the central nervous system drugs may include (i) trans-cranial drug delivery to the brain, i.e., drug delivery to the ventricles of the brain by injection through a trephine hole in the cranium (intracerebroventricular injection, intracerebral injection or convection-enhanced diffusion), (ii) trans-nasal drug delivery to the brain, a method that mimics the current *in vivo Acanthamoeba* encephalitis model in mice (lipid-soluble small molecules may be instilled nasally, cross the nasal mucosa and the arachnoid membrane and enter the olfactory cerebrospinal fluid with the result similar to intracerebroventricular injection), and (iii) transient blood-brain barrier disruption (arterial infusion of hyperosmotic solution or ultrasonic irradiation of the brain). All of the above methods may have complications due to side effects. Other approaches may include lipidization of small molecules to enhance transport across the blood-brain barrier, carrier-mediated transport of drugs, or administration of non-viral plasmid DNA encoding antisense RNA against the virulence genes of *Acanthamoeba*. Some of these approaches may hold promise in the successful prognosis of *Acanthamoeba* infections.

3. Future Research

At the International Symposium on *Acanthamoeba* in 1991, participants discussed plans for future research priorities and devised a "Program Planning for Research on *Acanthamoeba*" (Wilhelmus and Jones, 1991). Since then, there has been tremendous progress; however there is an urgent need for continued research in the development of strategies in the prevention, pathogenesis and therapy for *Acanthamoeba* infections.

3.1. Prevention

- Development of improved contact lenses that resist contamination and can be cared for easily and effectively.
- Development of new multipurpose contact lens disinfectants that can target, protozoa, fungi, viruses and bacteria.
- Improvement of educational efforts aimed at contact lens wearers to maintain proper hygiene in contact lens care.
- Identification of anti-*Acanthamoeba* compounds with minimal side effects that can be included in the highly active antiretroviral therapy

(HAART), which should prevent granulomatous amoebic encephalitis due to *Acanthamoeba* in the susceptible population.

3.2. Diagnosis

- Awareness aimed at ophthalmologist, neurologists, physicians, contact lens wearers.
- Continued development of rapid, cost-effective, antibody tests, antigen detection, and molecular-based assays in the early diagnosis of *Acanthamoeba* infections.
- Demonstrate the efficacy of diagnostic assays in clinical specimens (tissue biopsies, blood, cerebrospinal fluid), contact lens and storage case solution, as well as environmental samples.
- Determine the usefulness of sputum samples as a diagnostic tool for disseminated acanthamoebiasis and granulomatous amoebic encephalitis due to *Acanthamoeba*.

3.3. Basic Research

- Epidemiology studies to determine the true burden of *Acanthamoeba* infections, both in humans and animals.
- Do disease patterns differ around the globe?
- Effect of environmental factors (high temperature, low humidity, dust etc), demographic factors, socioeconomic factors, and malnutrition on *Acanthamoeba* infections.
- Who are the most likely carriers of *Acanthamoeba* and their physiological state?
- Continued assessment of host factors of *Acanthamoeba* patients, including hygiene, immunity, nutrition, occupation and underlying disease.
- Type and net state of immune-suppression (as well as type of transplant) in *Acanthamoeba* granulomatous encephalitis.
- For serious infections, corneal and neural tissues are the primary targets.
- What is the role of biofilm in *Acanthamoeba* keratitis? Survey contact lens and cases of *Acanthamoeba* keratitis patients for the presence of biofilms?
- Continuation of investigation of the disease processes *in vivo* and *in vitro*.
- Elucidation of the stages of *Acanthamoeba* pathogenesis.
- Understanding of mechanisms of encystation and excystation.
- Understanding of metabolic pathways in *Acanthamoeba*.
- Determination of the roles of inflammation and immune responses in *Acanthamoeba* infections.
- Determination of the roles of non-immunological defence mechanisms in *Acanthamoeba* infections.

3.4. Host for microbial pathogens

- Does *Acanthamoeba* act as a host for other microbial pathogens?
- Survey of *Acanthamoeba* in the environment and determination of endosymbionts.

- Understanding *Acanthamoeba* and endosymbionts interactions and their evolutionary relationships.
- How do endosymbionts act in clinical and non-clinical isolates of *Acanthamoeba* as well as in different genotypes?
- Do endosymbionts contribute to *Acanthamoeba* pathogenesis and vice versa?
- What are the complications of concurrent infections due to *Acanthamoeba* and the endosymbionts?
- Do *Acanthamoeba* contribute to the evolution microbial resistance to drugs and vice versa?
- How do parasite-parasite interactions affect their ecological distribution and transmission to the susceptible people?

4. Treatment

- Identification of novel antimicrobial agents for both therapeutic and prophylactic effects from natural products and rational development of therapeutic approaches against specific targets.
- Identification of possible vaccine targets.
- Development of targeted anti-*Acanthamoeba* chemo-therapeutic and immuno-therapeutic approaches.

From: *Acanthamoeba*: Biology and Pathogenesis. By: Naveed Khan.

Section I

Bibliography and Further Reading

Abd H, Johansson T, Golovliov I, Sandström G and Forsman M. (2003). Survival and growth of Francisella tularensis in *Acanthamoeba castellanii*. Appl Environ Microbiol. 69(1): 600-6.

Abedin A, Mohammed I, Hopkinson A and Dua HS. (2008). A novel antimicrobial peptide on the ocular surface shows decreased expression in inflammation and infection. Invest Ophthalmol Vis Sci. 49: 28-33.

Achar SB and Weisman RA. 1980. Adenylate cyclase activity during growth and encystment of *Acanthamoeba castellanii*. Biochim Biophys Acta. 629(2): 225-234.

Admoni MM, Bartolomei A, Qureshi MN, Bottone EJ and Asbell PA. (1993). Disinfection efficacy in an integrated ultraviolet light contact lens care system. CLAO J. 20: 246-248.

Aebi, U, Smith PR, Isenberg G and Pollard TD. 1980. Structure of crystalline actin sheets. Nature. 288(5788): 296-298.

Aebi, U, Fowler WE, Isenberg G, Pollard TD and Smith PR. 1981. Crystalline actin sheets: their structure and polymorphism. J Cell Biol. 340-351.

Aksozek A, McClellan K, Howard K, Niederkorn JY and Alizadeh H. (2002). Resistance of *Acanthamoeba castellanii* cysts to physical, chemical, and radiological conditions. J Parasitol. 88: 621-623.

Albanesi JP, Lynch TJ, Fujisaki H and Korn ED. (1986). Regulation of the actin-activated ATPase activity of *Acanthamoeba* myosin I by cross-linking actin filaments. J Biol Chem. 261: 10445-10449.

Albanesi JP, Lynch TJ, Fujisaki H, Bowers B and Korn ED. (1987). Purification and characterization of an ATP-sensitive actin gelation protein from *Acanthamoeba castellanii*. J Biol Chem. 262: 3404-3408.

Alfieri SC, Correia CE, Motegi SA and Pral EM. (2000). Proteinase activities in total extracts and in medium conditioned by *Acanthamoeba polyphaga* trophozoites. J Parasitol. 86(2): 220-227.

Alizadeh H, Pidherney MS, McCulley JP and Niederkorn JY. (1994). Apoptosis as a mechanism of cytolysis of tumor cells by a pathogenic free-living amoeba. Infect Immun. 62(4): 1298-1303.

Alizadeh H, He Y, McCulley JP, Ma D, Stewart GL, Via M, Haehling E and Niederkorn JY. (1995). Successful immunization against *Acanthamoeba* keratitis in a pig model. Cornea. 14: 180-186.

Alizadeh H, Apte S, El-Agha MS, Li L, Hurt M, Howard K, Cavanagh HD, McCulley JP and Niederkorn JY. (2001). Tear IgA and serum IgG antibodies against *Acanthamoeba* in patients with *Acanthamoeba* keratitis. Cornea. 20(6): 622-627.

Alizadeh H, Neelam S and Niederkorn JY. (2007). Role of activated macrophages in *Acanthamoeba* keratitis. J Parasitol. 93: 1114-1120.

Alizadeh H, Neelam S and Niederkorn JY. (2007). Effect of immunization with the mannose-induced *Acanthamoeba* protein and *Acanthamoeba* plasminogen activator in mitigating *Acanthamoeba* keratitis. Invest Ophthalmol Vis Sci. 48: 5597-5604.

Allen PG and Dawidowicz EA. (1990). Phagocytosis in *Acanthamoeba*: 1. A mannose receptor is responsible for the binding and phagocytosis of yeast. J Cell Physiol. 145: 508-513.

Allen PG and Dawidowicz EA. (1990a). Phagocytosis in *Acanthamoeba*: II. Soluble and insoluble mannose-rich ligands stimulate phosphoinositide metabolism. J Cell Physiol. 145: 514-521.

Alsam S, Kim KS, Stins M, Rivas AO, Sissons J and Khan NA. (2003). *Acanthamoeba* interactions with human brain microvascular endothelial cells. Microb Pathogen 35(6): 235-241.

Alsam S, Sissons J, Dudley R and Khan NA. (2005a). Mechanisms associated with *Acanthamoeba castellanii.* (T4) phagocytosis. Parasitol Res. 96(6): 402 – 409.

Alsam S, Sissons J, Jayasekera S and Khan NA. (2005b). Extracellular proteases of *Acanthamoeba castellanii*. (encephalitis isolate belonging to T1 genotype) contribute to increased permeability in an *in vitro* model of the human blood-brain barrier. J Infect. 51(2): 150-156.

Alsam S, Jeong SR, Sissons J, Dudley R, Kim KS and Khan NA. (2006). *Escherichia coli* interactions with *Acanthamoeba*: a symbiosis with environmental and clinical implications. J Med Microbiol. 55: 689-694

Alvord L, Court J, Davis T, Morgan CF, Schindhelm K, Vogt J and Winterton L. (1998). Oxygen permeability of a new type of high risk soft contact lens material. Optom Vis Sci. 75: 30-36.

Ampe C, Sato M, Pollard TD and Vandekerckhove J. (1988). The primary structure of the basic isoform of *Acanthamoeba* profilin. Eur J Biochem. 170: 597-601.

Anand CM, Skinner AR, Malic A, Kurtz JB. (1983). Interaction of *L. pneumophila* and a free-living amoeba. (*Acanthamoeba palestinensis*). J Hyg. (Lond). 91: 167-178.

Anderson IJ, Watkins RF, Samuelson J, Spencer DF, Majoros WH, Gray MW and Loftus BJ. (2005). Gene discovery in the *Acanthamoeba castellanii* genome. Protist. 156: 203-214.

Badenoch PR, Adams M and Coster DJ. (1994). Corneal virulence, cytopathic effect on human keratocytes and genetic characterization of *Acanthamoeba*. Int J Parasitol. 25: 229-239.

Baily CF and Bowers B. (1981). Localization of lipophosphonoglycan in membranes of *Acanthamoeba* by using specific antibodies. Mol Cell Biol. 1(4): 358-369.

Baines IC and Korn ED. (1990). Localization of myosin IC and myosin II in *Acanthamoeba castellanii* by indirect immunofluorescence and immunogold electron microscopy. J Cell Biol. 111: 1895-1904.

Baines IC, Brzeska H, Korn ED. (1992). Differential localization of *Acanthamoeba* myosin I isoform. J Cell Biol. 119: 1193-1203.

Baines IC, Corigliano-Murphy A and Korn ED. (1994). Quantification and localization of phosphorylated myosin I isoforms in *Acanthamoeba castellanii*. J Cell Biol. 130: 591-603.

Baldo ET, Moon EK, Kong HH and Chung DI. (2005). *Acanthamoeba healyi*: molecular cloning and characterization of a coronin homologue, an actin-related protein. Exp Parasitol. 110: 114-122.

Baldwin KM and Bowers B. (1995). Isolation of N-acetyl-beta-hexosaminidase from *Acanthamoeba castellanii*. J Eukaryot Microbiol. 42: 237-342.

Band RN and Mohrlok S. (1973). The cell cycle and induced amitosis in *Acanthamoeba*. J Protozool. 20(5): 654-657.

Barker J, Brown MW, Collier PJ, Farrell I and Gilbert P. (1992). Relationship between *Legionella pneumophila* and *Acanthamoeba polyphaga*: physiological status and susceptibility to chemical inactivation. Appl Environ Microbiol. 58: 2420-2425.

Barker J, Lambert PA and Brown MW. (1993). Influence of intra-amoebic and other growth conditions on the surface properties of *Legionella pneumophila*. Infect Immun. 61: 3503-3510.

Barker J, Humphrey TJ and Brown MW. (1999). Survival of *Escherichia coli* O157 in a soil protozoan: implications for disease. FEMS Microbiol Lett. 173(2): 291-295.

Baron D, Danglot C and Vilagines R. (1980). Role of a free-living amoeba from water, *Acanthamoeba castellanii*, in transport of naked or enveloped animal viruses. C R Seances Acad Sci D. 291(7): 629-632.

Barr, J. (1998). The 1997 report on contact lenses. Contact Lens Spec. 1: 23-33.

Barrett, RA and Alexander M. (1977). Resistance of cysts of amoebae to microbial decomposition. Appl Environ Microbiol. 33(3): 670-674.

Batzri S and Korn ED. (1975). Interaction of phospholipid vesicles with cells. Endocytosis and fusion as alternate mechanisms for the uptake of lipid-soluble and water-soluble molecules. J Cell Biol. 66: 621-634.

Bauer H. (1967). Ultrastruktur and Zellwand-bildug von *Acanthamoeba* sp. Vierteljahresschr. Naturforsch. Ges. Zurich 112: 173.

Baumann O and Murphy DB. (1995). Microtubule-associated movement of mitochondria and small particles in Acanthamoeba castellanii. Cell Motil Cytoskeleton. 32(4): 305-17.

Beattie TK, Tomlinson A and Seal DV. (2002). Anti-*Acanthamoeba* efficacy in contact-lens disinfeting systems. Br J Ophthalmol. 86: 1319-1320.

Beattie TK, Tomlinson A, McFadyen AK, Seal DV and Grimason AM. (2003). Enhanced attachment of *Acanthamoeba* to extended-wear silicone hydrogel contact lenses: a new risk factor for infection? Ophthalmol 110(4): 765-771.

Beattie TK, Seal DV, Tomlinson A, McFadyen AK, Grimason AM. (2003). Determination of amoebicidal activities of multipurpose contact lens solutions by using a most probable number enumeration technique. J Clin Microbiol. 41: 2992-3000.

Benedetto N and Auriault C. (2001). Prolactin-cytokine network in the defence against *Acanthamoeba castellanii* in murine microglia [corrected]. Eur Cytokine Netw. 13: 447-455.

Benedetto N and Auriault C. (2002). Complex network of cytokines activating murine microglial cell activity against *Acanthamoeba castellanii*. Eur Cytokine Netw. 13: 351-357.

Benedetto N, Rossano F, Gorga F, Folgore A, Rao M, Romano and Carratelli C. (2003). Defense mechanisms of IFN-gamma and LPS-primed murine microglia against *Acanthamoeba castellanii* infection. Int Immunopharmacol. 3: 825-834.

Binder PS. (1989). Cryotherapy for medically unresponsive *Acanthamoeba* keratitis. Cornea. 8: 106-114.

Biron, D., Libros, P., Sagi, D., Mirelman, D. and Moses, E. (2001). 'Midwives' assist dividing amoebae. Nature. 410: 430.

Bose K, Ghosh DK and Bhattacharya A. (1989). Membrane carbohydrate characterization of *Acanthamoeba astronyxis*, *A. castellanii* and *Naegleria fowleri* by fluorescein-conjugated lectins. Int J Parasitol. 19: 737-741.

Bottone EJ, Madayag RM and Qureshi MN. (1992). *Acanthamoeba* keratitis: synergy between amebic and bacterial cocontaminants in contact lens care systems as a prelude to infection. J Clin Microbiol. 30: 2447-2450.

Bottone EJ, Perez AA, Gordon RE and Qureshi MN. (1994). Differential binding capacity and internalisation of bacterial substrates as factors in growth rate of *Acanthamoeba* spp. J Med Microbiol. 40(2): 148-154.

Bouyer S, Imbert C, Daniault G, Cateau E and Rodier MH. (2007). Effect of caspofungin on trophozoites and cysts of three species of *Acanthamoeba*. J Antimicrob Chemother. 59: 122-124.

Bowers B and Korn ED. (1968). The fine structure of *Acanthamoeba castellanii*. I. The trophozoite. J Cell Biol. 39(1): 95-111.

Bowers B and Korn ED. (1969). The fine structure of *Acanthamoeba castellanii*. II. Encystment. J Cell Biol. 41: 786-805.

Bowers B and Olszewski TE. (1972). Pinocytosis in *Acanthamoeba castellanii*, kinetics and morphology. J Cell Biol. 53(3): 681-694

Bowers B and Korn ED. (1973). Cytochemical identification of phosphatase activity in the contractile vacuole of *Acanthamoeba castellanii*. J Cell Biol. 59: 784-791.

Bowers B and Korn ED. (1974). Localization of lipophosphonoglycan on both sides of *Acanthamoeba* plasma membrane. J Cell Biol. 62(2): 533-540.

Bowers B. (1980). A morphological study of plasma and phagosome membranes during endocytosis in *Acanthamoeba*. J Cell Biol. 84: 246-260.

Bowers B, Olszewski TE and Hyde J. (1981). Morphometric analysis of volumes and surface areas in membrane compartments during endocytosis in *Acanthamoeba*. J Cell Biol. 88: 509-515.

Bowers B and Olszewski TE. (1983). *Acanthamoeba* discriminates internally between digestible and indigestible particles. J Cell Biol. 97: 317

Brandt FH, Ware DA and Visvesvara GS. (1989). Viability of *Acanthamoeba* cysts in ophthalmic solutions. Appl Environ Microbiol. 55: 1144-1146.

Brasseur G, Favennec L, Perrine D, Chenu JP and Brasseur P. (1994). Successful treatment of *Acanthamoeba* keratitis by hexamidine. Cornea. 13: 459-462.

Brennan NA. (2002). Is there a question of safety with continuous wear? Clin Exp Optom 85(3): 127-140.

Brooks SE and Schneider DL. 1985. Oxidative metabolism associated with phagocytosis in *Acanthamoeba castellanii*. J Protozool. 32: 330-333.

Brown TJ and Cursons RTM. (1977). Pathgenic free-living amoebae from frozen swimming areas in Oslo, Norway. Scand J Infect Dis. 9: 237-240.

Brown TJ and Cursons RTM, Keys EA. (1982). Amoebae from Antarctica soil and water. Appl Environ Microbiol 44(2): 491-493.

Brown MRW and Barker J. (1999). Unexplored reservoirs of pathogenic bacteria: protozoa and biofilms. Trends Microbiol. 7: 46-50.

Bryant RJ, Woods LE, Coleman DC, Farbanks BC, McClellan JF and Cole CV. (1982). Interactions of bacterial and amoebal populations in soil microcosms with fluctuating moisture contnt. Appl Environ Microbiol. 43(4): 747-752.

Bryant AE and Stevens DL. (1996). Phospholipase C and perfringolysin O from *Clostridium perfringens* upregulate endothelial cell-leukocyte adherence molecule 1 and intercellular leukocyte adherence molecule 1 expression and induce interleukin-8 synthesis in cultured human umbilical vein endothelial cells. Infect Immun. 64(1): 358-362.

Brzeska H, Lynch TJ and Korn ED. (1988). Localization of the actin-binding sites of *Acanthamoeba* myosin IB and effect of limited proteolysis on its actin-activated Mg^{2+}-ATPase activity. J Biol Chem. 263: 427-435.

Brzeska H, Lynch TJ, Martin B and Korn ED. (1989). The localization and sequence of the phosphorylation sites of *Acanthamoeba* myosins I. J Biol Chem. 264: 19340-19348.

Bunting LA, Neilson JB and Bulmer GS. (1979). *Cryptococcus neoformans*: gastronomic delight of a soil amoebae. Sabouraudia. 17(3): 225-232.

Bunting M, Lorant DE, Bryant AE, Zimmerman GA, McIntyre TM, Stevens DL and Prescott SM. (1997). Alpha toxin from *Clostridium perfringens* induces proinflammatory changes in endothelial cells. J Clin Invest. 100(3): 565-574.

Burger RM, Franco RJ and Drlica K. (1994). Killing *Acanthamoeba*e with polyaminopropyl biguanide: quantitation and kinetics. Antimicrob. Agents Chemother. 38: 886-888.

Burger G, Plante I, Lonergan KM and Gray MW. (1995). The mitochondrial DNA of the amoeboid protozoon, *Acanthamoeba castellanii*: complete sequence, gene content and genome organization. J Mol Biol. 245: 522-537.

Byers TJ, Rudick VL and Rudick MJ. (1969). Cell size, macromolecule composition, nuclear number, oxygen consumption and cyst formation during two growth phases in unagitated cultures of *Acanthamoeba castellanii*. J Protozool. 16: 693-9.

Byers TJ. (1979). Growth, reproduction, and differentiation in *Acanthamoeba*. Int Rev Cytol 61: 283-338

Byers TJ, Hugo ER and Stewart VJ. (1990). Genes of *Acanthamoeba*: DNA, RNA and protein sequences. (a review). J Protozool. 37(4): 17S-25S.

Byers TJ, Kim BG, King LE and Hugo ER. (1991). Molecular aspects of the cell cycle and encystment of *Acanthamoeba*. Rev Infect Dis. 13: S373-S384.

Cabral GA and Marciano-Cabral F. (2004). Cannabinoid-mediated exacerbation of brain infection by opportunistic amebae. J Neuroimmunol. 147: 127-130.

Cabral GA, Raborn ES, Griffin L, Dennis J and Marciano-Cabral F. (2008). CB2 receptors in the brain: role in central immune function. Br J Pharmacol. 153: 240-251.

Cao Z, Jefferson MD and Panjwani N. (1998). Role of carbohydrate-mediated adherence in cytopathogenic mechanisms of *Acanthamoeba*. J Biol Chem. 273: 15838-15845.

Carter, RF, Cullity GJ, Ojeda VJ, Silberstein P and Willaert E. (1981). A fata case of meningoencephalitis due to free-living amoeba of uncertain indentity-probably *Acanthamoeba* sp. Pathology. 13(1): 51-68.

Carter WW, Gompf SG, Toney JF, Greene JN and Cutolo EP. (2004). Disseminated *Acanthamoeba* sinusitis in a patient with AIDS: a possible role for early antiretroviral therapy. AIDS Read. 14(1): 41-49

Casemore DP. (1970). Sensitivity of Hartmanella. (*Acanthamoeba*) to 5-fluorocytosine, hydroxystilbamidine and other substances. J Clin Pathol. 23: 649-652.

Castellani A. (1930). An amoeba found in culture of yeast: preliminary note. J Trop Med Hyg p.160.

Center for Disease Control. (1986). *Acanthamoeba* keratitis associated with contact lenses-United States. MMWR, 35: 405-408.

Cerva L. (1981). Spontaneous occurance of antibodies against pathogenic amoebae of the limax group in domestic animals. Folia Parasitol. (Praha). 28(2): 105-108.

Cerva L. (1989). *Acanthamoeba culbertoni* and *Naegleria fowleri*: occurrence of antibodies in man. J Hyg, Epidem, Microbiol Immunol. 33: 99-103

Chapman-Andersen C. (1965). The induction of pinocytosis in amoebae. Arch Biol 76: 189.

Chang PC and Soong HK. (1991). *Acanthamoeba* keratitis in non-contact lens wearers. Arch Ophthalmol. 109(4): 463-464.

Chappell CL, Wright JA, Coletta M and Newsome AL. (2001). Standardized method of measuring *Acanthamoeba* antibodies in sera from healthy human subjects. Clin Diagn Lab Immunol. 8(4): 724-730.

Cheng KH, Leung SL, Hoekman HW, Beekhuis WH, Mulder PGH, Geerards AJM and Kijlstra A. (1999). Incidence of contact-lensassociated microbial keratitis and its related morbidity. Lancet 354(9174): 181-185.

Chew SJ, Beuerman RW, Assouline M, Kaufman HE, Barron BA and Hill JM. (1992). Early diagnosis of infectious keratitis with *in vivo* real time confocal microscopy. CLAO J. 18: 197-201.

Chisholm GE and Vaughan MH. (1979). Isolation and characterization of a cycloheximide-resistant mutant of *Acanthamoeba castellanii* Neff. J Biol Chem. 138(1): 280-283.

Chlapowski FJ and Band RN. (1971). Assembly of lipids into membranes in *Acanthamoeba palestinensis*. II. The origins and fat of glycerol-^{3}H-labeled phospholipids of cellulare membranes. J Cell Biol 50: 634-651.

Chlapowski FJ and Butcher RW. (1986). Activation of adenylate cyclase in *Acanthamoeba palestinensis*. Life Sci. 38: 849-859.

Cho JH, Na BK, Kim TS and Song CY. (2000). Purification and characterization of an extracellular serine proteinase from *Acanthamoeba castellanii*. IUBMB Life. 50: 209-214.

Choi DH, Na BK, Seo MS, Song HR and Song CY. (2000). Purification and characterization of iron superoxide dismutase and copper-zinc superoxide dismutase from *Acanthamoeba castellanii*. J Parasitol. 86: 899-907.

Chu DM, Miles H, Toney D, Ngyuen C and Marciano-Cabral F. (1998). Amebicidal activity of plant extracts from Southeast Asia on *Acanthamoeba* spp. Parasitol Res. 84: 746-752.

Cirillo JD, Falkow S and Tompkins LS. (1994). Growth of Legionella pneumophila in *Acanthamoeba castellanii* enhances invasion. Infect Immun. 62: 3254-3261.

Cirillo JD, Falkow S, Tompkins LS and Bermudez LE. (1997). Interaction of *Mycobacterium avium* with environmental amoebae enhances virulence. Infect Immun. 65(9): 3759-3767.

Cirillo JD, Cirillo SL, Yan L, Bermudez LE, Falkow S and Tompkins LS. (1999). Intracellular growth in *Acanthamoeba castellanii* affects monocyte entry mechanisms and enhances virulence of *Legionella pneumophila*. Infect Immun. 67: 4427-4434.

Clarholm M. (1981). Protozoan grazing of bacteria in soil – impact and importance. Microb Ecol. 7: 343-350.

Clarke DW, Alizadeh H and Niederkorn JY. (2005). Failure of *Acanthamoeba castellanii* to produce intraocular infections. Invest Ophthalmol Vis Sci. 46: 2472-2478.

Cohn ZA. (1971). Endocytosis and the vacuolar systems. *In* Cell Membranes. Biological and Pathological aspects. Goetz W. Richter and Dante G. Scarpelli, editors. Williams Wilkins Co., Baltimore.

Cohen EJ, Parlato CJ, Arentsen JJ, Genvert GI, Eagle RC, Wieland MR and Laibson PR. (1987). Medical and surgical treatment of *Acanthamoeba* keratitis. Am J Ophthalmol. 103: 615-625.

Connor CG, Blocker Y and Pitts DG. (1989). *Acanthamoeba culbertsoni* and contact lens disinfection systems. Optom Vis Sci. 66: 690-693.

Connor CG, Hopkins SL, Salisbury RD. (1991). Effectivity of contact lens disinfection systems against *Acanthamoeba culbertsoni*. Optom Vis Sci. 68: 138-141.

Cooper JA, Blum JD, Williams RC Jr and Pollard TD. (1986). Purification and characterization of actophorin, a new 15,000-dalton actin-binding protein from *Acanthamoeba castellanii*. J Biol Chem. 261: 477 – 485.

Cordingley JS, Wills RA and Villemez CL. (1996). Osmolarity is an independent trigger of *Acanthamoeba castellanii* differentiation. J Cell Biochem. 61: 167-171.

Couso R, Lang L, Roberts RM and Kornfeld S. (1986). Phosphorylation of the oligosaccharide of uteroferrin by UDP-GlcNAc: glycoprotein N-acetylglucosamine-1-phosphotransferases from rat liver, *Acanthamoeba castellanii,* and *Dictyostelium discoideum* requires α1,2-linked mannose residues. J Biol Chem. 261: 6326-6331.

Culbertson CG, Smith JW and Miner JR. (1958). *Acanthamoeba*: Observation on animal pathogenicity. Science. 127: 1506.

Culbertson CG, Smith JW, Cohen HK and Miner JR. (1959). Experimental infection of mice and monkeys by *Acanthamoeba*. Am J Pathol. 35: 185-197.

Cursons RTM, Brown TJ and Keys EA. (1978). Virulence of pathogenic free-living amoebae. J Parasitol. 64: 744-745.

Cursons RT, Brown TJ and Keys EA. (1980). Effect of disinfectants on pathogenic free-living amoebae: in axenic conditions. Appl Environ Microbiol. 40(1): 62-66.

Cursons RT, Brown TJ, Keys EA, Moriarty KM and Till D. (1980). Immunity to pathogenic free-living amoebae: role of humoral antibody. Infect Immun. 29(2): 401-407.

Czarna M and Jarmuszkiewicz W. (2005). Activation of alternative oxidase and uncoupling protein lowers hydrogen peroxide formation in amoeba *Acanthamoeba castellanii* mitochondria. FEBS Lett. 579: 3136-3140.

Czarna M, Sluse FE and Jarmuszkiewicz W. (2007). Mitochondrial function plasticity in *Acanthamoeba castellanii* during growth in batch culture. J Bioenerg Biomembr. 39: 149-157.

D'Alessio JM, Spindler SR. and Paule MR. (1979). DNA-dependent RNA polymerase II from *Acanthamoeba castellanii*. J Biol Chem. 254(10): 4085-4091.

D'Alessio JM, Harris GH, Perna PJ and Paule MR. (1981). Ribosomal ribonucleic acid repeat unit of *Acanthamoeba castellanii*: cloning and restriction endonuclease map. Biochem. 20(13): 3822-3827.

Danes L and Cerva L. (1981). Survival of polioviruses and echoviruses in *Acanthamoeba castellanii* cultivated *in vitro*. J Hyg Epidemiol Microbiol Immunol. 25(2): 169-174.

Dart JKG, Stapleton F and Minassian D. (1991). Contact lenses and other risk factors in microbial keratitis. Lancet. 338: 650-653.

Das SR, Asiri S, el-Soofi A and Baer HP. (1991). Protective and curative effects of rifampicin in *Acanthamoeba* meningitis of the mouse. J Infect Dis. 163: 916-917.

David JT. (1993). Opportunistically pathogenic free-living amoebae. Parasitic Protozoa., 2nd Ed. Vol 3, Academic Press Inc., San Diego.

Davis D, Edwards JE, Mitchell AP and Ibrahim AS. (2000). *Candida albicans RIM101* pH response pathway is required for host pathogen interactions. Infect Immun. 68: 5953-5959.

Dearborn DG and Korn ED. (1974). Lipophophonoglycan of the plasma membrane of *Acanthamoeba castellanii*, fatty acid composition. J Biol Chem. 249(11): 3342-3346.

Dearborn DG, Smith S and Korn ED. (1976). Lipophophonoglycan of the plasma membrane of *Acanthamoeba castellanii*. Inositol and phytosphingosine content and general structural features. J Biol Chem. 251(10): 2976-2982.

De Bernardis F, Muhlschlegel FA, Cassone A and Fonzi WA. (1998). The pH of the host niche controls gene expression in and virulence of *Candida albicans*. Infect Immun. 66: 3317-3325.

De Jonckheere J and De Voorde HV. (1976). Differences in destruction of cysts of pathogenic and non-pathogenic *Naegleria* and *Acanthamoeba* by chlorine. Appl Environ Microbiol. 31(2).: 294-297.

De Jonckheere JF. (1980). Growth characteristics, cytopathic effect in cell culture and virulence in mice of 36 type strains belonging to 19 different *Acanthamoeba* spp. Appl Environ Microbiol. 39: 681-685.

De Jonckheere JF. (1981). Pathogenic and non-pathogenic *Acanthamoeba* spp. In thermally polluted discharges and surface waters. J Protozool. 28(1): 56-59.

De Jonckheere J. (1987). Taxonomy. In: Rondanelli, E.G., ed. Amphizoic amoebae human pathology. Padua, Italy: Piccin Nuova Libraria, 25-48.

De Jonckheere JF. (1991). Ecology of *Acanthamoeba*. Rev Infect Dis. 13. (suppl. 5): S385-S387.

De Jonckheere JF. (2002). A century of research on the Amoeboflagellate Genus *Naegleria*. Acta Protozool. 41: 309-342.

Dennis EA, Rhee SG, Billah MM and Hannun YA. (1991). Role of phospholipase in generating lipid second messengers in signal transduction. FASEB J. 5(7): 2068-2077.

Deslauriers R, Jarrell HC, Byrd RA and Smith ICP. (1980). Observation by ^{13}C NMR of metabolites in differentiating amoeba. FEBS Lett. 118(2): 185-190.

Detke, S and Paule, M. R. (1978). DNA-dependent RNA polymerases from *Acanthamoeba castellanii*. Multiple forms of the class III enzyme and levels of activity of the polymerase classes during encystment. Biochim. Biophys Acta. 520(1): 131-138.

Detke, S and Paule, M. R. (1978). DNA-dependent RNA polymerase from *Acanthamoeba castellanii*. Comparison of the catalytic properties and subunit architectures of the trophozoites and cyst enzymes. Biochim. Biophys Acta. 520(2): 376-392.

Detke, S and Paule, M. R. (1979). DNA-dependent RNA polymerase III from *Acanthamoeba castellanii*. Comparison of the catalytic properties of the trophozoites and cyst enzymes. J Protozool. 26(2): 319-323.

Devonshire P, Munro FA, Abernethy C and Clark BJ. (1993). Microbial contamination of contact lens cases in the west of Scotland. Br J Ophthalmol. 77(1): 41-45.

Diaz J, Osuna A, Rosales MJ, Cifuentes J and Mascaro C. (1991). Sucker-like structures in two strains of *Acanthamoeba*: scanning electron microscopy study. Int J Parasitol. 21: 365-367.

Dierickx PJ, Almar MM and De Jonckheere JF. (1990). Glutathione transferase activity in some flagellates and amoebae, and purification of the soluble glutathione transferases from *Acanthamoeba*. Biochem Int. 22: 593-600.

Doberstein SK, Baines IC, Wiegand G, Korn ED and Pollard TD. (1993). Inhibition of contractile vacuole function *in vivo* by antibodies against myosin-I. Nature. 365: 841-843.

Dolman PJ and Dobrogowski MJ. (1989). Contact lens disinfection by ultraviolet light. Am J Ophthalmol. 1989 Dec 15;108(6): 665-669.

Dolphin WD. (1970). Photoinhibition of growth in *Acanthamoeba castellanii* cultures. J Bacteriol 103(3): 755-760.

Donzis PB, Mondino BJ, Weissman BA and Bruckner DA. (1989). Microbial analysis of contact lens care systems contaminated with *Acanthamoeba*. Am J Ophthalmol. 108: 53-56.

Douglas M. (1930). Notes on the classification of the amoeba found by Castellani on culture of a yeast-like fungus. J Trop Med Hyg. 33: 258-259.

Drozanski W. (1956). Fatal bacterial infection in soil amoebae. Acta Microbiol Pol. 5: 315–317.

Dudley R, Matin A, Alsam S, Sissons J, Mahsood AH and Khan NA. (2005). *Acanthamoeba* isolates belonging to T1, T2, T3, T4 but not T7 encyst in response to increased osmolarity and cysts do not bind to human corneal epithelial cells. Acta Tropica. 95: 100-108.

Duguid IG, Dart JK, Morlet N, Allan BD, Matheson M, Ficker L and Tuft S. (1997). Outcome of *Acanthamoeba* keratitis treated with polyhexamethyl biguanide and propamidine. Ophthalmol. 104: 1587-1592.

Duma RJ and Finley R. (1976). *In vitro* susceptibility of pathogenic naegleria and *Acanthamoeba* species to a variety of therapeutic agents. Antimicrob Agents Chemother. 10(2): 370-376.

Edwards SW, Chagla AH, Griffiths AJ and Lloyd D. (1977). The cytochromes of *Acanthamoeba castellanii*. Biochem J. 168: 113-121.

Edwards SW and Lloyd, D. (1978a). Properties of mitochondria isolated from cyanide- stimulted and -sensitive cultures of *Acanthamoeba castellanii*. Biochem J. 174: 203-211.

Edwards SW and Lloyd, D., . (1978b). in function of Alternative Terminal Oxidases. (Degn, H., Lloyd, D and Hill GC., eds) pp: 31-37. Pergamon Press, Oxford.

Elloway EA, Armstrong RA, Bird RA, Kelly SL and Smith SN. (2004). Analysis of *Acanthamoeba* polyphaga surface carbohydrate exposure by FITC-lectin binding and fluorescence evaluation. J Appl Microbiol. 97: 1319-1325.

Essig A, Heinemann M, Simnacher U and Marre R. (1997). Infection of *Acanthamoeba castellanii* by *Chlamydia pneumoniae*. Appl Environ Microbiol. 63(4): 1396-1399.

Epstein RJ, Wilson LA, Visvesvara GS and Plourde Jr EG. (1986). Rapid diagnosis of *Acanthamoeba* keratitis from corneal scrapings using indirect fluorescent antibody staining. Arch Ophthalmol. 104(9): 1318-1321.

Ferrante A and Allison AC. (1983). Alternative pathway activation of complement by African trypanosomes lacking a glycoprotein coat. Parasite Immunol. 5(5): 491-498.

Ferrante A and Rowan-Kelly B. (1983). Activation of the alternative pathway of complement by *Acanthamoeba culbertsoni*. Clin Exp Immunol. 54(2): 477-485.

Ferrante A, Rowan-Kelly B and Thong YH. (1984). *In vitro* sensitivity of virulent *Acanthamoeba culbertsoni* to a variety of drugs and antibiotics. Int J Parasitol. 14: 53-56.

Ferrante A and Bates EJ. (1988). Elastase in the pathogenic free-living amoebae, *Naegleria* and *Acanthamoeba* spp. Infect Immun. 56: 3320-3321.

Ferrante A and Abell TJ. (1986). Conditioned medium from stimulated mononuclear leukocytes augments human neutrophils-mediated killing of a virulent *Acanthamoeba* sp. Infect Immun. 51: 607-617.

Ferrante A and Bates EJ. (1988). Elastase in the pathogenic free-living amoebae *Naegleria* and *Acanthamoeba* spp. Infect Immun. 56: 3320-3321.

Ferrante A. (1991). Immunity to *Acanthamoeba*. Rev Infect Dis. 13: S403-S409.

Ferro S, Coppellotti O, Roncucci G, Ben Amor T, Jori G. (2006). Photosensitized inactivation of *Acanthamoeba palestinensis* in the cystic stage. J Appl Microbiol. 101: 206-212.

Ficker L. (1988). *Acanthamoeba* keratitis – The quest for a better prognosis. Eye. Suppl 2: S37-S45.

Ficker L, Seal D, Warhurst D and Wright P. (1990). *Acanthamoeba* keratitis: resistance to medical therapy. Eye. 4: 835-838.

Fiori PL, Mattana A, Dessì D, Conti S, Magliani W and Polonelli L. (2006). *In vitro* acanthamoebicidal activity of a killer monoclonal antibody and a synthetic peptide. J Antimicrob Chemother. 57: 891-898.

Fleiszig SM, Zaidi TS, Ramphal R and Pier GB. (1994). Modulation of *Pseudomonas aeruginosa* adherence to the corneal surface by mucus. Infect Immun. 62(5): 1799-1804.

Fritsche TR, Gautom RK, Seyedirashti S, Bergeron DL and Lindquist TD. (1993). Occurance of bacterial endosynbionts in *Acanthamoeba* spp. isolated from corneal and environmental specimens and contact lenses. J Clin Microbiol. 31: 1122-1126.

Fritsche TR, Sobek D and Gautom RK. (1998). Enhancement of *in vitro* cytopathogenicity by *Acanthamoeba* spp. following acquisition of bacterial endosymbionts. FEMS Microbiol Lett. 166: 231-236.

Fritsche TR, Horn M, Seyedirashti S, Gautom RK, Schleifer KH and Wagner M. (1999). *In situ* detection of novel bacterial endosymbionts of *Acanthamoeba* spp. phylogenetically related to members of the order Rickettsiales. Appl Environ Microbiol. 65: 206-212.

Furlong ST, Leary JA, Costello CE and Dawidowicz EA. (1986). Isolation and identification of 1(3),2-diacylglyceryl-(3)-O-4'-(N,N,N-trimethyl) homoserine from the soil amoeba, *Acanthamoeba castellanii*. J Lipid Res. 27: 1182-1189.

Fürnkranz U, Nagl M, Gottardi W, Köhsler M, Aspöck H and Walochnik J. (2008). Cytotoxic activity of N-chlorotaurine on *Acanthamoeba* spp. Antimicrob Agents Chemother. 52: 470-476.

Gadasi H, and Korn ED. (1980). Evidence for differential intracellular localization of the *Acanthamoeba* myosin isoenzymes. Nature. 286: 452-455.

Gao LY, Harb OS, Abu Kwaik Y. (1997). Utilization of similar mechanisms by *Legionella pneumophila* to parasitize two evolutionarily distant host cells, mammalian macrophages and protozoa. Infect Immun. 65: 4738-4746.

Garate M, Cao Z, Bateman E and Panjwani N. (2004). Cloning and characterization of a novel mannose-binding protein of *Acanthamoeba*. J Biol Chem. 279(28): 29849-29856.

Garate M, Cubillos I, Marchant J and Panjwani N. (2005). Biochemical characterization and functional studies of *Acanthamoeba* mannose-binding protein. Infect Immun. 73: 5775-5781.

Garate M, Marchant J, Cubillos I, Cao Z, Khan NA and Panjwani N. (2006). Pathogenicity of *Acanthamoeba* is associated with the expression of the mannose-binding protein. Invest Ophthalmol Vis Sci 47(3): 1056-1062.

Garate M, Alizadeh H, Neelam S, Niederkorn JY and Panjwani N. (2006). Oral immunization with *Acanthamoeba castellanii* mannose-binding protein ameliorates amoebic keratitis. Infect Immun. 74: 7032-7034.

Garner A. (1993). Pathogenesis of acanthamoebic keratitis: hypothesis based on a histological analysis of 30 cases. Br J Ophthalmol. 77(6): 366-370.

Gelman BB, Rauf SJ, Nader R, Popov V, Borkowski J, Chaljub G, Nauta HW and Visvesvara GS. (2001). Amoebic encephalitis due to *Sappinia diploidea*. JAMA. 285: 2450-2451.

Georges P and Perrine D. (1989). Possibility of the survival of free-living amoebae which cause keratitis in decontamination solutions used for the maintenance of contact lenses. Bull Acad Natl Med. 173: 421-424.

Giese MJ and Weissman BA. (2002). Contact lens associated corneal infections. Where do we go from here? Clin Exp Optom. 85(3): 141-148.

Ghajar M, Houlsby RD, and Chavez G. (1989). Microbiological evaluation of MiraFlow. J Am Optom Assoc. 60: 592-595.

Ghedin E, and Claverie JM. (2005). Mimivirus relatives in the Sargasso sea. Virol J. 2: 62.

Gooi P, Lee-Wing M, Brownstein S, El-Defrawy S, Jackson WB, and Mintsioulis G. (2008). *Acanthamoeba* keratitis: persistent organisms without inflammation after 1 year of topical chlorhexidine. Cornea. 27: 246-248.

Gordon DJ, eisenberg E and Korn ED. (1976). Characterization of cytoplasmic actin isolated from *Acanthamoeba castellanii* by a new method. J Biol Chem. 251(15): 4778-4786.

Gordon VR, Asem EK, Vodkin MH and McLaughlin GL. (1993). *Acanthamoeba* binds to extracellular matrix proteins *in vitro*. Invest Ophthalmol Vis Sci. 34: 658-662.

Górnik K and Kuźna-Grygiel W. (2005). Histological studies of selected organs of mice experimentally infected with *Acanthamoeba* spp. Folia Morphol. (Warsz). 64: 161-167.

Gómez-Couso H, Paniagua-Crespo E and Ares-Mazás E. (2007). *Acanthamoeba* as a temporal vehicle of *Cryptosporidium*. Parasitol Res. 100: 1151-1154.

Gould WD, Coleman DC and Rubink AJ. (1979). Effects of bacteria and amoebae on rhizosphere phosphatase activity. Appl Environ Microbiol. 37(5): 943-946.

Gray TB, Cursons RT, Sherwan JF and Rose PR. (1995). *Acanthamoeba*, bacterial, and fungal contamination of contact lens storage cases. Br J Ophthalmol. 79(6): 601-605.

Greub G and Raoult D. (2002). *Parachlamydiaceae*: Potential Emerging Pathogens. Emerg Infect Dis. 8(6): 625-630

Greub G and Raoult D. (2004). Microorganisms resistant to free-living amoebae. Clin Microbiol Rev. 17(2): 413-433.

Grillot R. and Ambroise-Thomas P. (1981). Comparative study of the pathogenicity of free-living amoebae in mice and their cytopathogenic effect *in vitro*. C R Seances Soc Biol Fil. 175(2): 187-193.

Gritz DC, Lee TY, McDonnell PJ, Shih K and Baron N. (1990). Ultraviolet radiation for the sterilization of contact lenses. CLAO J. 16: 294-298.

Gudmundsson OG, Woodward DF, Fowler SA and Allansmith MR. (1985). Identification of proteins in contact lens surface deposits by immunofluorescence microscopy. Arch Ophthalmol. 103(2): 196-197.

Gullett J, Mills J, Hadley K, Podemski B, Pitts L and Gelber R. (1979). Disseminated granulomatous *Acanthamoeba* infection presenting as an unusual skin lesion. Am J Med. 67(5): 891-896.

Gunderson JH and Sogin ML. (1986). Length variation in eukaryotic rRNAs: small subunit rRNAs from the protists *Acanthamoeba castellanii* and *Euglena gracilis*. Gene. 44: 63-70.

Gupta S, Shukla OP and Walter RD. (1987). Putrescine-activated S-adenosylmethionine decarboxylase from *Acanthamoeba culbertsoni*. Mol Biochem Parasitol. 23: 247-252.

Gupta S, Kishore P and Shukla OP. (1987). Polyamine metabolism as a target for growth inhibition of *Acanthamoeba culbertsoni*. Am J Trop Med Hyg. 37: 550-555.

Hadas E and Mazur T. (1993). Biochemical markers of pathogenicity and virulence of *Acanthamoeba* sp. strains. Trop Med Parasitol. 44: 197-200.

Hadas E and Mazur T. (1997). Biosynthesis of prostaglandins in pathogenic and nonpathogenic strains of *Acanthamoeba* spp. Parasitol Res. 83: 296-299.

Hamburg and De Jonckheere, 1980. Amoebic keratitis. Ophthalmologica, 181(2): 74-80.

Han KL, Lee J, Kim DS, Park SJ, Im KI and Yong TS. (2006). Identification of differentially expressed cDNAs in *Acanthamoeba culbertsoni* after mouse brain passage. Korean J Parasitol. 44: 15-20.

Hanel H, Kirsch R, Schmidts HL and Kottmann H. (1995). New systematically active antimycotics from the beta-blocker category. Mycoses. 38(7-8): 251-264.

Harrington KA and Hormaeche CE. (1986). Expression of the innate resistance gene Ity in mouse Kupffer cells infected with *Salmonella typhimurium in vitro*. Microb Pathogen. 1: 269-274.

Harwood CR, Rich GE, McAleer R and Cherian G. (1988). Isolation of *Acanthamoeba* from a cerebral abscess. Med J Aust. 148: 47-49.

Hassan HF and Coombs GH. (1987). Purine salvage by *Acanthamoeba castellanii*. Int J Parasitol. 17: 1267-1272.

Hax WMA, Demel RA, Spies F, Vossenberg JBJ and Linnemans WAM. (1974). Increased phospholipase A activity and formation of communicative contacts between *Acanthamoeba castellanii* cells: effects of 3',5'-cyclic AMP. Exp Cell Res. 89(2): 311-9.

Hay J, Kirkness CM, Seal DV and Wright P. (1994). Drug resistance and *Acanthamoeba* keratitis: the quest for alternative protozoal chemotherapy. Eye. 8: 555-563.

He YG, Neiderkorn JY, McCulley JP, Stewart GL, Meyer DR, Silvany R and Doughtery J. (1990). *In vivo* and *in vitro* collagenolytic activity of *Acanthamoeba castellanii*. Invest Opthal Visual Sci. 31: 2235-2240.

He YG, McCulley JP, Alizadeh H, Pidherney M, Mellon J, Ubelaker JE, Stewart GL, Silvany RE and Niederkorn JY. (1992). A pig model of *Acanthamoeba* keratitis: transmission via contaminated contact lenses. Invest Ophthalmol Vis Sci. 33: 126-133.

Heath KV, Frank O, Montaner JS, O'Shaughnessy MV, Schechter MT and Hogg RS. (1998). Human immunodeficiency virus. (HIV)/acquired immunodeficiency syndrome. (AIDS) mortality in industrialized nations, 1987-1991. Int J Epidemiol. 27(4): 685-690.

Helton J, Loveless M and White CR Jr. (1993). Cutaneous *Acanthamoeba* infection associated with leukocytoclastic vasculitis in an AIDS patient. Am J Dermatopathol. 15: 146-149.

Henke M and Seidel KM. (1986). Association between Legionella pneumophila and amoebae in water. Isr J Med Sci. 22: 690-695.

Hettiarachchy NS and Jones G. (1974). Isolation and characterization of mitochondrial deoxyribonucleic acid of Acanthamoebe castellanii. Biochem J. 141: 159-164.

Hill, G. C.,. (1978). In functions of Alternative terminal oxidases. (Degn, H., Lloyd, D. and Hill, GC., eds.,), pp: 67-77, Pergamon Press, Oxford.

Hirukawa Y, Nakato H, Izumi S, Tsuruhara T and Tomino S. (1998). Structure and expression of a cyst specific protein of *Acanthamoeba castellanii*. Biochim Biophys Acta. 1398: 47-56.

Hiti K, Walochnik J, Faschinger C, Haller-Schober EM and Aspöck H. (2001). Microwave treatment of contact lens cases contaminated with *Acanthamoeba*. Cornea. 20: 467-470.

Hiti K, Walochnik J, Haller-Schober EM, Faschinger C and Aspöck H. (2002). Viability of *Acanthamoeba* after exposure to a multipurpose disinfecting contact lens solution and two hydrogen peroxide systems. Br J Ophthalmol. 86: 144-146.

Hiti K, Walochnik J, Faschinger C, Haller-Schober EM and Aspöck H. (2005). One- and two-step hydrogen peroxide contact lens disinfection solutions against *Acanthamoeba*: how effective are they? Eye. 19: 1301-1305.

Hohman TC and Bowers B. (1993). Hydrolase compartmentalization limits rate of digestion in *Acanthamoeba*. J Eukaryot Microbiol. 40: 589-593.

Holden B. (1990). A report card on hydrogen peroxide for contact lens disinfection. CLAO J. 16: S61-S64.

Hong YC, Hwang MY, Yun HC, Yu HS, Kong HH, Yong TS and Chung DI. (2002). Isolation and characterization of a cDNA encoding a mammalian cathepsin L-like cysteine proteinase from *Acanthamoeba healyi*. Korean J Parasitol. 40(1): 17-24.

Hong YC, Lee WM, Kong HH, Jeong HJ and Chung DI. (2004). Molecular cloning and characterization of a cDNA encoding a laminin-binding protein. (AhLBP). from *Acanthamoeba healyi*. Exp Parasitol. 106(3-4): 95-102.

Horn M, Fritsche TR, Gautom RK, Schleifer KH and Wagner M. (1999). Novel bacterial endosymbionts of *Acanthamoeba* spp. related to the *Paramecium caudatum* symbiont *Caedibacter caryophilus*. Environ Microbiol. 1: 357-367.

Horn M, Harzenetter MD, Linner T, Schmid EN, Müller KD, Michel R and Wagner M. (2001). Members of the Cytophaga-Flavobacterium-Bacteroides phylum as intracellular bacteria of *Acanthamoeba*e: proposal of '*Candidatus Amoebophilus asiaticus*'. Environ Microbiol. 3: 440-449.

Horsburgh B, Hirst LW, Carey T, Whitby RM, Nimmo G, Stallard K. (1991). Steroid sensitive *Acanthamoeba* keratitis. Aust N Z J Ophthalmol. 19: 349-350.

Hryniewiecka, L., Janek J., Michejda JW. 1980. Necessity of iron for the alternative respiratory pathway in *Acanthamoeba castellanii*. Biochem Biophys Res Commun. 93(1): 141-148.

Hsiao CH, Lin HC, Chen YF, Ma DH, Yeh LK, Tan HY, Huang SC and Lin KK. (2005). Infectious keratitis related to overnight orthokeratology. Cornea. 24: 783-788.

Huang ZH, Ferrante A and Carter R. (1999). Serum antibodies to *Balamuthia mandrillaris*, a free-living amoeba recently demonstrated to cause granulomatous amoebic encephalitis. J Infect Dis 179: 1305-1308.

Hughes R, Dart J, Kilvington S. (2003). Activity of the amidoamine myristamidopropyl dimethylamine against keratitis pathogens. J Antimicrob Chemother. 51: 1415-1418.

Hugo ER and Byers TJ. (1993). S-adenosyl-L-methionine decarboxylase of *Acanthamoeba castellanii*. (Neff): purification and properties. Biochem J 295: 203-209.

Hundt MJ and Ruffolo CG. (2005). Interaction of *Pasteurella multocida* with free-living amoebae. Appl Environ Microbiol. 71: 5458-5464.

Hurt M, Apte S, Leher H, Howard K, Niederkorn J and Alizadeh H. (2001). Exacerbation of *Acanthamoeba* keratitis in animals treated with anti-macrophage inflammatory protein 2 or antineutrophil antibodies. Infect Immun. 69(5): 2988-2995.

Hurt M, Proy V, Niederkorn JY and Alizadeh H. (2003). The interaction of *Acanthamoeba castellanii* cysts with macrophages and neutrophils. J Parasitol. 89(3): 565-572.

Huws SA, Smith AW, Enright MC, Wood PJ and Brown MR. (2006). Amoebae promote persistence of epidemic strains of MRSA. Environ Microbiol. 8: 1130-1133.

Hyun DK, Shin CO and Im K. (1992). Natural killer cell activity in mice infected with *Acanthamoeba* culbertsoni. Kisaengchunghak Chapchi. 30: 101-112.

Im K. (1990). Pathogenic free-living amoebae. Korean J Parasitol. 28: 29-39.

Imbert-Bouyer S, Merlaud A, Imbert C, Daniault G and Rodier MH. (2004). A mannose binding protein is involved in the adherence of *Acanthamoeba* species to inert surfaces. FEMS Microbiol Lett. 238: 207-211.

Ingalls CS and Brent MM. (1983). Defined minimal growth medium for *Acanthamoeba polyphaga*. J Protozool. 30: 606-608.

Isenberg G, Aebi U, Pollard TD. (1980). An actin-binding protein from *Acanthamoeba* regulates actin filament polymerization and interactions. Nature. 288(5790): 455-459.

Ito S, Chang RS and Pollard TD. (1969). Cytoplasmic distribution of DNA in a strain of hartmannellid amoeba. J Protozool. 16(4): 638-645.

Jager BV and Stamm WP. (1972). Brain abscesses caused by free-living amoeba probably of the genus *Hartmannella* in a patient with Hodgkin's disease. Lancet. ii: 1343-1345.

Jahnes WG, Fullmer HM and Li CP. (1957). Free-living amoebae as contaminants in monkey kidney tissue culture. Proc Soc Exp Biol Med. 96: 484-488.

Jaison PL, Cao Z and Panjwani N. (1998). Binding of *Acanthamoeba* to [corrected] mannose-glycoproteins of corneal epithelium: effect of injury. Curr Eye Res. 17(8): 770-776.

Jantzen H. (1981). Ribosomal phosphoproteins in *Acanthamoeba castellanii*. Eur J Biochem, 119(2): 347-352.

Jantzen H and Schulze I. (1988). Relationship between the timing of DNA replication and the developmental competence in *Acanthamoeba castellanii*. J Cell Science. 91: 389-399.

Jarmuszkiewicz W, Sluse-Goffart CM, Hryniewiecka L, Michejda J and Sluse FE. (1998). Electron partitioning between the two branching quinol-oxidizing pathways in *Acanthamoeba castellanii* mitochondria during steady-state state 3 respiration. J Biol Chem. 273: 10174-10180.

Jarmuszkiewicz W, Fraczyk O and Hryniewiecka L. (2001). Effect of growth at low temperature on the alternative pathway respiration in *Acanthamoeba castellanii* mitochondria. Acta Biochim Pol. 48: 729-737.

Jarmuszkiewicz W, Sluse FE, Hryniewiecka L and Sluse-Goffart CM. (2002). Interactions between the cytochrome pathway and the alternative oxidase in isolated *Acanthamoeba castellanii* mitochondria. J Bioenerg Biomembr. 34: 31-40.

Jeong HJ, Jang ES, Han BI, Lee KH, Ock MS, Kong HH, Chung DI, Seol SY, Cho DT and Yu HS. (2007). *Acanthamoeba*: could it be an environmental host of Shigella? Exp Parasitol. 115: 181-186.

Jimenez C, Portela RA, Mellado M, Rodriguez-Frade JM, Collard J, Serrano A, Martinez AC, Avila J and Carrera AC. (2000). Role of the PI3K regulatory subunit in the control of actin organization and cell migration. J Cell Biol. 151(2): 249-262.

John DT and John RA. (1994). Cryopreservation of pathogenic free-living amebae. *Folia Parasitol* 41: 110-114.

John T, Desai D and Sahm D. (1989). Adherence of *Acanthamoeba castellanii* cysts and trophozoites to unworn soft contact lenses. Am J Ophthalmol. 108: 658-664.

John T, Desai D and Sahm D. (1991). Adherence of *Acanthamoeba castellanii* to new daily wear, extended wear, and disposable soft contact lenses. CLAO J. 17: 109-113.

Johns KJ, O'Day DM, Head WS, Neff RJ and Elliott JH. (1987). Herpes simplex masquerade syndrome: *Acanthamoeba* keratitis. Curr Eye Res. 6: 207-212.

Johns KJ, Head WS and O'Day DM. (1988). Corneal toxicity of propamidine. Arch Ophthalmol. 106: 68-69.

Johnson HA and Thomas NR. (2002). Polyhydroxylated azepanes as new motifs for DNA minor groove binding agents. Bioorg Med Chem Lett. 12(2): 237-241.

Jones AL, Lloyd D, Hann AC, Harwood JL. (1991). Biochem Soc Trans 19: 318S.

Jones AL, Hann AC, Harwood JL, Lloyd D. (1993). Temperature-induced membrane-lipid adaptation in *Acanthamoeba castellanii*. Biochem J. 290: 273-278.

Joos L and Gicquaud C. 1987. Effect of phalloidin and viroisin on *Acanthamoeba castellanii* after permeabilization of the cell. Biochem Cell Biol. 65: 261-270.

Jung G, Korn ED and Hammer JA III. (1987). The heavy chain of *Acanthamoeba* myosin IB is a fusion of myosin-like and non-myosin-like sequences. PNAS. 84: 6720-6724.

Kachar B, Albanesi JP, Fujisaki H and Korn ED. (1987). Extensive purification from *Acanthamoeba castellanii* of a microtubule-dependent translocator with microtubule-activated Mg^{2+}-ATPase activity. J Biol Chem. 262: 16180-16185.

Kahane S, Dvoskin B, Mathias M and Friedman MG. (2001). Infection of *Acanthamoeba polyphaga* with *Simkania negevensis* and *S. negevensis* survival within amoebal cysts. Appl Environ Microbiol. 67(10): 4789-4795.

Kameyama S, Sato H and Murata R. (1975). The role of alpha-toxin of *Clostridium perfringens* in experimental gas gangrene in guinea pigs. Jpn J Med Sci Biol. 25: 200.

Kassab K, Dei D, Roncucci G, Jori G and Coppellotti O. (2003). Phthalocyanine-photosensitized inactivation of a pathogenic protozoan, *Acanthamoeba palestinensis*. Photochem Photobiol Sci. 2: 668-672.

Kelleher JF, Atkinson SJ and Pollard TD. (1994). Sequence, structural models, and cellular localization of the actin-related proteins Arp2 and Arp3 from *Acanthamoeba*. J Cell Biol. 131: 385-397.

Kelley PS, Dossey AP, Patel D, Whitson JT, Hogan RN and Cavanagh HD. (2006). Secondary glaucoma associated with advanced *Acanthamoeba* keratitis. Eye Contact Lens. 32: 178-182.

Kennett MJ, Hook RR Jr, Franklin CL and Riley LK. (1999). *Acanthamoeba castellanii*: characterization of an adhesin molecule. Exp Parasitol. 92: 161-169.

Ketcham CM, Kornfeld S. (1992). Purification of UDP-N-acetylglucosamine: glycoprotein N-acetylglucosamine-1-phosphotransferase from *Acanthamoeba castellanii* and identification of a subunit of the enzyme. J Biol Chem. 267: 11645-11653.

Ketcham CM and Kornfeld S. (1992). Characterization of UDP-N-acetylglucosamine: glycoprotein N-acetylglucosamine-1-phosphotransferase from *Acanthamoeba castellanii*. J Biol Chem. 267: 11654-11659.

Khan M, La Scola B, Lepidi H and Raoult D. (2007). Pneumonia in mice inoculated experimentally with *Acanthamoeba* polyphaga mimivirus. Microb Pathog. 42: 56-61.

Khan NA, Greenman J, Topping KP, Hough VC, Temple GS and Paget TA. (2000). Isolation of *Acanthamoeba*-specific antibodies from a bacteriophage display library. J Clin Microbiol. 38: 2374-2377.

Khan NA, Jarroll EL, Panjwani N, Cao Z and Paget TA. (2000). Proteases as markers of differentiation of pathogenic and non-pathogenic *Acanthamoeba*. J Clin Microbiol. 38: 2858-2861.

Khan NA. (2001). Pathogenicity, morphology and differentiation of *Acanthamoeba*. Curr Microbiol 43: 391-395.

Khan NA, Jarroll EL and Paget TA. (2001). *Acanthamoeba* can be differentiated by the polymerase chain reaction and simple plating assays. Curr Microbiol. 43(3): 204-208.

Khan NA and Paget TA. (2002). Molecular tools for speciation and epidemiological studies of *Acanthamoeba*. *Curr Microbiol* 44(6): 444-449.

Khan NA, Jarroll EL and Paget TA. (2002). Molecular and physiological differentiation between pathogenic and non-pathogenic *Acanthamoeba*. Curr Microbiol. 45(3): 197-202.

Khan NA. (2003). Pathogenesis of *Acanthamoeba* infections. Microb Pathogen 34(6): 277-285.

Khan NA. (2006). Protozoa, In: Instant Notes in Microbiology. (3nd edn., edited by B. D. Hames), Taylor and Francis, Oxford, UK, pp. 241 – 269.

Kicinska A, Swida A, Bednarczyk P, Koszela-Piotrowska I, Choma K, Dolowy K, Szewczyk A and Jarmuszkiewicz W. (2007). ATP-sensitive potassium channel in mitochondria of the eukaryotic microorganism *Acanthamoeba castellanii*. J Biol Chem. 282: 17433-17441.

Kilvington S and Price J. (1990). Survival of *Legionella pneumophila* within cysts of *Acanthamoeba polyphaga* following chlorine exposure. J Appl Bacteriol. 68: 519-525.

Kilvington S and Larkin DF. (1990). *Acanthamoeba* adherence to contact lenses and removal by cleaning agents. Eye. 4(Pt 4): 589-593.

Kilvington S, Larkin DFP, White DG and Beeching JR. (1990). Laboratory investigation of *Acanthamoeba* leratitis. J Clin Microbiol. 28: 2722-2725.

Kilvington S, Anger C. (2001). A comparison of cyst age and assay method of the efficacy of contact lens disinfectants against *Acanthamoeba*. Br J Ophthalmol. 85: 336-340.

Kilvington S, Hughes R, Byas J and Dart J. (2002). Activities of therapeutic agents and myristamidopropyl dimethylamine against *Acanthamoeba* isolates. Antimicrob Agents Chemother. 46: 2007-2009.

Kim BG, McCann PP, and Byers TJ. (1987). Inhibition of multiplication in *Acanthamoeba castellanii* by specific inhibitors of ornithine decarboxylase. J Protozool. 34: 264-266.

Kim BG, Sobota A, Bitonti AJ, McCann PP and Byers TJ. (1987). Polyamine metabolism in *Acanthamoeba*: polyamine content and synthesis of ornithine, putrescine, and diaminopropane. J Protozool. 34: 278-284.

Kim KH, Shin CO and Im K. (1993). Natural killer cell activity in mice infected with free-living amoeba with reference to their pathogenicity. Korean J Parasitol. 31(3): 239-248.

Kim MJ, Shin CO and Im KI. (1990). Cell-mediated immunity in mice infected with *Acanthamoeba culbertsoni*. Kisaengchunghak Chapchi. 28: 143-154.

Kim WT, Kong HH, Ha YR, Hong YC, Jeong HJ, Yu HS, Chung DI. (2006). Comparison of specific activity and cytopathic effects of purified 33 kDa serine proteinase from *Acanthamoeba* strains with different degree of virulence. Korean J Parasitol. 44: 321-330.

King CH, Schotts EB Jr, Wooley RE and Porter KG. (1988). Survival of coliforms and bacterial pathogens within protozoa during chlorination. Appl Environ Microbiol. 54: 3023-3033.

Kishore P, Gupta S, Srivastava DK and Shukla OP. (1990). Action of methylglyoxal bis. (guanyl hydrazone) and related antiprotozoals on *Acanthamoeba culbertsoni*. Indian J Exp Biol. 28: 1174-1179.

Kishore P, Wittich RM, Walter RD and Shukla OP. (1993). Polyamine metabolism in *Acanthamoeba culbertsoni*. Microbios. 73: 7-21.

Klotz SI, Misra RP and Butrus SI. (1987). Carbohydrate deposits on the surfaces of worn extended-wear soft contact lenses. Arch Ophthalmol. 105: 974-977.

Kong HH, Kim TH and Chung DI. (2000). Purification and characterization of a secretory proteinase of *Acanthamoeba healyi* isolates from GAE. J Parasitol. 86(1): 12-17.

Kopańska K, Najda A, Zebrowska J, Chomicz L, Piekarczyk J, Myjak P and Bretner M. (2004). Synthesis and activity of 1H-benzimidazole and 1H-benzotriazole derivatives as inhibitors of *Acanthamoeba castellanii*. Bioorg Med Chem. 12: 2617-2624.

Korn ED. (1963). Fatty acids of *Acanthamoeba* sp. *Journal of Biological Chem* 238(11): 3584-3587.

Korn ED. (1964). Biosynthesis of unsaturated fatty acids in *Acanthamoeba* sp. J Biol Chem. 239(2): 396-400.

Korn ED and Weisman RA. (1967). Phagocytosis of latex beads by *Acanthamoeba*. II. Electron microscopic study of the initial events. J Cell Biol. 34: 219 – 227.

Korn ED and Olivecrona T. (1971). Composition of an amoeba plasma membrane. Biochem Biophys Res Commun. 45(1): 90-97.

Korn ED, Dearborn DG and Wright PL. (1974). Lipophosphonoglycan of the plasma membrance of *Acanthamoeba castellanii*. Isolation from whole amoebae and identification of the water-soluble products of acid hydrolysis. J Biol Chem. 249(11): 3335-3341.

Kotting J, Berger MR, Unger C and Eibl H. (1992). Alkylphosphocholines: influence of structural variations on biodistribution of antineoplastically active concentrations. Cancer Chemother Pharmacol. 30: 105-112.

Krishna-Prasad BN and Gupta SK. (1978). Preliminary report on engulfment and retention of mycobacteria by trophozoites of axenically grown *Acanthamoeba castellanii* Douglas. *Curr Sci* 47: 245-247.

La Scola B, Audic S, Robert C, Jungang L, de Lamballerie X, Drancourt M, Birtles R, Claverie JM and Raoult D. (2003). A giant virus in amoebae. Science. 299(5615): 2033.

Lalitha MK, Anandi V, Srivastava A, Thomas K, Cherian AM and Chandi SM. (1985). Isolation of *Acanthamoeba culbertsoni* from a patient with meningitis. J Clin Microbiol. 21: 666-667.

Lam DSC, Houang E, Fan SP, Lyon D, Seal D, Wong E and the Hong Kong Microbial keratitis study group. (2002). Incidence and risk factors for microbial keratitis in Hong Kong: comparison with Europe and North America. Eye. 16: 608-618.

Lamothe J, Thyssen S and Valvano MA. (2004). Burkholderia cepacia complex isolates survive intracellularly without replication within acidic vacuoles of *Acanthamoeba polyphaga*. Cell Microbiol. 6: 1127-1138.

Lambooy PK and Korn ED. (1986). Purification and characterization of actobindin, a new actin monomer-binding protein from *Acanthamoeba castellanii*. J Biol Chem. 261: 17150-17155.

Landers P, Kerr KG, Rowbotham TJ, Tipper JL, Keig PM, Ingham E and Denton M. (2000). Survival and growth of *Burkholderia cepacia* within the free-living amoeba *Acanthamoeba polyphaga*. Eur J Clin Microbiol Infect Dis. 19(2): 121-123.

Lang L, Couso R and Kornfeld S. (1986). Glycoprotein phosphorylation in simple eukaryotic organisms. J Biol Chem. 261: 6320-6325.

Larkin DF, Kilvington S and Easty DL. (1990). Contamination of contact lens storage cases by *Acanthamoeba* and bacteria. Br J Ophthalmol. 74(3): 133-135.

Larkin DFP, Kilvington S and Dart JKG. (1992). Treatment of *Acanthamoeba* keratitis with polyhexamethylene biguanide. Ophthalmol. 99: 185-191.

La Scola B and Raoult D. (2001). Survival of *Coxiella burnetii* within free-living amoeba *Acanthamoeba castellanii*. Clin Microbiol Infect. 7(2): 75-79.

Leher H, Zaragoza F, Taherzadeh S, Alizadeh H and Niederkorn JY. (1999). Monoclonal IgA antibodies protect against *Acanthamoeba* keratitis. Exp Eye Res. 69(1): 75-84.

Lehmann OJ, Green SM, Morlet N, Kilvington S, Keys MF, Matheson MM, Dart JK, McGill JI and Watt PJ. (1998). Polymerase chain reaction analysis of corneal epithelial and tear samples in the diagnosis of *Acanthamoeba* keratitis. Invest Ophthalmol Vis Sci. 39(7): 1261-1265.

Li L, Sun X. (2008). Impaired innate immunity of ocular surface is the key bridge between extended contact lens wearing and occurrence of *Acanthamoeba* keratitis. Med Hypotheses. 70: 260-264.

Lim L, Coster DJ and Badenoch PR. (2000). Antimicrobial susceptibility of 19 Australian corneal isolates of *Acanthamoeba. Clin Exp* Ophthalmol. 28: 119-124.

Lindquist TD, Fritsche TR and Grutzmacher RD. (1990). Scleral ectasia secondary to *Acanthamoeba* keratitis. Cornea. 9: 74-76.

Lloyd D. and Edwards S. 1979. The effect of inhibitor on the oxygen kinetics of terminal oxidases of *Acanthamoeba castellanii*. Biochem J. 182: 11-15.

Lloyd D, Edwards S, Kristensen B and Degn H. (1979). The effect of inhibitors on the oxygen kinetics of terminal oxidases of *Acanthamoeba castellanii*. Biochem J. 182: 11-15.

Lloyd D, Edwards SW and Chance B. (1981). Carbon monoxide- and oxygen-reacting haemoproteins in the mitochondrial fraction from the soil amoeba *Acanthamoeba castellanii*. Biochem J. 200: 337-342.

Lloyd D, Turner NA, Khunkitti W, Hann AC, Furr JR and Russell AD. (2001). Encystation in *Acanthamoeba castellanii*: development of biocide resistance. J Eukaryot Microbiol. 48(1): 11-16.

Lorenzo-Morales J, Ortega-Rivas A, Foronda P, Abreu-Acosta N, Ballart D, Martinez E and Valladares B. (2005). RNA interference. (RNAi) for the silencing of extracellular serine proteases genes in *Acanthamoeba*: Molecular analysis and effect on pathogenecity. Mol Biochem Parasitol. 144(1): 10-15.

Lorenzo-Morales J, Coronado-Alvarez N, Martínez-Carretero E, Maciver SK and Valladares B. (2007). Detection of four adenovirus serotypes within water-isolated strains of *Acanthamoeba* in the Canary Islands, Spain. Am J Trop Med Hyg. 77: 753-756.

Lorenzo-Morales J, Kliescikova J, Martinez-Carretero E, De Pablos LM, Profotova B, Nohynkova E, Osuna A and Valladares B. (2008). Glycogen phosphorylase in *Acanthamoeba* spp.: determining the role of the enzyme during the encystment process using RNA interference. Eukaryot Cell. 7: 509-517.

Ly TM and Muller HE. (1990). Ingested *Listeria monocyto*genes survive and multiply in protozoa. J Med Microbiol 33(1): 51-54.

Ma P, Willaert E, Juechter KB and Stevens AR. (1981). A case of keratitis due to *Acanthamoeba* in New York, New York, and features of 10 cases. J Infect Dis. 143(5): 662-667.

Machesky LM, Goldschmidt-Clermont PJ, Pollard TD. (1990). The affinities of human platelet and *Acanthamoeba* profilin isoforms for phosphoinositides account for their relative abilities to inhibit phospholipase C. Cell Reg. 1: 937-950.

Machesky LM, Atkinson SJ, Ampe C, Vandekerckhove J and Pollard TD. (1994). Purification of a cortical complex containing two unconventional actins from *Acanthamoeba* by affinity chromatography on profilin-agarose. J Cell Biol. 127: 107-115.

Maciver SK, Zot HG and Pollard TD. (1991). Charactrization of actin filament severing by actophorin from *Acanthamoeba castellanii*. J Cell Biol. 115: 1611-1620.

Maciver SK, Wachsstock DH, Schwarz WH and Pollard TD. (1991a). The actin filament severing protein actophorin promotes the formation of rigid bundles of actin filaments crosslinked with α-actinin. J Cell Biol. 115: 1621-1628.

Maciver SK and Weeds AG. (1994). Actophorin preferentially binds to monomeric ADP-ctin over ATP-bound actin: consequences for cell locomotion. FEBS Lett. 347: 251-256.

Mackay, D. J. and Hall, A. (1998). Rho GTPases. J Biol Chem. 273(33): 20685-20688.

MacLean-Fletcher, SD and Pollard TD. (1980). Viscometric analysis of the gelation of *Acanthamoeba* extracts and purification of two gelation factors. J Cell Biol. 85: 414-428.

Maghsood AH, Sissons J, Rezaian M, Nolder D, Warhurst D and Khan NA. (2005). *Acanthamoeba* genotype T4 from the UK and Iran and isolation of the T2 genotype from clinical isolates. J Med Microbiol. 54(8): 755-759.

Marciano-Cabral F and Toney DM. (1998). The interaction of *Acanthamoeba* spp. with activated macrophages and with macrophage cell lines. J Eukaryot Microbiol. 45(4): 452-458.

Marciano-Cabral F, Ferguson T, Bradley SG and Cabral G. (2001). Delta-9-tetrahydrocannabinol. (THC), the major psychoactive component of marijuana, exacerbates brain infection by *Acanthamoeba*. J Eukaryot Microbiol. suppl: 4S-5S.

Marciano-Cabral F and Cabral G. (2003). *Acanthamoeba* spp. as agents of disease in humans. Clin Microbiol Rev. 16(2): 273-307.

Marciano-Cabral F, Ludwick C, Puffenbarger RA and Cabral GA. (2004). Differential stimulation of microglial pro-inflammatory cytokines by *Acanthamoeba culbertsoni* versus *Acanthamoeba castellanii*. J Eukaryot Microbiol. 51: 472-479.

Margo, C.E. (1987). *Acanthamoeba* keratitis. Arch. Pathol. Lab. Med. 111: 759-760.

Markowitz, SM., Sobieski T., Martinez, AJ. and Duma RJ. (1978). Experimental *Acanthamoeba* infections in mice pretreated with methylprednisolone or tetracycline. Am J Pathol. 92(3): 733-743.

Marolda CL, Hauroder B, John MA, Michel R and Valvano MA. (1999). Intracellular survival and saprophytic growth of isolates from the *Burkholderia cepacia* complex in free-living amoebae. Microbiol. 145(7): 1509-1517.

Martinez AJ, Garcia CA, Halks-Miller M and Arce-Vela R. (1980). Granulomatous amebic encephalitis presenting as a cerebral mass lesion. Acta Neuropathol. (Berl). 51(2): 85-91.

Martinez AJ. (1980). Is *Acanthamoeba* encephalitis an opportunistic infection. Neurology. 30(6): 567-574.

Martinez AJ. (1985). Free-living Amebas: Natural History, Prevention, Diagnosis, Pathology and treatment of disease. CRC Press, Boca Raton, Florida. p.156.

Martinez AJ and Visvesvara GS. (1991). Laboratory diagnosis of pathogenic free-living amoebas: *Naegleria*, *Acanthamoeba*, and Leptomyxid. Clin Lab Med. 11(4): 861-872.

Martinez AJ. (1991). Infections of the central nervous system due to *Acanthamoeba*. Rev Infect Dis 13: S399-S402.

Martinez AJ and Visvesvara GS. (1997). Free-living, amphizoic and opportunistic amebas. Brain Pathol. 7(1): 583-598.

Martinez MS, Gonzalez-Mediero G, Santiago P, Rodriguez de Lope A, Diz J, Conde C and Visvesvara GS. (2000). Granulomatous Amebic Encephalitis in a Patient with AIDS: Isolation of *Acanthamoeba* sp. Group II from Brain Tissue and Successful Treatment with Sulfadiazine and Fluconazole. J Clin Microbiol. 38: 3892-3895.

Maruta H and Korn ED. (1977). *Acanthamoeba* cofactor protein is a heavy chain kinase required for actin activation of the Mg^{2+}-ATPase activity of *Acanthamoeba* myosin I. J Biol Chem. 252(23): 8329-8332.

Maruta H and Korn ED. (1977). Purification from *Acanthamoeba castellanii* of proteins that induce gelation and syneresis of F-actin. J Biol Chem. 252: 399-402.

Maruta H, Gadasi H, Collins JH and Korn ED. (1978). The isolated heavy chain of an *Acanthamoeba* myosin contains full enzymatic activity. J Biol Chem. 253(18): 6297-6300.

Maruta H and Korn ED. (1981). Proteolytic separation of the actin-activeable ATPase site from the phosphorylation site on the heavy chain of *Acanthamoeba* myosin IA. J Biol Chem. 256(1): 503-506.

Maruta H, Gadasi H, Collins JH and Korn ED. (1979). Multiple forms of *Acanthamoeba* myosin I. J Biol Chem. 254(9): 3624-3630.

Matoba AY, Pare PD, Le TD and Osato MS. 1989. The effects of freezing and antibiotics on the viability of *Acanthamoeba* cysts. Arch Ophthalmol. 107: 439-440.

Mattana A, Bennardini F, Usai S, Fiori PL, Franconi F and Cappuccinelli P. (1997). *Acanthamoeba castellanii* metabolites increase the intracellular calcium level and cause cytotoxicity in wish cells. Microb Pathog. 23(2): 85-93.

Mattana A, Cappai V, Alberti L, Serra C, Fiori PL and Cappuccinelli P. (2002). ADP and other metabolites released from *Acanthamoeba castellanii* lead to human monocytic cell death through apoptosis and stimulate the secretion of proinflammatory cytokines. Infect Immun. 70(8): 4424-4432.

Mattana A, Biancu G, Alberti L, Accardo A, Delogu G, Fiori PL and Cappuccinelli P. (2004). *In vitro* evaluation of the effectiveness of the macrolide rokitamycin and chlorpromazine against *Acanthamoeba castellanii*. Antimicrob Agents Chemother. 48: 4520-4527.

Mattana A, Serra C, Mariotti E, Delogu G, Fiori PL and Cappuccinelli P. (2006). *Acanthamoeba castellanii* promotion of *in vitro* survival and transmission of coxsackie b3 viruses. Eukaryot Cell. 5: 665-671.

Matthews TD, Frazer DG, Minassian DC, Radford CF and Dart JK. (1992). Risks of keratitis and patterns of use with disposable contact lenses. Arch Ophthalmol. 110: 1559-1562.

Matura, H. and Korn ED. (1977). *Acanthamoeba* myosin II. J Biol Chem. 252(18): 6501-6509.

Mazur T and Zóźwiak M. 1989. Value of the color test in assessing the viability of cysts of *Acanthamoeba* sp. Wiad Parazytol. 35(1): 11-17.

Mazur T and Jóźwiak M. (1993). Extracerebral infections of *Acanthamoeba* spp. in mice. Wiad Parazytol. 39: 357-366.

Mazur T and Hadas E. (1994). The effect of passages of *Acanthamoeba* strains through mice tissues on their virulence and its biochemical markers. Parasitol Res. 80: 431-434.

Mazur T, Hadas E and Iwanicka I. (1995). The duration of the cyst stage and the viability and virulence of *Acanthamoeba* isolates. Trop Med Parasitol. 46: 106-108.

McBride J, Mullen AB, Carter KC and Roberts CW. (2007). Differential cytotoxicity of phospholipid analogues to pathogenic *Acanthamoeba* species and mammalian cells. J Antimicrob Chemother. 60: 521-525.

McClellan K, Howard K, Niederkorn JY and Alizadeh H. (2001). Effect of steroids on *Acanthamoeba* cysts and trophozoites. Invest Ophthalmol Vis Sci. 42(12): 2885-2893.

McClellan K, Howard K, Mayhew E, Niederkorn J and Alizadeh H. (2002). Adaptive immune responses to *Acanthamoeba* cysts. Exp Eye Res. 75(3): 285-293.

McLaughlin GL, Stimac JE, Luke JM, Kuhlenschmidt MS, Vernon RA, Morton LD, Visvesvara GS and Whiteley HE. (1991). Development of *Acanthamoeba*-cornea conincubation assays. Rev Infect Dis. 13. (suppl 3): S397-398.

Mehdi H and Garg NK. (1987). Changes in the lipid composition and activities of isocitrate dehydrogenase and isocitrate lyase during encystation of *Acanthamoeba culbertsoni* strain A-1. Trans R Soc Trop Med Hyg. 81: 633-636.

Mehdi H, Garg HS, Garg NK and Bhakuni DS. (1988). Sterols of *Acanthamoeba culbertsoni* strain A-1. Steroids. 51(5-6): 551-558.

Mehlotra RK and Shukla OP. (1993). *In vitro* susceptibility of *Acanthamoeba culbertsoni* to inhibitors of folate biosynthesis. J Eukaryot Microbiol. 40: 14-17.

Meisler DM, Ludwig IH, Rutherford I, Bican FE, Langston RH and Visvesvara GS. (1986). Susceptibility of *Acanthamoeba* to cryotherapeutic method. Arch Ophthalmol. 104: 130-131.

Mery-Drugeon E, Crouse EJ, Schmitt JM, Bohnert HJ and Bernardi G. (1981). The mitochondrial genomes of *Ustilago cynodontis* and *Acanthamoeba castellanii*. Eur J Biochem. 114(3): 577-583.

Michel R and Menn T. (1991). *Acanthamoeba*, Naegleria and invertebrates in wet areas of physiotherapy equipment in hospitals. Zentralbl Hyg Umweltmed. 191: 423-437.

Michel R, Burghardt H and Bergmann H. (1995). *Acanthamoeba*, naturally intracellularly infected with *Pseudomonas aeruginosa*, after their isolation from a microbiologically contaminated drinking water system in a hospital. Zentralbl Hyg Umweltmed. 196(6): 532-544.

Michel R and Hauröder B. (1997). Isolation of an *Acanthamoeba* strain with intracellular *Burkholderia pickettii* infection. Zentralbl Bakteriol. 285: 541-557.

Misra SK, Sharma AK, Mehdi H and Garg NK. (1986). Effect of cholesterol and alpha-p-chlorophenoxyisobutyrate passage on virulence in *Acanthamoeba culbertsoni* strain A-1 and C-7. Int J Parasitol. 16: 191-196.

Mitra MM, Alizadeh H, Gerard RD and Niederkorn JY. (1995). Characterization of a plasminogen activator produced by *Acanthamoeba castellanii*. Mol Biochem Parasitol. 73(1-2): 157-164.

Mitro K, Bhagavathiammai A, Zhou OM, Bobbett G, McKerrow JH, Chokshi R, Chokshi B and James ER. (1994). Partial characterization of the proteolytic secretions of *Acanthamoeba polyphaga*. Exp Parasitol. 78(4): 377-385.

Miyata H, Bowers B and Korn ED. (1989). Plasma membrane association of *Acanthamoeba* myosin I. J Cell Biol. 109: 1519-1528.

Mockrin SC and Korn ED. (1980). *Acanthamoeba* profilin interacts with G-actin to increase the rate of exchange of actin-bound adenosine 5'-triphosphate. Biochem. 19(23): 5359-5362.

Moon EK, Chung DI, Hong YC and Kong HH. (2007). Differentially expressed genes of *Acanthamoeba castellanii* during encystation. Korean J Parasitol. 45: 283-285.

Moon EK, Chung DI, Hong YC, Ahn TI and Kong HH. (2008). *Acanthamoeba castellanii*: Gene profile of encystation by ESTs analysis and KOG assignment. Exp Parasitol. 119: 111-116.

Moore MB and McCulley JP. (1989). *Acanthamoeba* keratitis associated with contact lenses: six consecutive cases of successful management. Br J Ophthalmol. 73: 271-275.

Morton LD, McLaughlin GL and Whiteley HE. (1991). Effect of temperature, amebic strain and carbohydrates on *Acanthamoeba* adherence to corneal epithelium *in vitro*. Infect Immun. 59: 3819-3822.

Mounier J, Laurent V, Hall A, Fort P, Carlier MF, Sansonetti PJ and Egile C. (1999). Rho family GTPases control entry of *Shigella flexneri* into epithelial cells but not intracellular motility. J Cell Sci. 112(13): 2069-2080.

Muldrow LL, Tyndall RL and Fliermans CB. (1982). Application of flow cytometry to studies of pathogenic free-living amoebae. Appl Environ Microbiol. 44(6): 1258-1269.

Muller M. (1969). Lysosomal hydrolases in *Acanthamoeba* sp. J Protozool. 16: 428-431.

Muller M and Moller KM. (1969). Urate oxidase and its association with peroxisomes in *Acanthamoeba* sp. Eur J Biochem. 9: 424-430.

Müller KD, Schmid EN and Michel R. (1999). Intracellular bacteria of *Acanthamoeba*e resembling Legionella spp. turned out to be Cytophaga sp. Zentralbl Bakteriol. 289: 389-397.

Murdoch D, Gray TB, Cursons R and Parr D. (1998). *Acanthamoeba* keratitis in New Zealand, including two cases with *in vivo*resistance to polyhexamethylene biguanide. Aust N Z J Ophthalmol. 26: 231-236.

Na BK, Kim JC and Song CY. (2001). Characterization and pathogenetic role of proteinase from *Acanthamoeba castellanii*. Microb Pathog. 30(1): 39-48.

Na BK, Cho JH, Song CY and Kim TS. (2002). Degradation of immunoglobulins, protease inhibitors and interleukin-1 by a secretory proteinase of *Acanthamoeba castellanii*. Korean J Parasitol. 40(2): 93-99.

Nacapunchai D, Phadungkul K and Kaewcharus S. (2001). *In vitro* effect of artesunate against *Acanthamoeba* spp. Southeast Asian J Trop Med Public Health. 33 Suppl 3: 49-52.

Nagington J, Watson PG, Playfair TJ, McGill J, Jones BR and Steele AD. (1974). Amoebic infections of the eye. Lancet. ii: 1537-1540.

Nagington J and Richards JE. (1976). Chemotherapeutic compounds and *Acanthamoeba*e from eye infections. J Clin Pathol. 29: 648-651.

Nassif KF. (1996). Ocular surface defense mechanisms. Infections of the eye. Little, Brown and Company, New York, NY. USA.

Neff RJ and Benton WF. (1962). Localization of cellulose in the cysts of *Acanthamoeba* sp. J Protozool. 9: suppl., 11.

Neff RJ and Neff RH. (1969). The biochemistry of amoebic encystment. Symp Soc Exp Biol. 23: 51-81.

Nero LC, Tarver MG and Hedrick LR. (1964). Growth of *Acanthamoeba castellani* with the yeast Torulopsis famata. J Bacteriol. 87(1): 220-225.

Niederkorn JY, Ubelaker JE, McCulley JP, Stewart GL, Meyer DR, Mellon JA, Silvany RE, He YG, Pidherney M, Martin JH, et al. (1992). Susceptibility of corneas from various animal species to *in vitro* binding and invasion by *Acanthamoeba castellanii* [corrected] Invest Ophthalmol Vis Sci. 33: 104-112.

Niederkorn JY, Alizadeh H, Leher H and McCulley JP. (1999). The pathogenesis of *Acanthamoeba* keratitis. Microbes Infect. 1: 437-443.

Nilsson SEG and Montan PG. (1994). The annualized incidence of contact lens induced keratitis in Sweden and its relation to lens type and wear schedule: results of a three-month prospective study. CLAO J 20: 225-230.

Oates PJ and Touster O. (1976). *In vitro* fusion of *Acanthamoeba* phagolysosomes. I. Demonstration and quantitation of vacuole fusion in *Acanthamoeba* homogenates. J Cell Biol. 68: 319-338.

Oates PJ and Touster O. (1978). *In vitro* fusion of *Acanthamoeba* phagolysosomes. II. Quantitative characterization of *in vitro* vacuole fusion by improved electron microscope and new light microscope technique. J Cell Biol. 79: 217-234.

Oates PJ and Touster O. (1980). *In vitro* fusion of *Acanthamoeba* phagolysosomes. III. Evidence that cyclic nucleotides and vacuole subpopulations respectively control the rate and the extent of vacuole fusion in *Acanthamoeba* homogenates. J Cell Biol. 85: 804-810.

Odom A, Del Poeta M, Perfect J and Heitman J. (1997a). The immunosuppressant FK506 and its nonimmunosuppressive analog L-685,818 are toxic to *Cryptococcus neoformans* by inhibition of a common target protein. Antimicrob Agents Chemother. 41: 156-161.

Odom A, Muir S, Lim E, Toffaletti DL, Perfect J and Heitman J. (1997b). Calcineurin is required for virulence of *Cryptococcus neoformans*. EMBO J. 16: 2576-2589.

Ofori-Kwakye SK, Sidebottom DG, Herbert J, Fischer EG and Visvesvara GS. (1986). Granulomatous brain tumor caused by *Acanthamoeba*. J Neurosurg. 64: 505-509.

Ogbunude PO and Baer HP. (1993). *De novo* purine biosynthesis in a free living protozoan parasite, *Acanthamoeba polyphaga* [correction of polypha] Trop Med Parasitol. 44: 19-22.

Oishi K, Raynor RL, Charp PA and Kuo JF. (1988). Regulation of protein kinase C by lysophospholipids. Potential role in signal transduction. J Biol Chem. 263(14): 6865-6871.

Ondarza RN, Iturbe A and Hernández E. (2006). *In vitro* antiproliferative effects of neuroleptics, antimycotics and antibiotics on the human pathogens *Acanthamoeba polyphaga* and *Naegleria fowleri*. Arch Med Res. 37: 723-729.

Ondarza RN, Iturbe A and Hernández E. (2007). The effects by neuroleptics, antimycotics and antibiotics on disulfide reducing enzymes from the human pathogens *Acanthamoeba polyphaga* and *Naegleria fowleri*. Exp Parasitol. 115: 41-47.

Osato MS, Robinson NM, Wilhelmus KR and Jones DB. (1991). *In vitro* evaluation of antimicrobial compounds for cysticidal activity against *Acanthamoeba*. Rev Infect Dis. 13 Suppl 5: S431-S435.

Paatero G, Isomaa B, Johansson M and Kiiskinen G. (1981). Effect of surface-active alkyltrimethylammonium salts on concanavalin A-mediated agglutination in *Acanthamoeba castellanii*. Eur J Cell Biol. 25(1): 66-70.

Page FC. (1967a). Taxonomic criteria for limax amoebae, with descriptions of 3 new species of *Hartmannella* and 3 of *Vahlkampfia*. J Protozool. 14: 499-521.

Page FC. (1967b). Re-definition of the genus *Acanthamoeba* with descriptions of three species. J Protozool. 14: 709-724.

Page FC. (1988). A new key to fresh water and soil amoebae. Freshwater Biological Association Scientific Publications, Cumbria, England. 1-122.

Paget T, Khan N, Temple G, Hough V and Greenman J. (2000). Use of phage antibodies to distinguish closely related species of protozoan parasites. Dis Markers. 16: 83-90.

Parmar DN, Awwad ST, Petroll WM, Bowman RW, McCulley JP and Cavanagh HD. (2006). Tandem scanning confocal corneal microscopy in the diagnosis of suspected *Acanthamoeba* keratitis. Ophthalmol. 113: 538-547.

Panjwani N, Zhao Z, Baum J, Pereira M and Zaidi T. (1992). *Acanthamoeba* bind to glycolipids of rabbit corneal epitelium. Infect Immun. 60: 3460-3463.

Park DH, Palay DA, Daya SM, Stulting RD, Krachmer JH and Holland EJ. (1997). The role of topical corticosteroids in the management of *Acanthamoeba* keratitis. Cornea. 16: 277-283.

Pellegrin JL, Ortega-Barria E, Barza M, Baum J and Pereira ME. (1991). Neuraminidase activity in *Acanthamoeba* species trophozoites and cysts. Invest Ophthalmol Vis Sci. 32: 3061-3066.

Pellegrin JL, Ortega-Barria E, Prioli RP, Meijia JS, Pereira ME. (1992). The neuraminidases of Trypanosoma cruzi and *Acanthamoeba castellanii* are immunologically related. Trop Med Parasitol. 43: 33-37.

Peng Z, Omaruddin R and Bateman E. (2005). Stable transfection of *Acanthamoeba castellanii*. Biochim Biophys Acta. 1743(1-2): 93-100.

Penley CA, Willis SW, Sickler SG. (1989). Comparative antimicrobial efficacy of soft and rigid gas permeable contact lens solutions against *Acanthamoeba*. CLAO J. 15: 257-260.

Perez-Santonja JJ, Kilvington S, Hughes R, Tufail A, Metheson M and Dart JKG. (2003). Persistently culture positive *Acanthamoeba* keratitis; *in vivo* resistance and *in vitro* sensitivity. Ophthalmol. 110(8): 1593-1600.

Pérez-Serrano J, Martínez J, Pérez B, Bernadina WE and Rodríguez-Caabeiro F. (2000). *In vitro* shock response to different stressors in free living and pathogenic *Acanthamoeba*. Int J Parasitol. 30: 829-835.

Perrine D, Chenu JP, Georges P, Lancelot JC, Saturnino C and Robba M. (1995). Amoebicidal efficiencies of various diamidines against two strains of *Acanthamoeba polyphaga*. Antimicrob Agents Chemother. 39(2): 339-342.

Pettit DA, Williamson J, Cabral GA and Marciano-Cabral F. (1996). *In vitro* destruction of nerve cell cultures by *Acanthamoeba* spp.: a transmission and scanning electron microscopy study. J Parasitol. 82(5): 769-777.

Pickup ZL, Pickup R, and Parry JD. (2007). A comparison of the growth and starvation responses of *Acanthamoeba castellanii* and *Hartmannella vermiformis* in the presence of suspended and attached *Escherichia coli* K12. FEMS Microbiol Ecol. 59: 556-563.

Pigon A. (1981). Growth-affecting activity of cellular microexudate of *Acanthamoeba*. Cytobios. 32(125): 15-28.

Pigon, A. (1986). Limited life-span and growth of solitary cells of *Acanthamoeba* and stimulatory effect of cellular microexudate, Cytobios. 46: 93-105.

Pigon, A. (1987). *Acanthamoeba* microexudate: a cell detoxification system. Cytobios. 50: 133-144.

Poggio EC, Glynn RJ, Schein OD, Seddon JM, Shannon MJ, Scardino VA and Kenyon KR. (1989). The incidence of ulcerative keratitis among users of daily-wear and extended-wear soft contact lenses. N Engl J Med. 321: 779-783.

Polat ZA, Tepe B and Vural A. (2007). *In vitro* effectiveness of Thymus sipyleus subsp. sipyleus var. sipyleus on *Acanthamoeba castellanii* and its cytotoxic potential on corneal cells. Parasitol Res. 101: 1551-1555.

Polat ZA, Vural A, Ozan F, Tepe B, Ozcelik S and Cetin A. (2007). *In vitro* evaluation of the amoebicidal activity of garlic. (Allium sativum). extract on *Acanthamoeba castellanii* and its cytotoxic potential on corneal cells. J Ocul Pharmacol Ther. 24: 8-14.

Pollard TD and Korn ED. (1973). *Acanthamoeba* myosin. I. Isolation from *Acanthamoeba castellanii* of an enzyme similar to muscle myosin. J Biol Chem. 248(13): 4682-4690.

Pollard, TD. (1976). The role of actin in the temperature-dependent gelation and contraction of extracts of *Acanthamoeba*. J Cell Biol. 68: 579-601.

Pollard, TD, Stafford WF and Porter ME. (1978). Characterization of a second myosin from *Acanthamoeba castellanii*. J Biol Chem. 253(13): 4798-4808.

Pollard TD. (1981). Purification of a calcium-sensitive actin gelation protein from *Acanthamoeba*. J Biol Chem. 256(14): 7666-7670.

Pollard TD. (1982). Structure and polymerization of *Acanthamoeba* myosin-II filaments. J Cell Biol. 95: 816-825.

Pollard TD, Tseng PC, Rimm DL, Bichell DP, Williams RC Jr, Sinard J, and Sato M. (1986). Characterization of alpha-actinin from *Acanthamoeba*. Cell Motil Cytoskeleton. 6: 649-661.

Poulin R, Larochelle J and Nadeau P. (1984). Polyamines in *Acanthamoeba castellanii*: presence of an unusually high, osmotically sensitive pool of 1,3-diaminopropane. Biochem Biophys Res Commun. 122: 388-393.

Pozio E and Morales MA. (2005). The impact of HIV-protease inhibitors on opportunistic parasites. Trends Parasitol. 21(2): 58-63.

Preston TM and King CA. (1984). Amoeboid locomotion of *Acanthamoeba castellanii* with special reference to cell-substratum interactions. J Gen Microbiol. 130: 2317-2323.

Preston TM, Richard H and Wotton RS. (2001). Locomotion and feeding of *Acanthamoeba* at the water-air interface of ponds. FEMS Lett. 194: 143-147.

Proca-Ciobanu M, Lupascu GH, Petrovici A and Ionescu MD. (1975). Electron microscopic study of a pathogenic *Acanthamoeba castellani* strain: the presence of bacterial endosymbionts. Int J Parasitol. 5(1): 49-56.

Proszenyák A, Charnock C, Hedner E, Larsson R, Bohlin L and Gundersen LL. (2007). Synthesis, antimicrobial and antineoplastic activities for agelasine and agelasimine analogs with a beta-cyclocitral derived substituent. Arch Pharm. (Weinheim). 340: 625-634.

Przelecka A and Sobota A. (1982). Growth phase dependent alterations in the surface coat of *Acanthamoeba castellanii*. Acta Histochem. 71: 219-229.

Pussard M. (1966). Le genre *Acanthamoeba* Volkonsky 1931. (Hartmannellidae-Amoebida). Protistologica. 2: 71-93.

Pussard M and Pons R. (1977). Morphologies de la paroi kystique et taxonomie du genre *Acanthamoeba*. (Protozoa, Amoebida). Protistologica. 13: 557–610.

Qu, XD and Leher RL. (1998). Secretory Phospholipase A2 is the principal bactericide for *Staphylococci* and other gram positive bacteria in human tears. Infect Immun. 6(66): 2791-2797.

Raali E, Vaahtoranta-Lehtonen HH and Lehtonen OP. (2001). Detachment of trophozoites of *Acanthamoeba* species from soft contact lenses with BEN22 detergent, BioSoak and Renu multi-purpose solutions. CLAO J. 27: 155-158.

Radford CF, Bacon AS, Dart JK and Minassian DC. (1995). Risk factors for *Acanthamoeba* keratitis in contact lens users: a case-control study. Br M J. 310: 1567-1570.

Radford CF, Lehmann OJ and Dart JK. (1998). *Acanthamoeba* keratitis: multicentre survey in England 1992-6. National *Acanthamoeba* Keratitis Study Group. Br J Ophthalmol. 82(12): 1387-1392.

Radford CF, Minassian DC and Dart JK. (2002). *Acanthamoeba* keratitis in England and Wales: incidence, outcome, and risk factors. Br J Ophthalmol. 86(5): 536-542.

Rahman M, Abd H, Romling U, Sandstrom G and Möllby R. (2008). Aeromonas-*Acanthamoeba* interaction and early shift to a viable but nonculturable state of Aeromonas by *Acanthamoeba*. J Appl Microbiol. 104: 1449-1457.

Raoult D, Audic S, Robert C, Abergel C, Renesto P, Ogata H, La Scola B, Suzan M and Claverie JM. (2004). The 1.2-megabase genome sequence of Mimivirus. Science. 306(5700): 1344-1350.

Reichstein E and Korn ED. 1979. *Acanthamoeba* profilin. A protein of low molecular weight from *Acanthamoeba castellanii* that inhibits actin nucleation. J Biol Chem. 254(13): 6174-6179.

Riddick DH. (1968). Contractile vacuole in the amoeba, *Pelomyxa carolinensis*. Am J Physiol. 215: 736.

Rimm DL, Pollard TD and Hieter P. (1988). Resolution of *Acanthamoeba castellanii* chromosomes by pulsed field gel electrophoresis and construction of the initial linkage map. Chromosoma. 97(3): 219-223.

Robertson AF and Lands WEM. (1964). Metabolismof phospholipids in normal and spherocytic human erythrocytes. J Lipid Res. 5: 88-93.

Robin JB, Chan R and Andersen BR. (1988). Rapid visualization of *Acanthamoeba* using fluorescein-conjugated lectins. Arch Ophthalmol. 106(9): 1273-1276.

Rodríguez-Zaragoza S, Ordaz C, Avila G, Muñoz JL, Arciniegas A and Romo de Vivar A. (1999). *In vitro* evaluation of the amebicidal activity of Buddleia cordata. (Loganiaceae, H.B.K.) on several strains of *Acanthamoeba*. J Ethnopharmacol. 66: 327-334.

Rosenthal S, Reed EJ and Weisman RA. (1969). Effect of lytic enzymes of *Acanthamoeba castellanii* on bacterial cell walls. J Bacteriol. 98(1): 182 – 189.

Rosenthal RA, Buck S, McAnally C, Abshire R, and Schlech B. (1999). Antimicrobial comparison of a new multi-purpose disinfecting solution to a 3% hydrogen peroxide system. CLAO J. 25: 213-217.

Rosenthal RA, McAnally CL, McNamee LS, Buck SL, Schlitzer RL, and Stone RP. (2000). Broad spectrum antimicrobial activity of a new multi-purpose disinfecting solution. CLAO J. 26: 120-126.

Roti Roti LW and Stevens AR. (1974). Effect of 5-bromodeoxyuridine on growth, encystment and excystment of *Acanthamoeba castellanii*. J Cell Biol. 61: 233-237.

Roti Roti LW and Stevens AR. (1975). DNA synthesis in growth and encystment of *Acanthamoeba castellanii*. J Cell Sci 17: 503-515.

Rowbotham TJ. (1980). Preliminary report on the pathogenicity of *Legionella pneumophila* for freshwater and soil amoebae. J Clin Pathol. 33(12): 1179-1183.

Rowan-Kelly B, Ferrante A and Thong YH. (1982). The chemotherapeutic value of sulphadiazine in treatment of *Acanthamoeba* meningoencephalitis in mice. Trans R Soc Trop Med Hyg. 76: 636-638.

Rubin R. W., Hill, M. C., Hepworth P. and Boehmer, J. (1976). J Cell Biol., 68: 740-751.

Rudick VL, Weisman RA. (1973). Effects of cycloheximide on RNA polymerase specific activity and encystment in *Acanthamoeba castellanii*. Nature. 244(5413): 220-2.

Rudick VL and Weisman RA. (1973). DNA-dependent RNA polymerase from trophozoites and cysts of *Acanthamoeba castellanii*. Biochim Biophys Acta. 299(1): 91-102.

Rudick VL and Weisman RA. (1974). Uridine diphosphate glucose pyrophosphorylase of *Acanthamoeba castellanii*. Purification, kinetic, and developmental studies. J Biol Chem. 249: 7832-40.

Russell AD and Chopra I. (1996). Understanding antibacterial action and resistance. 2nd edition, London: Ellis Horwood p.113-118.

Saunders PP, Proctor EM, Rollins DF and Richards JS. (1992). Enhanced killing of *Acanthamoeba* cysts *in vitro* using dimethylsulfoxide. Ophthalmology. 99: 1197-1200.

Sawyer T. (1971). *Acanthamoeba griffini*, a new species of marine amoeba. J Protozool 18: 650-654.

Sawyer TK and Griffin JL. (1975). A proposed new family, *Acanthamoebidae*. (order Amoebida), for certain cyst-forming filose amoebae. Trans Am Microsc Soc. 94: 93-98.

Sayanova O, Haslam R, Guschina I, Lloyd D, Christie WW, Harwood JL and Napier JA. (2006). A bifunctional Delta12,Delta15-desaturase from *Acanthamoeba castellanii* directs the synthesis of highly unusual n-1 series unsaturated fatty acids. J Biol Chem. 281: 36533-36541.

Schein OD, Ormerod LD, Barraquer E, Alfonso E, Egan KM, Paton BG and Kenyon KR. (1989). Microbiology of contact lens-related keratitis. Cornea. 8: 281-285.

Schroeder JM, Booton GC, Hay J, Niszl IA, Seal DV, Markus MB, Fuerst PA and Byers TJ. (2001). Use of subgenic 18S ribosomal DNA PCR and sequencing for genus and genotype identification of *Acanthamoebae* from humans with keratitis and from sewage sludge. J Clin Microbiol. 39(5): 1903-1911.

Schuster FL and Mandel N. (1984). Phenothiazine compounds inhibit *in vitro* growth of pathogenic free-living amoebae. Antimicrob. Agents Chemothr. 25: 109-112.

Schuster FL and Jacob LS. (1992). Effects of magainins on amoeba and cyst stages of *Acanthamoeba polyphaga*. Anti Microb Agents Chemothr. 36: 1263-1271.

Seal DV, Hay J and Kirkness CM. (1995). Chlorhexidine or polyhexamethylene biguanide for *Acanthamoeba* keratitis. Lancet 345(8942): 136.

Seal DV, Kirkness CM and Bennett HGB. (1999). Population-based cohort study of microbial keratitis in Scotland: incidence and features. Cont lens Ant Eye. 22: 49-57.

Sesma MJ and Ramos LZ. (1989). Isolation of free-living amoebas from the intestinal contents of reptiles. J Parasitol. 75(2): 322-324.

Schulze I and Jantzen H. (1982). Coordinate regulation of synthesis of ribosomal proteins during encystation of *Acanthamoeba castellanii*. Eur J Biochem. 126: 285-292.

Schuster FL and Levandowsky M. (1996). Chemosensory responses of *Acanthamoeba castellanii*: visual analysis of random movement and responses to chemical signals. J Eukaryot Microbiol. 43: 150-158.

Schuster FL and Visvesvara GS. (2004). Free-living amoebae as opportunistic and non-opportunistic pathogens of humans and animals. Int J Parasitol. 34(9): 1-27.

Seijo Martinez M, Gonzalez-Mediero G, Santiago P, De Lope Rodriguez A, Diz J, Conde C and Visvesvara GS. (2000). Granulomatous amebic encephalitis in a patient with AIDS: isolation of *Acanthamoeba* sp. Group II from brain tissue and successful treatment with sulfadiazine and fluconazole. J Clin Microbiol. 38(10): 3892-3895.

Seilhamer JJ and Byers TJ. (1978). Mutants of *Acanthamoeba castellanii* resistant to erythromycin, chloramphenicol, and oligomycin. J Protozool. 25(4): 486-489.

Seilhamer JJ and Byers TJ. (1982). ATPase-associated oligomycin resistance in *Acanthamoeba castellanii*. J Protozool. 29: 394-397.

Serhan CN, Haeggstrom JZ and Leslie CC. (1996). Lipid mediator networks in cell signaling: update and impact of cytokines. FASEB J. 10(10): 1147-1158.

Sharma S, Srinivasan M and George C. (1990). *Acanthamoeba* keratitis in non-contact lens wearers. Arch Ophthalmol. 108(5): 676-678.

Shin HJ, Cho MS, Jung SY, Kim HI, Park S, Seo JH, Yoo JC and Im KI. (2001). Cytopathic changes in rat microglial cells induced by pathogenic *Acanthamoeba culbertsoni*: morphology and cytokine release. Clin Diagn Lab Immunol. 8(4): 837-840.

Shukla OP, Kaul SM and Mehlotra RK. (1990). Nutritional studies on *Acanthamoeba culbertsoni* and development of chemically defined medium. J Protozool. 37: 237-242.

Shukla OP, Müller S and Walter RD. (1992). Polyamine oxidase from *Acanthamoeba culbertsoni* specific for N8-acetylspermidine. Mol Biochem Parasitol. 51: 91-98.

Silvany RE, Dougherty JM, McCulley JP, Wood TS, Bowman RW and Moore MB. (1990). The effect of currently available contact lens disinfection systems on *Acanthamoeba castellanii* and *Acanthamoeba polyphaga*. Ophthalmology. 97: 286-290.

Silvany RE, Dougherty JM and McCulley JP. (1991). Effect of contact lens preservatives on *Acanthamoeba*. Ophthalmology. 98: 854-857.

Simmons PA, Tomlinson A and Seal DV. (1998). The role of *Pseudomonas aeruginosa* biofilm in the attachment of *Acanthamoeba* to four types of hydrogel contact lens materials. Optom Vis Sci 75(12): 860-866.

Sinard JH and Pollard TD. (1989). The effect of heavy chain phosphorylation and solution conditions on the assembly of *Acanthamoeba* myosin-II. J Cell Biol. 109: 1529-1536.

Sinard JH and Pollard TD. 1989. Microinjection into *Acanthamoeba castellanii* of monoclonal antibodies to myosin-II slows but does not stop cell locomotion. Cell Motil Cytoskeleton. 12: 42-52.

Sinclair JL, McClellan JF and Coleman DC. (1981). Nitrogen mneralization by *Acanthamoeba polyphaga* in grazed *Pseudomonas paucimobilis* populations. Appl Environ Microbiol. 42(4): 667-671.

Singh BN. (1950). Nuclear division in nine species of small free-living amoebae and its bearing on the classification of the order Amoebida. Philos. Trans R Soc Lond [Biol.]. 236: 405-461.

Singh BN and Das SR. (1970). Studies on pathogenic and non-pathogenic small free-living amoebae and the bearing of nuclear division on the classification of the order Amoebida. Philos. Trans R Soc Lond [Biol.]. 259: 435-476.

Sissons J, Alsam S, Jayasekera S, Kim KS, Stins M and Khan NA. (2004). *Acanthamoeba* induces cell-cycle arrest in the host cells. J Med Microbiol. 53(8): 711-717.

Sissons J, Alsam S, Jayasekera S and Khan NA. (2004). Ecto-ATPases of clinical and non-clinical isolates of *Acanthamoeba*. Microb Pathogen. 37(5): 231-239.

Sissons J, Kim KS, Stins M, Jayasekera S, Alsam S and Khan NA. (2005a). *Acanthamoeba castellanii* induces host cell death via a phosphatidylinositol 3-kinase-dependent mechanism. Infect Immun. 73(5): 2704-2708.

Sissons J, Alsam S, Goldsworthy G, Lightfoot M, Jarroll EL and Khan NA. (2006). Identification and properties of proteases from an *Acanthamoeba* isolate capable of producing granulomatous encephalitis. BMC Microbiol. 6: 42.

Sissons J, Alsam S, Stins M, Rivas AO, Morales JL, Faull J and Khan NA. (2006). Use of in vitro assays to determine effects of human serum on biological characteristics of *Acanthamoeba castellanii*. J Clin Microbiol. 44: 2595-600.

Smith FR and Korn ED. (1968). 7-Dehydrostigmasterol and ergosterol: the major sterol of an amoebae. J Lipid Res. 9: 405 – 408.

Smith JD and Law JH. (1970). Phosphatidylcholine biosynthesis in *Tetrahymena pyriforms*. Biochim Biophys Acta. 202: 141-152.

Snelling WJ, McKenna JP, Lecky DM, and Dooley JS. (2005). Survival of *Campylobacter jejuni* in waterborne protozoa. Appl Environ Microbiol. 71: 5560-5571.

Sobota A and Przelecka, A. (1981). Visualization of calcium-binding sites at plasma membrane of shock-frozen *Acanthamoeba* cells. Acta Histochem. 68(1): 125-129.

Sobota A and Przelecka, A. (1981a). Develpmental changes in the localization of calcium-binding sites in *Acanthamoeba castellanii*. Histochemistry 71(1): 135-144.

Sobota A, Pogorelov AG and Burovina IV. (1981b). Hetergenous distribution of potassium and phosphorous in *Acanthamoeba castellanii*. Cell Biol Int Rep 5(3): 221-227.

Sobota A, Burovina IV, Pogorelov AG and Solus AA. (1984). Correlation between potassium and phosphorus content and their nonuniform distribution in *Acanthamoeba castellanii*. Histochemistry. 81: 201-204.

Soulides DA, and Allison FE. (1961). Effect of drying and freezing soils on carbon dioxide production, available mineral nutrients, aggregation and bacterial population. Soil Sci 91: 192-298.

Spanakos G, Tzanetou K, Miltsakakis D, Patsoula E, Malamou-Lada E and Vakalis NC. (2006). Genotyping of pathogenic *Acanthamoeba*e isolated from clinical samples in Greece--report of a clinical isolate presenting T5 genotype. Parasitol Int. 55: 147-149.

Spindler SR, Duester GL, D'Alessio JM and Paule MR. (1978). A rapid and facile procedure for the preparation of RNA polymerase I from *Acanthamoeba castellanii*: purification and subunit structure. J Biol Chem 253(13): 4669-4675.

Spindler SR, D'Alessio JM, Duester GL and Paule MR. (1978). DNA-dependent RNA polymerase III from *Acanthamoeba castellanii*: a rapid procedure for the large scale preparation of homogeneous enzyme. J Biol Chem 253(17): 6242-6248.

Steenbergen JN, Shuman HA and Casadevall A. (2001). *Cryptococcus neoformans* interactions with amoebae suggest an explanation for its virulence and intracellular pathogenic strategy in macrophages. Proc Natl Acad Sci U S A. 98: 15245-15250.

Stehr-Green JK, Baily TM and Visvesvara GS. (1989). The epidemiology of *Acanthamoeba* keratitis in the United States. Am J Ophthalmol 107: 331-336.

Steinert M, Birkness KK, White E, Fields B and Quinn F. (1998). *Mycobacterium avium* bacilli grow saprozoically in coculture with *Acanthamoeba polyphaga* and survive within cyst walls. Appl Environ Microbiol 64(6): 2256-2261.

Stevens AR and Pachler PF. (1973). RNA synthesis and turnover during density-inhibited growth and encystment of *Acanthamoeba castellanii*. J Cell Biol 57: 525-537.

Stevens AR and O'Dell WD. (1974). *In vitro* and *in vivo* activity of 5-fluorocytosine on *Acanthamoeba*. Antimicrob Agen Chemother. 6(3): 282-289.

Stewart GL, Kim I, Shupe K, Alizadeh H, Silvany R, McCulley JP and Niederkorn JY. (1992). Chemotactic response of macrophages to *Acanthamoeba castellanii* antigen and antibody-dependent macrophage-mediated killing of the parasite. J Parasitol. 78(5): 849-55.

Stewart GL, Shupe KK, Kim I, Silvany RE, Alizadeh H, McCulley JP and Niederkorn JY. (1994). Antibody-dependent neutrophil-mediated killing of *Acanthamoeba castellanii*. Int J Parasitol. 24(5): 739-42.

Stohr M, Bommert K, Schulze I and Jantzen H. (1987). The cell cycle and its relationship to development in *Acanthamoeba castellanii*. J Cell Sci. 88: 579-589.

Suhre K, Audic S and Claverie JM. (2005). Mimivirus gene promoters exhibit an unprecedented conservation among all eukaryotes. Proc Natl Acad Sci U S A. 102: 14689-14693.

Sujata S and Berk SG. (1993). Stimulatory effect of cooling tower biocides on amoebae. Appl. Environmental Microbiol. 59: 3245-3249.

Suzan-Monti M, La Scola B and Raoult D. (2006). Genomic and evolutionary aspects of Mimivirus. Virus Res. 117: 145-155.

Sykes DE and Band RN. (1985). Polyphenol oxidase produced during encystation of *Acanthamoeba castellanii*. J Protozool. 32: 512-517.

Tanaka Y, Suguri S, Harada M, Hayabara T, Suzumori K and Ohta N. (1994). *Acanthamoeba*-specific human T-cell clones isolated from healthy individuals. Parasitol Res. 80(7): 549-553.

Taylor WM, Pidherney MS, Alizadeh H and Niederkorn JY. (1995). *In vitro* characterization of *Acanthamoeba castellanii* cytopathic effect. J Parasitol. 81(4): 603-609.

Thakur A and Willcox DP. (2000). Contact lens wear alters the production of certain inflammatory mediators in tears. Exp Eye Res. 70: 255-259.

Thom S, Warhurst D and Drasar BS. (1992). Association of *Vibrio cholerae* with fresh water amoebae. J Med Microbiol. 36(5): 303-306.

Thyrell L, Hjortsberg L, Arulampalam V, Panaretakis T, Uhles S, Zhivotovsky B, Leibiger I, Grander D and Pokrovskaja K. (2004). Interferon-alpha induced apoptosis in tumor cells is mediated through PI3K/mTOR signaling pathway. J Biol Chem. 279(23): 24152-24162.

Tobacman LS and Korn ED. (1982). The regulation of actin polymerization and the inhibition of monomeric actin ATPase activity by *Acanthamoeba* profilin. J Biol Chem. 257(8): 4166-4170.

Tomita M, Shimmura S, Tsubota K and Shimazaki J. (2006). Dacryoadenitis associated with *Acanthamoeba* keratitis. Arch Ophthalmol. 124: 1239-1242.

Tomlinson G and Jones EA. (1962). Isolation of cellulose from the cyst wall of a soil amoeba. Biochim Biophys Acta. 63: 194-200.

Tomlinson A, Simmons PA, Seal DV and McFadyen AK. (2000). Salicylate inhibition of *Acanthamoeba* attachment to contact lenses: a model to reduce risk of infection. Ophthalmol.107(1): 112-117.

Toney DM and Marciano-Cabral F. (1998). Resistance of *Acanthamoeba* species to complement lysis. J Parasitol. 84(2): 338-344.

Topalkara A, Vural A, Polat Z, Toker MI, Arici MK, Ozan F and Cetin A. (2007). *In vitro* amoebicidal activity of propolis on *Acanthamoeba castellanii*. J Ocul Pharmacol Ther. 23: 40-45.

Torno MS Jr., Babapour R, Gurevitch A and Witt MD. (2000). Cutaneous acanthamoebiasis in AIDS. J Am Acad Dermatol. 42(2): 351-354.

Trebst, A., Horth, H. and Draber, W. (1970). On a new inhibitor of photosynthetic electron-transport in isolated chloroplasts. Z. Naturforsch. 25B, 1157-1159.

Tripathi RC and Tripathi BJ. (1984). Lens spoilage. In: Contact Lenses: The Contact Lens Assoc. Ophth. Guide to Basic Science and Clinical Practice, edited by O. Diabezies. pp: 45.1 – 45.33. Grune and Stratton. Orlando. FL. USA.

Trzyna WC, Legras XD and Cordingley JS. (2006). A type-1 metacaspase from *Acanthamoeba castellanii*. Microbiol Res. 163: 414-23.

Tseng PCH and Pollard TD. (1982). Mechanism of action of *Acanthamoeba* profilin: demonstration of actin species specificity and regulation by micromolar concentration of $MgCl^2$. J Cell Biol. 94: 213-218.

Tseng CH, Fong CF, Chen WL, Hou YC, Wang IJ and Hu FR. (2005). Overnight orthokeratology-associated microbial keratitis. Cornea. 24: 778-782.

Turner NA, Russell AD, Furr JR and Lloyd D. (2000). Emergence of resistance to biocides during differentiation of *Acanthamoeba castellanii*. J Antimicrob Chem. 46: 27-34.

Tyndall RL and Domingue EL. (1982). Cocultivation of *Legionella pneumophila* and free-living amoebae. Appl Environ Microbiol. 44(4): 954-959.

Ulsamer AG, Smith FR and Korn ED. (1969). Lipids of *Acanthamoeba castellanii*: composition and effects of phagocytosis on incorporation of radioactive precursors. J Cell Biol. 43: 105-114.

Ueno T and Korn ED. (1986). Isolation and partial characterization of a 10-kD dimmer actin-binding protein. J Cell Biol. 103: 621-630.

Ulsamer AG, Wright PL, Wetzel MG and Korn ED. (1971). Plasma and phagosome membranes of *Acanthamoeba castellanii*. J Cell Biol. 51: 193-215.

Vandekerckhove J, van Damme J, Vancompernolle K, Bubb MR, Lambooy PK and Korn ED. (1990). The covalent structure of *Acanthamoeba* actobindin. J Biol Chem. 265: 12801-12805.

Vandenesch F, Surgot M, Bornstein N, Paucod JC, Marmet D, Isoard P and Fleurette J. (1990). Relationship between free amoeba and *Legionella*: studies *in vitro* and *in vivo*. Zentralbl Bakteriol. 272: 265-275.

Van Klink F, Alizadeh H, Stewart GL, Pidherney MS, Silvany RE, He Y, McCulley JP and Niederkorn JY. (1992). Characterization and pathogenic potential of a soil isolate and an ocular isolate of *Acanthamoeba castellanii* in relation to *Acanthamoeba* keratitis. Curr Eye Res. 11(12): 1207-2012.

Van Klink F, Alizadeh H, He Y, Mellon JA, Silvany RE, McCulley JP and Niederkorn JY. (1993). The role of contact lenses, trauma, and Langerhans cells in a Chinese hamster model of *Acanthamoeba* keratitis. Invest Ophthalmol Vis Sci. 34(6): 1937-1944.

Van Klink F, Taylor WM, Alizadeh H, Jager MJ, van Rooijen N and Niederkorn JY. (1996). The role of macrophages in *Acanthamoeba* keratitis. Invest Ophthalmol Vis Sci. 37(7): 1271-1281.

Van Klink F, Leher H, Jager MJ, Alizadeh H, Taylor W and Niederkorn JY. (1997). Systemic immune response to *Acanthamoeba* keratitis in the Chinese hamster. Ocul Immunol Inflamm. 5(4): 235-244.

Vaquero Ruano M, Encinas Martín JL, Ramón Y Cajal S and Cajigal Morales C. (2002). *Acanthamoeba* keratitis, a possible dibromo propamidine isethionate resistance. Arch Soc Esp Oftalmol. 77: 279-281.

Verma AK and Raizada MK. (1975). The inhibitory effect of glucose on the differentiation of trophic *Hartmannella culbertsoni* into viable cysts. Cell Differen. 4: 167-177.

Vemuganti GK, Reddy K, Iftekhar G, Garg P and Sharma S. (2004). Keratocyte loss in corneal infection through apoptosis: a histologic study of 59 cases. BMC Ophthalmol. 4(1): 16.

Victoria EJ and Korn ED. (1975a). Plasma membrane and soluble lysophospholipases of *Acanthamoeba castellanii*. Arch Biochem Biophys. 171(1): 255-258.

Victoria EJ and Korn ED. (1975b). Enzymes of phospholipid metabolism in the plasma membrane of *Acanthamoeba castellanii*. J Lipid Re.s 16(1): 54-60.

Visvesvara GS and Healy GR. (1975). Comparative antigenic analysis of pathogenic and free-living *Naegleria* species by the gel diffusion and immunoelectrphoresis techniques. Infect Immn. 11(1) 95-108.

Visvesvara GS and Balamuth W. (1975). Comparative studies on related free-living and pathogenic amebae with special reference to *Acanthamoeba*. J Protozool. 22(2): 245-256.

Visvesvara GS, Martinez AJ, Schuster FL, Leitch GJ, Wallace SV, Sawyer TK and Anderson M. (1990). Leptomyxid ameba, a new agent of amebic meningoencephalitis in humans and animals. J Clin Microbiol. 28(12): 2750–2756.

Visvesvara GS. (1991). Classification of *Acanthamoeba*. Rev Infect Dis. 13(Suppl 5): 369-372.

Visvesvara GS, Schuster FL and Martinez AJ. (1993). *Balamuthia mandrillaris*, N. G., N. Sp., agent of amebic meningoencephalitis in humans and other animals. J Eukaryot Microbiol 40(4): 504-514.

Volkonsky M. (1931). *Hartmanella castellanii* Douglas, et classification des hartmannelles. *Arch Zool Exp Gen* 72: 317-339.

Wachsstock DH, Schwarz WH, and Pollard TD. (1993). Affinity of α-actinin for actin determines the structure and mechanical properties of actin filaments gels. Biophysical J 65: 205-214.

Wagner Y, Noack B, Hoffmann T, Jacobs E, Christian Lück P. (2006). Periodontopathogenic bacteria multiply in the environmental amoeba *Acanthamoeba castellanii*. Int J Hyg Environ Health. 209: 535-539.

Walochnik J, Obwaller A and Aspock H. (2000). Correlations between morphological, molecular biological, and physiological characteristics in clinical and nonclinical isolates of *Acanthamoeba* spp. Appl Environ Microbiol 66(10): 4408-4413.

Walochnik J, Obwaller A, Haller-Schober EM and Aspock H. (2001). Anti-*Acanthamoeba* IgG, IgM, and IgA immunoreactivities in correlation to strain pathogenicity. Parasitol Res 87(8): 651-658.

Walochnik J, Duchene M, Seifert K, Obwaller A, Hottkowitz T, Wiedermann G, Eibl H and Aspock H. (2002). Cytotoxic activities of alkylphosphocholines against clinical isolates of *Acanthamoeba* spp. Antimicrob Agents Chemother 46(3): 695-701.

Walochnik J, Aichelburg A, Assadian O, Steuer A, Visvesvara G, Vetter N, and Aspöck H. (2008). Granulomatous amoebic encephalitis caused by *Acanthamoeba* amoebae of genotype T2 in a human immunodeficiency virus-negative patient. J Clin Microbiol. 46: 338-340.

Wang L, Asem EK and McLaughlin GL. (1994). Calcium enhances *Acanthamoeba polyphaga* binding to extracellular matrix proteins. Invest Ophthalmol Vis Sci. 35: 2421-2426.

Watanabe R, Ishibashi Y, Hommura S and Ishii K. (1994). *Acanthamoeba* isolation from contact lens solution of contact lens wearers without keratitis. J Jpn Ophthalmol Soc 98(5): 477-480.

Weekers PHH, Bodelier PLE, Wijen JPH and Vogels GD. (1993). Effects of grazing by the free-living soil amoebae, *Acanthamoeba catellanii*, *Acanthamoeba polyphaga*, and *Hartmannella vermiformis* on various bacteria. Appl Environ Microbiol 59: 2317-2319.

Weihing RR and Korn ED. (1971). *Acanthamoeba* actin. Isolation and properties. Biochemistry 10: 590-600.

Weisman RA and Korn ED. (1967). Phagocytosis of latex beads by *Acanthamoeba*. I. Biochemical properties. *Biochem* 6(2): 485 – 497.

Weisman RA. (1976). Differentiation in *Acanthamoeba castellanii*. Annu Rev Microbiol. 30: 189-219.

Weisman B and Mondino B. (2002). Risk factors for contact lens associated microbial keratitis. *Cont Lens Ant* Eye 25: 3-9.

Wennström S, Hawkins P, Cooke F, Hara K, Yonezawa K, Kasuga M, Jackson T, Claesson-Welsh L and Stephens L. (1994). Activation of phosphoinositide 3-kinase is required for PDGF-stimulated membrane ruffling. Curr Biol. 4: 385-396.

Wilhelmus KR, Osato MS, Font RL, Robinson NM and Jones DB. (1986). Rapid diagnosis of *Acanthamoeba* keratitis using calcofluor white. Arch Ophthalmol. 104: 1309-1312.

Wilhelmus KR and Jones DB. (1991). Program Planning for research on *Acanthamoeba*. Rev Infect Dis. 13(suppl. 5): S446-S450.

Winchester K, Mathers WD, Sutphin JE and Daley TE. (1995). Diagnosis of *Acanthamoeba* keratitis *in vivo* with confocal microscopy. Cornea 14(1): 10-17.

Winiecka-Krusnell J, Wreiber K, von Euler A, Engstrand L and Linder E. (2002). Free-living amoebae promote growth and survival of *Helicobacter pylori*. *Scand* J Infect Dis 34(4): 253-256.

Wright P, Warhurst D and Jones BR. (1985). *Acanthamoeba* keratitis successfully treated medically. Br J Ophthalmol 69: 778-782.

Xu P, Mitchelhill KI, Kobe B, Kemp BE and Zot HG. (1997). The myosin-I-binding protein Acan125 binds the SH3 domain and belongs to the superfamily of leucine-rich repeat proteins. Proc Natl Acad Sci U S A. 94: 3685-3690.

Yang, Y., Gordon, DJ, Korn ED and Eisenberg E. (1977). Interaction between *Acanthamoeba* actin and rabbit skeletal muscle tropomyosin. J Biol Chem. 252(10): 3374-3378.

Yang, Y., Korn ED and Eisenberg E. (1979). Binding of tropomyosin to copolymers of *Acanthamoeba* actin and muscle actin. J Biol Chem. 254(6): 2084-2088.

Yang S, Villemez C. (1994). Cell surface control of differentiation in *Acanthamoeba*. J Cellular Biochem. 56: 592-596.

Yang Z, Cao Z and Panjwani N. (1997). Pathogenesis of *Acanthamoeba* keratitis: carbohydrate mediated host-parasite interactions. Infect Immun 65: 439-445.

Yepes N, Lee SB, Hill V, Ashenhurst M, Saunders PP and Slomovic AR. (2005). Infectious keratitis after overnight orthokeratology in Canada. Cornea. 24: 857-860.

Yin J and Henney HR Jr. (1997). Stable transfection of *Acanthamoeba*. *Can J Microbiol* 43(3): 239-244.

Yonemura S and Pollard TD. (1992). The localization of myosin I and myosin II in *Acanthamoeba* by fluorescence microscopy. J Cell Sci 102: 629-642.

Zaman V, Zaki M and Manzoor M. (1999). *Acanthamoeba* in human faeces from Karachi. Ann Trop Med Parasitol. 93: 189-191.

Zanetti S, Fiori PL, Pinna A, Usai S, Carta F and Fadda G. (1995). Susceptibiltiy of *Acanthamoeba castellanii* to contact lens disinfecting solutions. Antimicrob. Agents Chemother. 39: 1596-1598.

Zegans ME, Becker HI, Budzik J and O'Toole G. (2002). The role of bacterial biofilms in ocular infections. *DNA Cell Biol* 21(5-6): 415-420.

Zhu CM, Cumaraswamy A and Henney HR Jr. (1989). Comparison of polyamine and S-adenosylmethionine contents of growing and encysted *Acanthamoeba* isolates. Mol Cell Biochem. 90: 145-153.

Zot HG, Doberstein SK and Pollard TD. (1992). Myosin I moves actin filaments on a phospholipids substrate: implications for membrane targeting. J Cell Biol 116: 367-376.

Appendix 1

Glossary

Abscess

A collection of pus (dead immune cells) that has accumulated in a cavity formed by the tissue on the basis of an infectious process (usually caused by bacteria or parasites) or other foreign materials. It is a defensive reaction of the tissue to prevent the spread of infectious materials to other parts of the body.

Acanthopodia

Spine-like structures present on the surface of *Acanthamoeba* trophozoites.

Acridine orange

A nucleic acid selective fluorescent cationic dye useful for cell cycle determination. It is cell-permeable, and interacts with DNA and RNA by intercalation or electrostatic attractions.

Active transport

It is an energy-requiring process that moves material across a cell membrane and up the concentration gradient.

Adenosine-5'-triphosphate (ATP)

A multifunctional nucleotide that is most important as an intracellular energy transfer. ATP transports chemical energy within cells for metabolism. It is produced as an energy source during the processes of photosynthesis and cellular respiration and consumed by many enzymes and a multitude of cellular processes including biosynthetic reactions, motility and cell division.

Adrenal glands

The triangle-shaped endocrine glands that sit on top of the kidneys.

Amoeba

Protozoa that move and feed by pseudopodia and reproduce via binary fission.

Amoebiasis

A potentially fatal disease, caused by the amoeba *Entamoeba histolytica*.

Anabolism

The set of metabolic pathways that construct molecules from smaller units. These reactions require energy. Anabolism is powered by catabolism, where large molecules are broken down into smaller parts and then used up in respiration. Many anabolic processes are powered by adenosine triphosphate (ATP).

Anatomical seclusion

A mechanisms by which parasites can invade non-immune cells, replicate intracellularly and hide from the immune cells.

Antibiotic

Any substance produced by a micro-organism that is antagonistic to the growth of others.

Antibodies

Antibodies (also known as immunoglobulins) are gamma globulin proteins that are found in blood or other bodily fluids of vertebrates, and are used by the immune system to identify and neutralize foreign objects, such as microbial pathogens.

Antigenic variation

The process by which an infectious organism alters its surface proteins in order to evade a host immune response.

Antimicrobial

A substance that kills or inhibits the growth of microbes such as bacteria, fungi, viruses or protozoa. Antimicrobial drugs either kill microbes (microbicidal) or prevent the growth of microbes (microbistatic).

Antimicrobial peptides (also called host defence peptides)

An evolutionarily conserved component of the innate immune response and are found among all classes of life.

Aphasia

Inability to comprehend language.

Apoptosis

A cell's commitment to suicide.

Asexual reproduction

A form of reproduction which does not involve meiosis, ploidy reduction, or fertilization. Only one parent is involved in asexual reproduction.

Asthma

A disease process that is characterised by paradoxical narrowing of the bronchi (lung passageways) making breathing difficult.

Ataxia

Incoordination of muscle movements.

ATP synthase

A term for an enzyme that can synthesize adenosine triphosphate (ATP) from adenosine diphosphate (ADP) and inorganic phosphate by using some form of energy.

Axenic culture

A culture of a particular organism that is entirely free of all other "contaminating" organisms.

Basal ganglia

A group of nuclei in the brain interconnected with the cerebral cortex, thalamus and brainstem.

Basement membrane (Descemet's membrane)

Membrane lying between the corneal stroma, and the endothelial layer of the cornea.

B cells

Lymphocytes that play a large role in the humoral immune response (as opposed to the cell-mediated immune response, which is governed by T-cells). The principal function of B cells is to make antibodies against soluble antigens. B-cells are an essential component of the adaptive immune system.

Binary fission

A form of asexual reproduction, during which the organism splits directly into two equal-sized offspring, each with a copy of the parent's genetic material.

Biofilm

A structured community of microorganisms encapsulated within a self-developed polymeric matrix and adherent to a living or inert surface. Biofilms are also often characterized by surface attachment, structural heterogeneity, genetic diversity, complex community interactions, and an extracellular matrix of polymeric substances.

Biogeochemical cycle
: A circuit or pathway by which a chemical element or molecule moves through both biotic ("bio-") and abiotic ("geo-") compartments of an ecosystem.

Blood-brain barrier
: A membranic structure that acts primarily to protect the brain from chemicals in the blood, while still allowing essential metabolic function.

Bowman's membrane (Bowman's layer)
: A smooth layer in the eye. It is located between the front epithelium and the stroma in the cornea.

Brain stem
: The lower part of the brain, adjoining and structurally continuous with the spinal cord.

Budding
: Formation of a new organism by the protrusion of part of another organism.

Calcofluor white
: A dye that binds to beta (1, 4) glucan-like molecules.

Carbon cycle
: The biogeochemical cycle by which carbon is exchanged between the biosphere, geosphere, hydrosphere, and atmosphere of the Earth.

Catabolism
: The set of metabolic pathways which break down molecules into smaller units and release energy. In catabolism, large molecules such as polysaccharides, lipids, nucleic acids and proteins are broken down into smaller units such as monosaccharides, fatty acids, nucleotides and amino acids, respectively.

Cellulitis
: An acute, diffuse, spreading, oedematous, inflammation of the deep subcutaneous tissues and sometimes muscle, which may be associated with abscess formation.

Cellulose
: A polysaccharide consisting of a linear chain of several hundred to over ten thousand $\beta(1\rightarrow4)$ linked D-glucose units.

Central nervous system
: Contains the majority of the nervous system, and consists of the brain (invertebrates which have brains), and the spinal cord.

Cerebral hemisphere
: One of the two regions of the brain that are delineated by the body's median plane.

Cerebrospinal fluid
: A clear bodily fluid that occupies the subarachnoid space and the ventricular system around and inside the brain.

Chagas' disease
: A potentially fatal disease, caused by a protozoan named *Trypanosoma cruzi*, and is transmitted by kissing bug of the subfamily Triatominae (Family Reduviidae).

Chemo-organotroph
: A heterotroph, or chemo-organotroph is an organism that requires organic substrates to get its carbon for growth and development.

Chemotaxis
: The phenomenon in which cells direct their movements according to certain chemicals in their environment.

Chemotherapy
: Treatment of disease by chemicals that kill cells.

Chitin
: A long-chain polymer of N-acetylglucosamine, a derivative of glucose, and it is found in many places throughout the natural world.

Chloroplast
: Light harvesting organelle found in photosynthetic eukaryotes.

Cholera
: A form of infectious gastroenteritis (intestinal infection) that results in frequent watery stools, cramping abdominal pain, and eventual collapse from dehydration

Chromosome
: Organized structures of DNA and proteins.

Ciliary body
: The circumferential tissue inside the eye composed of the ciliary muscle and ciliary processes

Cisterna (plural cisternae)
: Comprises a flattened membrane disk that makes up the Golgi apparatus.

Collagen
: The main protein of connective tissue in animals and the most abundant protein in mammals, making up about 25% of the whole-body protein content.

Complement activation
: Complement is a set of proteins produced in part by the liver that circulate in the blood and are activated sequentially by proteolytic cleavage, a process called the complement activation.

Confocal microscopy
: An optical imaging technique used to increase micrograph contrast and/or to reconstruct three-dimensional images by using a spatial pinhole to eliminate out-of-focus light or flare in specimens that are thicker than the focal plane.

Conjunctiva
: A clear membrane that covers the sclera (white part of the eye) and lines the inside of the eyelids. It is made of lymphoid tissue.

Conjunctival hyperemia
: Engorgement conjunctival blood vessels or increased blood flow to the conjunctiva.

Conjunctivitis
: Infection or inflammation of the conjunctiva (the outermost layer of the eye and the inner surface of the eyelids).

Contractile vacuole
: A sub-cellular structure involved in osmoregulation. It pumps excess water out of a cell and is found prominently in freshwater protists.

Cornea
: The transparent front part of the eye that covers the iris, pupil, and anterior chamber. Together with the lens, the cornea refracts light, and as a result helps the eye to focus, accounting for approximately 80% of the eye's optical power

Corneal transplantation (corneal grafting or penetrating keratoplasty)
: A surgical procedure where a damaged or diseased cornea is replaced by donated corneal tissue.

Cryosurgery
: The application of extreme cold to destroy abnormal or diseased tissue.

Cyclic adenosine monophosphate (cAMP)
A second messenger that is important in many biological processes. cAMP is derived from adenosine triphosphate (ATP) and used for intracellular signal transduction in many different organisms.

Cyst
The inactive or resting stage of a protozoan, with minimal metabolic activity.

Cystcidal effects
Killing of the amoebae cysts.

Cystistatic effects
Inhibition of the emergence of the amoeba trophozoite from the cyst stage.

Cytochromes
Membrane-bound hemoproteins that contain heme groups and carry out electron transport.

Cytokines
A category of signalling proteins and glycoproteins that, like hormones and neurotransmitters, are used extensively in cellular communication.

Cytokinesis
The process whereby the cytoplasm of a single eukaryotic cell is divided to spawn two daughter cells. It usually initiates during the late stages of mitosis

Cytoskeleton
Internal network of fibres contributing to the basic shape of the cell.

Cytosol
The liquid portion of the cytoplasm.

Cytotoxic CD8 T-cell
Type of cells that are characterized by CD8 cell-surface glycoprotein that secretes perforins and granzyme, which destroy infected or abnormal body cells.

Dacryoadenitis
Inflammation of the lacrimal glands, i.e., the tear-producing glands.

Defensins
Small proteins that make holes in the microbial membranes by a charge or voltage-dependent mechanism

Deoxyribonucleic acid (DNA)
Nucleic acid consisting of nucleotides, made up of phosphate, a deoxyribose pentose sugar, and an arrangement of the bases adenine, guanine, cytosine and thymine.

Descemet's membrane (basement membrane)
A membrane lying between the corneal stroma, and the endothelial layer of the cornea.

Diabetes mellitus (diabetes)
A syndrome characterized by disordered metabolism and abnormally high blood sugar (hyperglycaemia) resulting from insufficient levels of the hormone insulin.

Diarrhea (diarrhoea)
Frequent loose or liquid bowel movements.

Duodenitis
Infection or inflammation of the duodenum

Ecology
The scientific study of the distribution and abundance of life and the interactions between organisms and their environment.

Ecosystem
: An ecosystem is a natural unit consisting of all plants, animals and micro-organisms (biotic factors) in an area functioning together with all of the non-living physical (abiotic) factors of the environment

Ecto-ATPases
: Enzymes that hydrolyze the extracellular ATP and other nucleoside. triphosphates

Elastases
: Enzymes that degrade a range of connective tissue proteins such as elastin, an elastic fiber, fibrinogen, collagen, and proteoglycans, which together determined the mechanical properties of the connective tissue.

Electron transport chain
: Electron transport chain associates electron carriers (such as NADH and $FADH_2$) and mediating biochemical reactions that produce ATP, which is a major energy intermediate in living organisms.

Encephalitis
: Infection or the inflammation of the brain tissue.

Encystation
: Transformation of a biochemically active stage of an organism during which the organism reproduces to a dormant state.

Endocarditis
: Exudative and proliferative inflammatory alterations of the endocardium, characterised by the presence of vegetations on the surface of the endocardium or in the endocardium itself and most commonly involving a heart valve, but sometimes affecting the inner lining of the cardiac chambers or the endocardium elsewhere. It may occur as a primary disorder or as a complication of or in association with another disease.

Endophthalmitis
: Inflammation of the internal coats of the eye.

Endoplasmic reticulum
: An organelle found in all eukaryotic cells that is an interconnected network of tubules, vesicles and cisternae and responsible for several specialized functions including protein translation, folding and transport of proteins.

Endosymbiont
: Any organism that lives within the body or cells of another organism, i.e. forming an endosymbiosis.

Epidemic
: In epidemiology, an epidemic is a classification of a disease that appears as new cases in a given human population, during a given period, at a rate that substantially exceeds what is “expected” based on recent experience.

Epithelial defects
: Damage to the outermost layer of the cornea.

Epidemiology
: The study of factors affecting the health and illness of populations, and serves as the foundation and logic of interventions made in the interest of public health and preventive medicine.

Etiological factor
: Causative agent of disease.

Etiology
: The study of the causation of disease.

Eukaryotes

Any organism made up of cells containing a nucleus composed of genetic material and enclosed within a membrane.

Excystation

Transformation of a dormant cyst stage of an organism to a biochemically active stage, during which the organism reproduces.

Exocytosis

A process by which a cell directs secretory vesicles out of the cell membrane.

Exponential phase (also called the log phase)

A period characterized by cell doubling.

Extracellular matrix (ECM)

The extracellular part of animal tissue that usually provides structural support to the cells in addition to performing various other important functions. The extracellular matrix is the defining feature of connective tissue in animals.

Ex vivo

Latin meaning "out of the living", which takes place outside an organism. *Ex vivo* refers to experimentation or measurements done in or on living tissue in an artificial environment outside the organism with the minimum alteration of the natural conditions.

Eyelid ptosis

Abnormally low position or drooping of the upper eyelid.

Fatty acid

A carboxylic acid often with a long unbranched aliphatic tail (chain), which is either saturated or unsaturated.

Fibroblast

A type of cell that synthesizes and maintains the extracellular matrix of many animal tissues.

Fibronectin

A high-molecular-weight extracellular matrix glycoprotein containing carbohydrate that binds to membrane spanning receptor proteins called such as integrins, as well as other extracellular matrix components such as collagen, fibrin and heparan sulfate.

Fibrous proteins

This class of proteins are insoluble in aqueous solutions.

Flow cytometry

A technique for counting, examining, and sorting microscopic particles suspended in a stream of fluid.

Fluorescent microscope

Type of light microscope that uses an ultraviolet light source to fluoresce objects.

Fungi

Eukaryotic organisms that have cell wall and obtain food from other organisms.

G1 phase

A period in the cell cycle during interphase, after cytokinesis and before the S phase. For many cells, this phase is the major period of cell growth during its life span. During this stage new organelles are being synthesized, so the cell requires both structural proteins and enzymes, resulting in great amount of protein synthesis.

G2 phase

During this phase, the cell undergoes a period of rapid growth to prepare for mitosis. It follows successful completion of DNA synthesis and chromosomal replication during the S phase.

Gamma-globulinaemia
Significant changes in the levels of immunoglobulins/antibodies.

Gastritis
Infection or inflammation of the stomach

Gastroenteritis
Inflammation of the mucous membrane of stomach and intestines.

Gene
A locatable region of genomic sequence, corresponding to a unit of inheritance.

Genotyping
The process of determining the genotype of an individual by the use of biological assays such as DNA sequencing.

Giemsa stain
It is a mixture of methylene blue and eosin. The stain binds to the phosphate groups of DNA and attaches itself to regions of DNA where there are high amounts of adenine-thymine bonding.

Glaucoma
Damage of the optic nerve, usually due to raised intraocular pressure

Globular proteins
These are more or less soluble in aqueous solutions (where they form colloidal solutions).

Glycogen
A polysaccharide of glucose, which functions as the primary short term energy storage in animal cells.

Glycolipids
Carbohydrate-attached lipids. Their role is to provide energy and also serve as markers for cellular recognition.

Glycolysis
The sequence of reactions that converts glucose into pyruvate with the concomitant production of a relatively small amount of ATP.

Glycoproteins
Proteins that contain oligosaccharide chains (glycans) covalently attached to their polypeptide side-chains.

Glycosidases
Glycoside hydrolases catalyze the hydrolysis of the glycosidic linkage to generate two smaller sugars.

Glycosylation
The enzymatic process that links saccharides to produce glycans, either free or attached to proteins and lipids.

Golgi bodies
A series of flattened, hollow sacks that are surrounded by phospholipid bilayers, and function to package molecules such as proteins and lipids for export in secretory vesicles.

Granulomatous
Formation of granuloma (granuloma is a nodule consisting mainly of inflammatory and immune cells as well as extracellular matrix; they are often surrounded by a lymphocyte cuff).

Guanosine-5'-triphosphate (GTP)
GTP is a purine nucleotide. One role is its use as a substrate for the synthesis of RNA during transcription. Its structure is similar to that of the guanine nucleoside, the only difference being that there are three extra phosphate groups added on to the 5' carbon. It also has the role of a source of energy or an activator of substrates

in metabolic reactions, like that of ATP, but more specific. It is used as a source of energy for protein synthesis. GTP is essential to signal transduction, particularly with G-proteins, where it is converted to GDP (guanosine diphosphate) through the action of GTPases.

Haematogenous dissemination

Dissemination through the bloodstream.

Haemolytic uraemic syndrome

A pathologic condition with involves the rupture (haemolysis) of red blood cells, subsequent anaemia, low platelet count and kidney failure.

Hematologic disorders

Diseases that affect the blood.

Hemiparesis

Partial paralysis of one side of body.

Herbivores

Organisms consume principally autotrophs such as plants, algae and photosynthesizing bacteria. By that definition, many fungi, some bacteria, many animals, some protists and a small number of parasitic plants can be considered herbivores.

Herniation

A potentially fatal side effect of very high intracranial pressure, occurs when the brain shifts across structures within the skull.

Heterotrophic

A heterotroph, or chemoorganotroph is an organism that requires organic substrates to get its carbon for growth and development.

Histopathology

The microscopic examination of tissue in order to study the manifestations of disease.

Hybridoma

Cells that have been engineered to produce a desired antibody in large amounts. To produce monoclonal antibodies, B-cells are removed from the spleen of an animal that has been challenged with the relevant antigen. These B-cells are then fused with myeloma tumor cells that can grow indefinitely in culture (myeloma is a B-cell cancer). This fusion is performed by making the cell membranes more permeable. The fused hybrid cells (called hybridomas), being cancer cells, will multiply rapidly and indefinitely and will produce large amounts of the desired antibodies.

Hydrogen peroxide (H_2O_2)

A colorless solution, that is a weak acid and has strong oxidizing properties.

Hypersensitivity

It refers to undesirable (damaging, discomfort-producing and sometimes fatal) reactions produced by the normal immune system.

Hyponatremia

The electrolyte disturbance hyponatremia exists in humans when the sodium concentration in the plasma falls below 135 mM per L. At lower levels water intoxication may result, an urgently dangerous condition. Hyponatremia is an abnormality that can occur in isolation or, as most often is the case, as a complication of other medical illnesses.

Immunocompromised

A state in which the immune system's ability to fight infectious disease is compromised or entirely absent.

Immunofluorescence assays
: The labeling of antibodies or antigens with fluorescent dyes. This technique is often used to visualize the subcellular distribution of biomolecules of interest. Immunofluorescent-labelled tissue sections or cultures are studied using a fluorescence microscope or by confocal microscopy.

Immunoglobulins
: Immunoglobulins (also known as antibodies) are gamma globulin proteins that are found in blood or other bodily fluids of vertebrates, and are used by the immune system to identify and neutralize foreign objects, such as microbial pathogens.

Inclusion bodies
: Nuclear or cytoplasmic aggregates of stainable substances, usually proteins. They typically represent sites of viral multiplication in a bacterium or a eukaryotic cell and usually consist of viral capsid proteins.

Inorganic compounds
: Inorganic compounds are considered to be of mineral, not biological, origin.

In silico
: An expression used to indicate "performed on a computer or via a computer simulation".

Intermediate filaments
: Components of the cytoskeleton.

Intracellular state
: A state in which pathogen remain inside the cell

Intracranial pressure (ICP)
: The pressure exerted by the cranium on the brain tissue, cerebrospinal fluid, and the brain's circulating blood volume.

Intravenous therapy
: Giving of liquid substances directly into a vein.

In vitro
: Refers to the technique of performing a given experiment in a controlled environment outside of a living organism; for example in a test tube.

In vivo
: Latin meaning "within the living", i.e., that taking place inside an organism. *In vivo* refers to experimentation done using the whole organism.

Iritis
: A form of anterior uveitis and refers to the inflammation of the iris of the eye.

Isoenzymes
: Enzymes that differ in amino acid sequence but catalyze the same chemical reaction.

Keratitis
: Infection or inflammation of the cornea.

Keratoneuritis
: Infection or inflammation of the corneal nerves.

Keratoplasty (corneal transplantation, corneal grafting or penetrating keratoplasty)
: A surgical procedure where a damaged or diseased cornea is replaced by donated corneal tissue.

Kinase
: An enzyme that attach phosphate groups to their substrates by using energetic molecules like ATP.

Krebs cycle
: Series of enzymatically catalyzed reactions that transform stored energy from acetyl-CoA to coenzymes NAD^+ and FAD.

Lactoferrin

Iron-binding protein that deprives microbes of iron

Laminin

A major non-collagenous component of the basal lamina, such as those on which cells of an epithelium sit. Basically, laminin is a protein found in the extracellular matrix.

Legionnaires' disease

A potentially fatal infection that initially exhibit symptoms such as pneumonia.

Leishmaniasis

A potentially fatal disease caused by protozoan parasite, *Leishmania* and is transmitted by the bite of certain species of sand fly.

Lesion

Any abnormal tissue found on or in an organism, usually damaged by disease or trauma.

Life cycle

A period involving 1 generation of an organism through means of reproduction, whether through asexual reproduction or sexual reproduction.

Linoleic acid

An unsaturated omega-6 fatty acid. Linoleic acid is a polyunsaturated fatty acid used in the biosynthesis of prostaglandins. It is found in the lipids of cell membranes. It is abundant in many vegetable oils, especially safflower and sunflower oils.

Lipids

Fat-soluble (lipophilic), naturally-occurring molecule, such as fats, oils, waxes, cholesterol, sterols, fat-soluble vitamins (such as vitamins A, D, E and K), monoglycerides, diglycerides, phospholipids, and others. The main biological functions of lipids include energy storage, acting as structural components of cell membranes, and participating as important signaling molecules.

Liposome

Tiny vesicles, made out of the same material as a cell membrane. Liposomes can be filled with drugs, and used to deliver drugs. Membranes are usually made of phospholipids, which are molecules that have a head and a tail. The head is attracted to water, and the tail, which is made of oil (hydrocarbon), is repelled by water.

Lymph node

An organ consisting of many types of cells, and is a part of the lymphatic system.

Lymphoproliferative disorders

It refers to several conditions in which lymphocytes are produced in excessive quantities. They typically occur in patients who have compromised immune systems.

Lysosomes

Organelles that contain digestive enzymes (acid hydrolases).

Lysozyme

An enzyme that digests cell walls of pathogens.

Macrophages

Immune cells that act both in the non-specific defence (or innate immunity) as well as specific defence (or cell-mediated immunity). Their role is to phagocytose (engulf and then digest) cellular debris and pathogens either as stationary or mobile cells, and to stimulate lymphocytes and other immune cells to respond to the pathogen.

Malaria

A potentially fatal disease caused by a protozoan parasite, *Plasmodium* spp. and transmitted by mosquito, *Anopheles* spp.

Mandible

Inferior maxillary bone that forms the lower jaw and holds the lower teeth in place.

Maxillary sinus

The largest of the paranasal sinuses, and is pyramidal in shape.

Metabolism

The sum of all chemical reactions, both anabolic and catabolic, within an organism.

Meningitis

Infection or inflammation of the meninges, i.e., membranes that enclose the brain and the spinal cord. Caused by viruses, bacteria, protozoa, and fungi.

Meningoencephalitis

A medical condition that simultaneously resembles both meningitis, which is an infection or inflammation of the meninges, and encephalitis, which is an infection or inflammation of the brain.

Messenger ribonucleic acid (mRNA)

A molecule of RNA encoding a chemical "blueprint" for a protein product. The mRNA is transcribed from a DNA template, and carries coding information to the sites of protein synthesis, i.e., the ribosomes.

Methicillin-resistant *Staphylococcus aureus* (MRSA)

A bacterium that is highly resistant to many available antibiotics. It may also be referred to as multiple-resistant *Staphylococcus aureus* or oxacillin-resistant *Staphylococcus aureus* (ORSA). MRSA are by definition strains of *Staphylococcus aureus* that are resistant to a large group of antibiotics called the beta-lactams, which include the penicillins and the cephalosporins.

Microbial loop

A term used to describe a trophic pathway in aquatic environments where dissolved organic carbon is reintroduced to the food web through the incorporation into bacteria. Bacteria are consumed mostly by protists, which in turn, are consumed by larger organisms.

Microfilaments

Components of the cytoskeleton.

Microscope

A microscope is an instrument for viewing objects that are too small to be seen by the naked or unaided eye.

Microscopy

The use of light or electrons to magnify objects.

Microtubules

Components of the cytoskeleton.

Midbrain

The middle of three vesicles that arise from the neural tube that forms the brain of developing animals.

Mineralization

The process where a substance is converted from an organic substance to an inorganic substance, thereby becoming mineralized.

Minimum cysticidal concentration

Lowest concentration of a test compound to which a suspension of cyst cultures gives no recoverable trophozoites.

Minimum cystistatic concentration
Lowest concentration of a test compound to which a suspension of cysts shows no emergence of trophozoites and/or growth, in the growth medium.

Minimum trophicidal concentration
Lowest concentration of a test compound to which a suspension of trophozoite cultures gives no recoverable trophozoites.

Minimum trophistatic concentration
Lowest concentration of a test compound to which a suspension of trophozoites shows no growth, in the growth medium.

Mitochondria
Elongated structure that is found in most eukaryotic cells, and produces most of the ATP in the cell.

Mitosis (M phase)
Nuclear division of a eukaryotic cell, resulting in two nuclei with the same ploidy as the original.

Monoclonal antibody
Identical antibodies secreted by cell lines originating from a single plasma cell.

Monophyletic group
In phylogenetics, monophyletic is a group that include all the descendants of the most recent common ancestor.

Monosaccharide
A monomer of carbohydrate, such as molecule of glucose.

Mortality
The condition of being mortal, or susceptible to death.

Mortality rate
A measure of the number of deaths in a given population.

Multiple drug resistance
Lack of sensitivity to three or more antimicrobials.

Mutation
A change in the nucleotide base sequence of a genome.

Myometrium
The middle layer of the uterine wall consisting of smooth muscle cells and supporting stromal and vascular tissue.

Myonecrosis
Necrosis of muscle.

NADH dehydrogenase
An enzyme located in the inner mitochondrial membrane that catalyzes the transfer of electrons from NADH to coenzyme Q.

Natural environment
The environment that comprises all living and non-living things that occur naturally on Earth or some region thereof.

Natural killer cells
Type of cytotoxic lymphocyte that constitute a major component of the innate immune system. NK cells play a major role in the rejection of tumors and cells infected by viruses. The cells kill by releasing small cytoplasmic granules of proteins called perforin and granzyme that cause the target cell to die by apoptosis or necrosis.

Nerve palsy
A disorder associated with dysfunction of nerve(s).

Neuritis
The general inflammation of the peripheral nervous system.

Neutropenia
: The decrease in white blood cells, chiefly in neutrophils.

Neutrophils
: The most abundant type of white blood cells in humans and form an essential part of the immune system.

Nitrogen cycle
: The biogeochemical cycle that describes the transformations of nitrogen and nitrogen-containing compounds in nature. It is a gaseous cycle.

Nucleolus
: A nuclear membraneless organelle in which the rDNA transcription takes place and in which the ribosomal subunits are assembled.

Nucleus
: A round to ovoid organelle within a eukaryotic cell that contains genetic material.

Nuclear envelope
: Double membrane composed of phospholipid bilayers surrounding a cell nucleus.

Nuclear lamina
: A major constituent of the nucleus that is associated with the inner face of the nuclear envelope whereas the outer face stays continuous with the endoplasmic reticulum. Besides providing mechanical support, the nuclear lamina regulates important cellular events such as DNA replication and cell division.

Nuclear pores
: Pores in the nuclear envelope that control the transport of substances through it.

Nuclease
: An enzyme capable of cleaving the phosphodiester bonds between the nucleotide subunits of nucleic acids.

Nucleoplasm
: Similar to the cytoplasm of a cell, the nucleus contains nucleoplasm or nuclear sap.

Open reading frame (ORF)
: A portion of an organism's genome which contains a sequence of gene(s) that could potentially encode a protein.

Opportunistic pathogens
: Microorganisms that produce disease when the immune system is suppressed or when introduced into an abnormal part of tissue or an area with limited immune system.

Optic atrophy
: The loss of some or most of the fibers of the optic nerve.

Organelles
: Organs of the cell.

Organic compound
: Any member of a large class of chemical compounds whose molecules contain carbon.

Orthokeratology
: The use of rigid gas-permeable contact lenses, normally worn only at night, to improve vision through the reshaping of the cornea.

Osmole
: A measurement that defines the number of moles of a chemical compound that contribute to a solution's osmotic pressure.

Osmolarity

A measure of the osmoles of solute per liter of solution, while the osmolality is a measure of the osmoles of solute per kilogram of solvent.

Osteomyelitis

An infection of bone or bone marrow. It can be sub-classified on the basis of the causative organism, the route, duration and anatomic location of the infection.

Ostiole

A small hole or opening in the cyst wall, through which *Acanthamoeba* trophozoite monitors environmental conditions and upon favourable conditions, trophozoite emerges.

Oxidation

The loss of electrons by a molecule, atom or ion.

Oxidative phosphorylation

A metabolic pathway that uses energy released by the oxidation of nutrients to produce ATP.

Pancreas

A gland organ in the digestive and endocrine system of vertebrates producing several important hormones, including insulin etc. that pass into the small intestine. These enzymes help in the further breakdown of the carbohydrates, protein, and fats.

Paraphyletic group

In phylogenetics, paraphyletic is a group that contains its most recent common ancestor but does not contain all the descendants of that ancestor.

Passive transport

Moving biochemicals and other atomic or molecular substances across membranes. However, unlike active transport, this process does not involve chemical energy.

Pathogenicity

The ability of an organism, a pathogen, to produce an infectious disease in another organism.

Pathogenesis of a disease

The pathogenesis of a disease is the mechanism by which an etiological factor causes the disease.

Pathophysiology

Pathophysiology is the study of the disturbance of normal mechanical, physical, and biochemical functions, either caused by a disease, or resulting from a disease or abnormal syndrome.

Perforation

The puncturing of a material to create a hole or aperture.

Periodic acid-Schiff (PAS)

A staining method used for staining structures containing a high proportion of carbohydrate macromolecules (glycogen, glycoprotein, proteoglycans). The reaction of periodic acid selectively oxidizes the glucose residues, creates aldehydes that react with the Schiff reagent and creates a purple-magenta color. A suitable basic stain is often used as a counterstain.

Peritonitis

Inflammation of the peritoneum, a condition marked by exudations in the peritoneum of serum, fibrin, cells and pus. It is attended by abdominal pain and tenderness, constipation, vomiting and moderate fever.

Peroxisomes

Ubiquitous organelles in eukaryotes that participate in the metabolism of fatty acids and other metabolites. Peroxisomes have enzymes that rid the cell of toxic peroxides.

Phagocytosis

A form of uptake in which solid particles are brought into the cell by the cell membrane to form an internal phagosome or food vacuole, which subsequently fuse with lysosomes and broken down.

Phagolysosome

A membrane-enclosed organelle which is formed from the fusing of a lysosome and a phagosome.

Phagosome

A vacuole formed around a particle absorbed by phagocytosis. The vacuole is formed by the fusion of the cell membrane around the particle.

Pharyngitis

Inflammation of the pharynx.

Pharynx

The part of the neck and throat situated immediately posterior to (behind) the mouth and nasal cavity, and cranial, or superior, to the esophagus, larynx, and trachea.

Phase contrast microscopy

An optical microscopy illumination technique in which small phase shifts in the light passing through a transparent specimen are converted into amplitude or contrast changes in the image.

Phosphatase

An enzyme that removes a phosphate group from its substrate by hydrolyzing phosphoric acid monoesters into a phosphate ion and a molecule with a free hydroxyl group.

Phospholipase

An enzyme that converts phospholipids into fatty acids and other lipophilic substances. There are four major classes, termed A, B, C and D distinguished by what type of reaction they catalyze.

Phosphorylase

An enzyme that attach phosphate groups to their substrates by using energetic molecules like ATP.

Photodynamic chemotherapy

Use of light to damage / kill the target cell.

Photophobia

Excessive sensitivity to the light.

Photosynthesis

Photosynthesis is a series of enzyme-catalysed steps for the conversion of light energy into chemical energy by living organisms.

Pinocytosis

A form of uptake in which fluid / small particles are brought into the cell suspended within small vesicles which subsequently fuse with lysosomes and hydrolyzed or breaken down.

Plasma

Acellular part of the blood.

Plasma membrane

A selectively permeable lipid bilayer found in all cells. It contains a wide variety of biological molecules, primarily proteins and lipids, which are involved in a vast array of cellular processes such as cell adhesion, ion channel conductance and cell signaling. The plasma membrane also serves as the attachment point for both the intracellular cytoskeleton and, if present, the cell wall.

Pneumonitis

Infection or inflammation of lung tissue.

Polyamines

Organic compounds having two or more primary amino groups - such as putrescine, cadaverine, spermidine, and spermine - that are growth factors in both eukaryotic and prokaryotic cells.

Polysaccharides

Complex carbohydrates that are polymers made up of many monosaccharides joined together by glycosidic bonds. They are therefore very large, often branched, macromolecules.

Polyunsaturated fatty acid

A fatty acid in which more than one double bond exists within the representative molecule. Polyunsaturated fatty acids can assume a *cis* or *trans* conformation depending on the geometry of the double bond.

Post-transcriptional regulation

The control of protein synthesis by genes after synthesis of RNA has begun.

Primary producers

Primary producers are those organisms in an ecosystem that produce biomass from inorganic compounds (autotrophs). In almost all cases these are photosynthetically active organisms (plants, cyanobacteria and a number of other unicellular organisms).

Prokaryote

Any unicellular organism that lacks a nucleus. Classification includes bacteria and archaea.

Prostaglandin

Released from the damaged cells and increase vascular permeability. Any member of a group of lipid compounds that are derived enzymatically from fatty acids and have important functions in the animal body. Every prostaglandin contains 20 carbon atoms, including a 5-carbon ring.

Prostate

A compound tubuloalveolar exocrine gland of the male mammalian reproductive system.

Protease

Enzyme that conducts proteolysis, that is, begins protein catabolism by hydrolysis of the peptide bonds that link amino acids together in the polypeptide chain.

Proteins

Large organic compounds made of amino acids arranged in a linear chain and joined together by peptide bonds between the carboxyl and amino groups of adjacent amino acid residues.

Protist

A group of single-celled eukaryotes, which includes protozoa, algae, and yeast (fungi).

Protozoa

Single-celled eukaryotes that lack a cell wall and are similar to animal cells in the nutritional requirements.

Q fever

Q stands for "query fever" because originally the causative agent was unknown). The Q fever involves respiratory symptoms.

Reduction

The gain of electrons by a molecule, atom or ion.

Rhinitis
: Irritation and inflammation of some internal areas of the nose.

Rho GTPases
: The major regulators of the actin cytoskeleton, which link external signals to the cytoskeleton.

Ribonucleic acid (RNA)
: Nucleic acid consisting of nucleotides made up of phosphate, a ribose pentose sugar, and an arrangement of the bases adenine, guanine, cytosine, and uracil.

Ribosomal RNA
: A form of RNA, which together with polypeptides, makes up the structure of ribosomes.

Ribosome
: Non-membranous organelle found made of RNA and polypeptides and functions to make proteins.

RNA interference (RNAi)
: A mechanism that inhibits gene expression at the stage of translation or by hindering the transcription of specific genes.

Rocky Mountain spotted fever
: The disease is characterised by sudden onset of headache, chills and fever which can persist for 2-3 weeks, muscle pain. A characteristic rash appears on the extremities and trunk about the 4th day of illness.

S phase
: A period in the cell cycle during interphase, between G1 phase and the G2 phase. Following G1, the cell enters the S stage, when DNA synthesis or replication occurs.

Scanning electron microscope (SEM)
: A type of electron microscope that images the sample surface by scanning it with a high-energy beam of electrons in a raster scan pattern. The electrons interact with the atoms that make up the sample producing signals that contain information about the sample's surface topography, composition and other properties such as electrical conductivity.

Sclera
: The white part of the eye.

Serum
: Acellular part of blood.

Septaecemia
: Systemic disease associated with the presence and persistence of pathogenic microorganisms or their toxins in the blood.

Sexual reproduction
: Two cells join, exchange genetic material and produce progeny by budding or fission.

Sleeping sickness
: A potentially fatal disease of people and animals, caused by protozoa of species *Trypanosoma brucei* and transmitted by the tsetse fly.

Speciation
: The evolutionary process by which new biological species arise.

Stationary phase
: During this phase, the growth rate slows as a result of nutrient depletion and accumulation of toxic products.

Stroma
: The connective, non-functional supportive framework of a biological cell, tissue, or organ.

Stroma of cornea

A fibrous, tough, unyielding, and perfectly transparent part of the eye.

Superoxide dismutase

Enzyme that catalyzes the dismutation of the superoxide into oxygen and hydrogen peroxide.

Systemic lupus erythematosus

Autoimmune diseases that most often harms the heart, joints, skin, lungs, blood vessels, liver, kidneys, and the nervous system.

Thalamus

A pair and symmetric part of the brain. It constitutes the main part of the diencephalon.

T helper cells (CD4 T-helper cells)

A sub-group of lymphocytes that plays an important role in establishing and maximizing the capabilities of the immune system.

Tissue plasminogen activator (tPA or PLAT)

A serine protease normally found on the surface of endothelial cells of veins, capillaries, and is secreted after vascular injury. It converts the proenzyme plasminogen to plasmin, a fibrinolytic enzyme.

Trans-cranial drug delivery to the brain

Drug delivery to the ventricles of the brain by injection through a trephine hole in the cranium.

Transcription

The process of copying DNA to RNA by an enzyme called RNA polymerase.

Transfection

The introduction of foreign material into eukaryotic cells using a virus vector or other means of transfer. The term "transfection" for non-viral methods is most often used in reference to mammalian cells, while the term "transformation" is preferred to describe non-viral DNA transfer in bacteria and non-animal eukaryotic cells such as fungi, algae and plants.

Transfer RNA (tRNA)

A small RNA (usually about 74-95 nucleotides) that transfers a specific amino acid to a growing polypeptide chain at the ribosomal site of protein synthesis during translation.

Trans-nasal drug delivery to the brain

A method in which lipid-soluble small molecules are instilled nasally, cross the nasal mucosa and the arachnoid membrane and enter the olfactory cerebrospinal fluid with the result similar to intracerebroventricular injection.

Transient blood-brain barrier disruption to deliver drug to the brain

Arterial infusion of hyperosmotic solution or ultrasonic irradiation of the brain.

Trophozoite

The motile, feeding stage of a protozoan.

Trophicidal effects

Killing of the amoebae trophozoites.

Trophistatic effects

Growth inhibitory effects on amoebae trophozoites.

Tularemia (rabbit fever)

A disease typically found in animals, especially rodents, rabbits, and hares. Typically, people become infected through the bite of infected insects (most commonly, ticks and deerflies), by handling infected sick or dead animals, by eating or drinking contaminated food or water, or by inhaling airborne bacteria. Possible symptoms include skin ulcers, swollen and painful lymph glands, inflamed eyes, sore throat, mouth sores, diarrhea or pneumonia. If the bacteria are inhaled, symptoms can include abrupt onset of fever, chills, headache, muscle aches, joint pain, dry cough, and progressive weakness. People with pneumonia can develop chest pain, difficulty breathing, bloody sputum, and respiratory failure. Tularemia can be fatal if the person is not treated with appropriate antibiotics.

Typhus

An acute infectious disease characterised by high fever, a skin eruption and severe headache, being spread by lice, ticks or fleas. The infecting organism is *Rickettsia prowazekii*, sensitive to sulpha drugs or tetracycline.

Ultraviolet (UV) light

Electromagnetic radiation with a wavelength shorter than that of visible light, but longer than X-rays.

Uridine diphosphate glucose (uracil-diphosphate glucose, UDP-glucose)

A nucleotide sugar. It is involved in glycosyltransferase reactions in metabolism.

Urokinase--type Plasminogen Activator (uPA)

A serine protease that targets plasminogen, which is an inactive zymogen and then forms plasmin. Activation of plasmin triggers a proteolysis cascade which, depending on the physiological environment participate in thrombolysis or extracellular matrix degradation.

Uveitis

Inflammation of the middle layer of the eye.

Vasodilator

A drug or chemical that relaxes the smooth muscle in blood vessels, which causes them to dilate. Dilation of arterial blood vessels (mainly arterioles) leads to a decrease in blood pressure.

Virulence

The degree of pathogenicity of a microbe, or the relative ability of a microbe to cause disease.

Virus

A virus (from the Latin *virus* meaning "toxin" or "poison"), is a sub-microscopic infectious agent that is unable to grow or reproduce outside a host cell.

Xenic cultivation

When one organism is cultured in association with another organism.

Zoonosis (zoonotic infection)

Zoonosis is any infectious disease that is able to be transmitted (by a vector) from other animals, both wild and domestic, to humans or from humans to animals.

www.ingramcontent.com/pod-product-compliance
Lightning Source LLC
LaVergne TN
LVHW011748200926
848484LV00003B/650
* 9 7 8 1 9 0 4 4 5 5 4 3 1 *